Lecture Notes on Multidisciplinary Industrial Engineering

Series editor

More information about this series at http://www.springer.com/series/15734

Uday S. Dixit · Ravi Kant
Editors

Simulations for Design and Manufacturing

Select Papers from AIMTDR 2016

Springer

Editors
Uday S. Dixit
Department of Mechanical Engineering
Indian Institute of Technology Guwahati
Guwahati, Assam
India

Ravi Kant
Indian Institute of Technology Ropar
Rupnagar, Punjab
India

ISSN 2522-5022 ISSN 2522-5030 (electronic)
Lecture Notes on Multidisciplinary Industrial Engineering
ISBN 978-981-10-8517-8 ISBN 978-981-10-8518-5 (eBook)
https://doi.org/10.1007/978-981-10-8518-5

Library of Congress Control Number: 2018934878

Printed on acid-free paper

This Springer imprint is published by the registered company Springer Nature Singapore Pte Ltd.
part of Springer Nature
The registered company address is: 152 Beach Road, #21-01/04 Gateway East, Singapore 189721, Singapore

Preface

Manufacturing technology is of great interest to industries where parts are shaped by forming, cutting or welding operations. It has attracted many researchers across the world in past several decades due to rapid development in new materials, processes, automation and control, microtechnology and computer technology. Manufacturing is becoming more important due to recent industrial requirements for the development of new products following sustainable manufacturing practices.

Initially, the manufacturing was considered as an art and was broadly dependent on skills of the worker. Research in manufacturing was focused on experimental investigations, and as a result, it suffered from a lack of understanding of the process. The development of computers and the interest of research community to understand the insight of the process changed the scenario. The development of process models and their simulations was started by a large number of researchers. Simulation results not only verified existing theories but also provided a nuanced understanding of different explanations. Simulations can predict the process response, which helps in the control of the process. A huge amount of time and cost associated with hit-and-trial technique of process design can be saved with effective simulation.

This book presents basic and advanced manufacturing processes such as drilling, milling, friction stir processing, TIG welding, incremental sheet forming, ECDM and selective sintering. The focus is on providing exposure to the recent research on the simulations of manufacturing processes. This book comprises 10 chapters, which are extended versions of peer-reviewed papers presented in 6th International and the 27th All India Manufacturing Technology, Design and Research (AIMTDR) Conference held at College of Engineering, Pune (COEP), during December 16–18, 2016. It covers various simulation techniques such as finite element method (FEM), multiphysics simulations, mathematical modeling, computational fluid dynamics (CFD), and molecular dynamics (MD) and simulation tools. The broad range of manufacturing processes and the tools and techniques for their simulations are covered in this book. This book will be useful to the students, teachers, research scholars, scientists and engineers working in the field of manufacturing and especially for those who are looking for a solution through process simulations.

Chapter "Study on the Delamination of GFRP Composites in Drilling: A Finite Element Model" presents finite element studies on delamination during drilling of glass fiber-reinforced plastic (GFRP) material. The variation in thrust force and torque with a change in speed and feed rate is studied by finite element simulations. Chapter "Computational Fluid Dynamics Analysis of MQL Spray Parameters and Its Influence on MQL Milling of SS304" focuses on the machining process using a limited amount of lubricant, known as minimum quantity lubrication (MQL). Computational fluid dynamics (CFD) simulations are carried out to replicate the mist formation in MQL milling. The significance of spray parameters like droplet size and velocity is evaluated with reference to a milling process.

Chapter "Finite Element Method and Experimental Study of Self-reacting Friction Stir Welding of Aluminium Alloy AA6061-T6" reports self-reacting friction stir welding, a variant of friction stir welding (FSW). FEM simulations and experimental investigations are carried out to analyze the effect of tool traverse speed on mechanical and metallurgical properties of welded joint. A regression relation is established between process parameters and average grain size in the welded region. Chapter "Thermomechanical Simulation of Friction Stir Welding Process Using Lagrangian Method" presents the modeling of FSW process in FEM-based DEFORM software using Lagrangian formulation. Three different solvers, viz conjugate gradient, sparse and combination of both, are compared based on computation time and solution accuracy. Shear friction model is defined to formulate the contact between tool and workpiece. Simulation results are used to explain the insight of process mechanism in terms of temperature distribution and material flow.

Chapter "Studies on Spring Back Effect of TIG Welded Ti-6Al-4V Sheets" reports forecasting of spring back effect in both base material and tungsten inert gas (TIG) welded Ti-6Al-4V alloy sheets through experimental and numerical analysis. The optimum parameters are identified to obtain good weld quality with appropriate depth of penetration. Chapter "Numerical Prediction of Fracture in Parts Formed with Incremental Sheet Forming Process" reports numerical investigation to predict the fracture depth in parts formed with incremental sheet forming (ISF) process. The plastic deformation and fracture are modeled through Johnson–Cook model and constant equivalent plastic strain criteria, respectively. The Lagrangian solver with explicit time integration and iterative configuration is used to handle the contact, material nonlinearity and geometric nonlinearity.

Chapter "Finite Element Analysis of V-Shape Incremental Equal Channel Angular Pressing" presents equal channel angular pressing (ECAP), which is a severe plastic deformation (SPD) process for improving the mechanical properties of the material. A new SPD process called V-shape incremental equal channel angular pressing is proposed. FEM simulations are carried out to see the feasibility of V-shape ECAP. Suitable tool geometry and process kinematics are established.

Chapter "Dynamic Analysis of Robotic Manipulator for Incremental Sheet Metal Forming" explains robotic assistance in an incremental sheet metal forming (ISF) to make the process flexible and efficient. Due to varying forces in ISF, the tool may deflect causing error in tool path, which in turn results in errors in geometry of the part. To compensate the deflection, accurate prediction of the stiffness of robot in all configurations and the forces acting on tool is required. Therefore, dynamic analysis is carried out using Lagrangian–Euler formulation for a manipulator to be used in incremental sheet metal forming. Joint torque and force are obtained by the inverse dynamic model for desired acceleration using state variables of the robot, i.e., position and velocity.

Chapter "FEM Analysis of Different Materials Based on Explicit Dynamics ANSYS in Electrochemical Discharge Machine" presents an advanced hybrid machining process, electrochemical discharge machining (ECDM), which is a combination of electrochemical (ECM) and electrodischarge (EDM) machining. A finite element analysis is carried out for ECDM machine structure by using explicit dynamic simulation in ANSYS software. The stress, strain and deformation of different materials (Al6061T6, magnesium alloy, nickel alloy, silicon carbide and titanium alloy) during ECDM are analyzed under different conditions. Chapter "Molecular Dynamics Simulation Study of Neck Growth in Micro-selective Laser Sintering of Copper Nanoparticles" presents molecular dynamics (MD) simulations for microselective laser sintering of copper nanoparticles. The mechanism of neck growth formation during selective laser sintering (SLS) of two spherical ultrafine copper powders is studied.

The chapters highlight the importance and capability of simulations in manufacturing. They also provide the directions for future work. We welcome the feedback from the readers about this book.

Guwahati, India — Uday S. Dixit
Rupnagar, India — Ravi Kant

About the AIMTDR Conference

AIMTDR Conference is a highly prestigious, biennial event organized in the field of mechanical and production engineering in India. The conference has a glorious history of organization since its inception. The first conference entitled 'All India Machine Tool Design and Research Conference' was held at Jadavpur University, Kolkata, in 1967. In early 1990s, it was thought appropriate to widen the scope of the conference to encompass areas related to different manufacturing process technologies and systems. Accordingly, it was renamed as 'All India Manufacturing Technology, Design and Research' Conference and the 16th in the series was organized at Central Machine Tool Institute, Bangalore, in 1994. It became an international event with the first international conference being held at Indian Institute of Technology Roorkee in 2006. The international conference aimed to bring together academicians, researchers and industry professionals working worldwide in the field of manufacturing to exchange and disseminate ideas.

The subsequent international AIMTDR Conferences were held at IIT Madras and Andhra University, Visakhapatnam, respectively. Jadavpur University hosted silver jubilee of the conference and organized 4th International and 25th AIMTDR Conference at Kolkata. The fifth international conference was organized at IIT Guwahati in December 2014, while the 6th one (AIMTDR 2016) was organized by the Department of Production Engineering and Industrial Management, College of Engineering, Pune (COEP), during December 16–18, 2016.

Mission, Vision, Challenges and Direction of AIMTDR Conference

(Excerpt from the address of Prof. Amitabha Ghosh, the Chief Guest of 26th AIMTDR Conference, held at IIT Guwahati during December 12–14, 2014)

Personally, I have a close association with AIMTDR Conference from the very first one for which I happened to be a humble and young member of the organizing team. That event was organized jointly by Late Prof. Amitabha Bhattacharyya and Prof. A. K. De at Jadavpur University in the year 1967. Keeping in view the growing industrialization of India, the AIMTDR Conference was planned along the line of MTDR Conference that used to be organized by Profs. Tobias and Koenigsberger at Cambridge and Birmingham alternately.

India being an emerging economy, the importance of 'manufacturing' was well recognized and one of the primary goals of AIMTDR Conference has been to bring the academicians, researchers and the engineers from the industry to a common platform for exchanging ideas and developing a deeper mutual understanding among all concerned. The organizers of the AIMTDR Conference in the past were eminently successful in this regard. With time, this event has gained maturity and has emerged as one of the most important international and national conferences held in India for all who are associated with the field of manufacturing.

Since the economic development of any country is very critically linked to the manufacturing sector, it is only very natural that the current political leadership of India has taken up 'Manufacturing in India' as one of its key objectives. In fact, 'manufacturing' should be a common objective for all the South Asian and the Southeast Asian countries to develop a good mutual understanding and cooperation to enhance the overall manufacturing capabilities of this region. Then only, this region of the world can become a powerhouse for economic growth and play a center-stage role in the world economy. This is essential if we have to eliminate the poverty that has plagued this region for a very long time.

From my long association with the evolution of AIMTDR Conference and my involvement in teaching and R&D in the field of manufacturing for almost half a century, placing a few observations before this august gathering may not be out of place.

Traditionally, the 'primary' manufacturing processes have remained in the domain of mechanical engineering and metallurgy. At the same time, the secondary and finishing processes along with the machine tools and systems involved in manufacturing have remained exclusively as part and parcel of mechanical engineering. This, in my personal opinion, has not always helped the manufacturing activities in India to take advantage of the progress made in physical and applied sciences. Barring some isolated cases, this has rendered the manufacturing activities in India to be largely devoid of major fundamental innovations. As a result, the manufacturing activities in India have remained mostly confined to the traditional lines without giving much attention to 'value addition.' Thus, for example, our earnings from the export of a couple of hundred 'made in India' cars to Europe can be offset by that through the sale of a single focused ion beam machine, measuring 1.5 m × 1 m × 1 m, by USA to India. This scenario must be changed in the coming days.

Over the years, advanced manufacturing has gradually developed into a multidisciplinary activity and real 'value addition' through 'manufacturing' can be achieved only when the advancement in physical, chemical and other sciences is used in innovating newer processes and possibilities of 'manufacturing.' Establishing a close link between the manufacturing and the related sciences has to be recognized as a necessary task. I believe that an event like AIMTDR can play a very significant role in this regard. Besides, revamping of the old-fashioned curricula for training engineers in manufacturing is essential to render them capable of facing the challenges from futuristic manufacturing; AIMTDR Conference can take a leading role in that direction by providing a separate session to discuss the issues involved in the matter. In the coming years, I am very hopeful that AIMTDR Conference will attract not only manufacturing engineers but also researchers from basic and applied sciences whose works are closely related to and important for innovations in manufacturing.

From my half-a-century-old teaching experience, I find that the current young generation is gradually becoming somewhat disinclined to take up careers in R&D related to manufacturing. This is particularly so for the brighter section of the student community; there is a feeling among them that there is not much intellectual challenge in the subjects related to manufacturing. There can be nothing farther from the truth. Perhaps, to a large extent, this is so as the curricula and syllabi have remained archaic in many universities and institutions. In fact, application of many advances in scientific principles to manufacturing is the key requirement to open the gate for the impending next Industrial Revolution. 'Manufacturing' needs the brilliant young minds to take up R&D careers in academia and industry with equal eagerness and enthusiasm. However, this can be possible only on receiving adequate and aggressive support from the industry houses—both financially and administratively.

Although the world has reaped the benefit of the 2nd Industrial Revolution which was triggered by the R&D in the Silicon Valley, California, except for some software-related activities, India (and many other countries in this region) had really nothing to do with the actual developments and related manufacturing. Remaining a good follower cannot take India to any leadership position though it may provide some financial relief. India and the countries in this region cannot afford to miss the opportunity to take up important position when the next Industrial Revolution comes.

Dear colleagues and friends, the silicon-based 2nd Industrial Revolution has reached a plateau and there are indications that 'carbon' may play a more important role and carbon-based devices will play the key role in ushering the world into the reign of the 3rd Industrial Revolution. Already enough indications are coming in that direction. If that be the case, India should play a major role and initiate well-planned pioneering activities so that manufacturing engineering becomes a multidisciplinary area involving relevant basic science and engineering subjects for the emergence and growth of 'carbon-based technology' in this region of the world. I am happy to notice that some of the leading world authorities on carbon devices and advanced fabrication are present in this conference. This event can be a great opportunity for the manufacturing community to take advantage of their presence and plan an appropriate course of action, to initiate planned activities, and to innovate carbon devices.

Many areas of manufacturing in the not-too-distant future will be very different from what we recognize as 'manufacturing' today; 'self-assembly,' 'self-regulation,' 'self-correction' and 'self-replication' will become the keywords in futuristic manufacturing. Obviously, it will be too drastic to think of redirecting all the R&D on manufacturing in this direction, but, at the same time, India should be well prepared to take active role in such areas of futuristic manufacturing (a name for that was coined a few years ago in a workshop at IIT Kanpur—'Fabrionics') as that will help the country to gain expertise for incorporating significant 'value addition' in our manufacturing activities.

AIMTDR Conference is one of the very few events that draw researchers and practitioners from the academia and the industry with equal enthusiasm. I am very hopeful that using this grand platform, we all can take India and the neighboring countries along the path of growing technological excellence and engineering marvel.

AIMTDR 2016 Conference: Objectives and Organization

Manufacturing has revolutionized itself from its contemporary form to its current digital access, more so in the era of Industry 4.0. With every Industrial Revolution, we have seen labor and asset productivity multiply and structural shifts emerge in the manufacturing world order. Several core technologies are driving Industry 4.0, be it Simulation, Autonomous Robots, Big Data and Analytics, Augmented Reality or Additive Manufacturing. If our economy needs to grow multifold to achieve sustainable development, the existing 15% share of manufacturing sector in India's GDP needs to scale to 25% in the immediate near future.

Manufacturing today seeks innovation to be ubiquitous by inventing ways to produce more with less inputs. In an era of integration where technologies complement one another, design and manufacturing face a daunting task with regard to the quality and cost-effectiveness of products. Concentrated efforts focusing on quality research need to be endorsed for improving the manufacturing processes, technology and systems to adopt world-class manufacturing technologies. The manufacturing education should also emphasize its importance to attract the talented young to this area and equip them with skills that embrace knowledge, information and techniques.

With this broad focus, College of Engineering, Pune (COEP), presented as a sequel to the AIMTDR Conferences of the past, the 6th International and 27th AIMTDR Conference in December 2016. The theme of the conference was 'Recontouring Manufacturing.' Several invited lectures and keynote addresses on cutting-edge technologies were presented in the conference by leading researchers from USA, Singapore and India. In all, 380 papers were selected for oral presentations and 80 for poster presentations. The organizing team brought out the proceedings on a CD covering all the papers presented. These papers will help to provide insights into the realistic exposure of current research and development trends in the field of manufacturing technology, design and research. During the exhibition in the conference, 40 leading companies participated by displaying new technology equipment, products and measuring equipment. The manufacturing community in India received this conference with appreciation and applause.

Through the deliberations in the conference, I hope to see the culmination of great thoughts and ideas that would introduce to develop technological solutions in the domain of manufacturing and design by the fusion of technologies straddling physical, digital and biological worlds.

B. B. Ahuja
Organizing Secretary, AIMTDR 2016
Professor of Production Engineering and Director
College of Engineering, Pune

Editorial Acknowledgements

We are thankful to Prof. B. B. Ahuja, Director of College of Engineering, Pune (COEP), who was chairman and organizing secretary of AIMTDR 2016, for allowing us to publish the book from extending versions of selected papers of the conference. We are also thankful to Dr. Rajiv B., who is the Head of the Department of Production Engineering and Industrial Management of College of Engineering, Pune (COEP), and his team comprising Prof. S. K. Basu, Prof. H. Bagchi, Prof. P. P. Chikate, Prof. (Mrs.) N. R. Rajhans, Prof. K. G. Patil, Dr. B. U. Sonawane, Dr. M. D. Jaybhaye, Dr. S. S. Anasane, Dr. (Mrs.) A. V. Muley, Dr. P. D. Pantawane, Dr. S. M. Patil, Dr. M. R. Dhanvijay, Mr. S. U. Ghunage, Mr. J. S. Karajagikar, Mr. Suleman Shaikh, Mrs. Vaishali Gaikwad and Mrs. Tripta Nair, for their motivation, encouragement, advice and support provided during organization of a very successful conference that led to the publication of this book.

We sincerely thank all the authors for contributing their valuable research work as extended chapters for this book. The authors submitted the chapters and responded for the revisions on time which could make it possible to publish this book on time. We sincerely thank the organizers and the National Advisory Committee (NAC) members of AIMTDR 2016 for their guidance and support. We are grateful to Prof. Amitabha Ghosh, Prof. V. K. Jain, Prof. V. Radhakrishnan and Prof. P. Radhakrishnan for their encouragement and suggestions.

We express our heartfelt gratitude to Prof. Paulo J. Davim from University of Aveiro, the Series Editor of this book for his encouragement and guidance. Last but not least, we express our sincere gratitude toward the staff members of Springer, Ms. Swati Meherishi, Ms. Aparajita Singh, Dr. Mayra Castro and Ms. Bhavana, for their dedicated support during publication process of this book.

Contents

About the Editors

Dr. Uday S. Dixit received his BE in Mechanical Engineering from the University of Roorkee (now the Indian Institute of Technology Roorkee) in 1987, MTech and PhD in Mechanical Engineering from the Indian Institute of Technology Kanpur (IIT Kanpur) in 1993 and 1998, respectively. He has worked for two companies, HMT, Pinjore; and Indomag Steel Technology, New Delhi, where his main responsibility was designing various machines. He joined the Department of Mechanical Engineering, Indian Institute of Technology Guwahati (IIT Guwahati), in 1998, where he is currently a Professor. He was also the Officiating Director of the Central Institute of Technology, Kokrajhar, from February 2014 to May 2015. He has been actively engaged in research in various areas of design and manufacturing for 25 years. He has authored/co-authored 75 journal papers, 91 conference papers, 20 chapters, and 6 books on mechanical engineering. He has also co-edited 3 books related to manufacturing. Of his 9 books, 6 have been published by Springer. He has guest-edited 10 special issues of journals. He is currently an Associate Editor of the Journal of the Institution of Engineers (India), Series C. He has investigated a number of sponsored projects and developed several courses. Presently, he is the Vice-President of the AIMTDR Conference.

Dr. Ravi Kant is an Assistant Professor at the Department of Mechanical Engineering, IIT Ropar. Prior to joining IIT Ropar, he was an Assistant Professor at the College of Engineering, Pune (COEP), which is one of the oldest and most respected institutes in India. He received his bachelor's degree in Mechanical Engineering from Maharshi Dayanand University, Rohtak (Haryana). He completed his MTech at the Department of Mechanical Engineering, IIT Guwahati, with a specialization in 'Computer-Assisted Manufacturing,' and worked on a project investigating the formability of adhesively bonded sheets. He also obtained his doctorate at IIT Guwahati, working on laser-forming processes, including experimental studies, numerical simulations, soft computing modeling and optimization. After his PhD, he worked on inverse estimation of unknown parameters and laser-assisted adhesive joining of metal to polymer sheets. His research interests include laser-based manufacturing, modeling and optimization of modern manufacturing processes, transmission welding, adhesive joining, inverse estimation, ultra-high-speed machining.

Study on the Delamination of GFRP Composites in Drilling: A Finite Element Model

Harshadeep S. Joshi, Surjya K. Pal and Goutam Chakraborty

1 Introduction

Fiber-reinforced plastic (FRP) composite materials are increasingly being used due to their higher strength-to-weight ratios and better control over strength properties. Machining of FRP composite has always been a major challenge for the machinists. Laminar nature of FRP composite material results in separation of the layers of FRP during machining which is called delamination. Delamination reduces the strength of the components considerably, making it unusable and causes rejection [1]. There are mainly two types of delamination, namely, pill-up delamination and pushout delamination. The uncut material ahead of the cutting edge of the twist drill tries to climb the helical surface of the twist drill resulting in pill-up delamination. The deflection of uncut laminate under the action of thrust force exerted by the twist drill is the main reason behind pushout delamination. Initially, as the uncut thickness of laminate under the action of axial thrust force is large, the laminate has sufficient stiffness. As the drill progresses, the uncut thickness under the drill reduces and thus the stiffness of the uncut laminate also decreases. Under the action of axial thrust force exerted by the drill, the uncut laminate undergoes bending. When this loading exceeds interlaminar bonding strength, delamination occurs away from the point of drilling.

H. S. Joshi (✉)
Department of Mechanical Engineering, Dr. Babasaheb Ambedkar Technological University, Lonere, Raigad 402103, Maharashtra, India
e-mail: harshadeepjoshi@gmail.com

S. K. Pal · G. Chakraborty
Department of Mechanical Engineering, Indian Institute of Technology Kharagpur, Kharagpur 721302, West Bengal, India
e-mail: skpal@mech.iitkgp.ernet.in

G. Chakraborty
e-mail: goutam@mech.iitkgp.ernet.in

U. S. Dixit and R. Kant (eds.), *Simulations for Design and Manufacturing*, Lecture Notes on Multidisciplinary Industrial Engineering,
https://doi.org/10.1007/978-981-10-8518-5_1

Various techniques are used for reducing the delamination during drilling of FRP composites. Few of the important methods includes the following:

- Optimization of the drilling parameters, i.e., speed and feed
- Optimization of drill geometry and use of special drill geometries such as core, saw, step, multi-fluted, drills, etc.
- Modification of drilling conditions such as, use of
  - pilot hole
  - backing plate
  - cryogenic drilling
  - vibratory drilling, etc.

Wide literature is available on experimental investigation of the drilling-induced delamination in FRP composites. Analytical models are also available in literature for the study of delamination during drilling of FRP composites. With the developments in the computational facilities, finite element method (FEM) provides a powerful tool for analyzing the complex problems. But the complex geometry of the twist drill, anisotropic heterogeneous nature of workpiece, etc., are some of the major difficulties in the modeling of drilling-induced delamination. Thrust force is mainly responsible for the delamination. There exists a critical value of thrust force below which delamination is not observed. Thrust force is further dependent on the drill geometry and machining parameters. Therefore, estimation of thrust force based on the drill geometry and machining parameters is inevitable.

In the present study, finite element model is developed for the study of delamination during drilling of FRP composites. Realistic geometry and boundary conditions are used for the simulations. The simulation results are compared with the experimental results from literature and also with the results from analytical models. The results were found in good agreement in both the cases. Drilling is one of the most commonly used processes for producing holes [2].

1.1 Mechanism of Drilling

Drilling is a complex oblique cutting process. The cutting mechanisms at the primary and secondary cutting edges are totally different. The cutting angle and cutting speeds change along the length of the primary cutting edge. While modeling of the drilling process, evaluating the cutting forces becomes a challenging job. A semi-analytical process is used for modeling the cutting forces during drilling. There are mainly two cutting forces in case of drilling: Thrust force and the torque.

1.1.1 Thrust Force

Thrust force is the axial force generated during the drilling process. Total thrust force is the sum of the components of cutting forces and frictional forces along the drill axis. Figure 1 shows different axial components of the cutting and the frictional forces.

The rake angle at the chisel edge is highly negative resulting in adverse cutting conditions. Moreover, the metal removal process in this region is more of extrusion rather than cutting. Obviously, the major component of the axial cutting force is the cutting force due to the chisel edge, F_{CCA}. Major material removal takes place at the primary cutting edges. Second major component of axial thrust force comes from the cutting action at both the lips. But the cutting force is not constant throughout the length of the cutting edge. The cutting angle and cutting velocity both vary over the lip [2]. The total axial cutting force due to both the lips is $2 \times F_{LCA}$.

Axial components of the frictional forces also contribute to the total thrust force. There are two types of frictional forces: due to friction between the drill and the cylindrical surface of the drilled hole and the other due to chip flowing over the helical groove of the flute. Axial component of the friction between the drill and the cylindrical surface of the drilled hole is F_{SFA} and the axial component of the frictional force between the helical groove and the chips is F_{GFA}. Hence, the total thrust force during drilling F_A is given by [3]

Fig. 1 Various components of thrust force during drilling

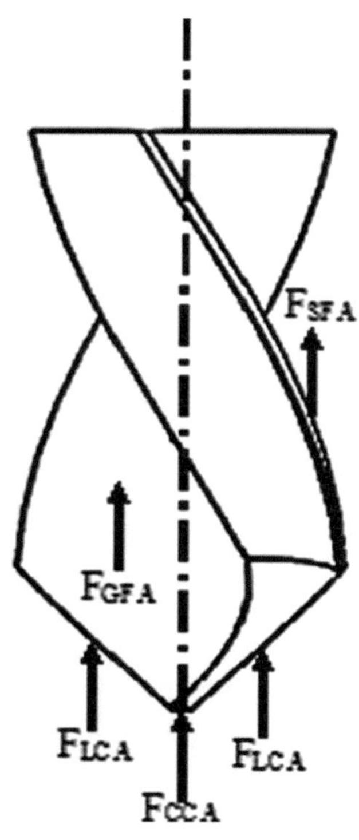

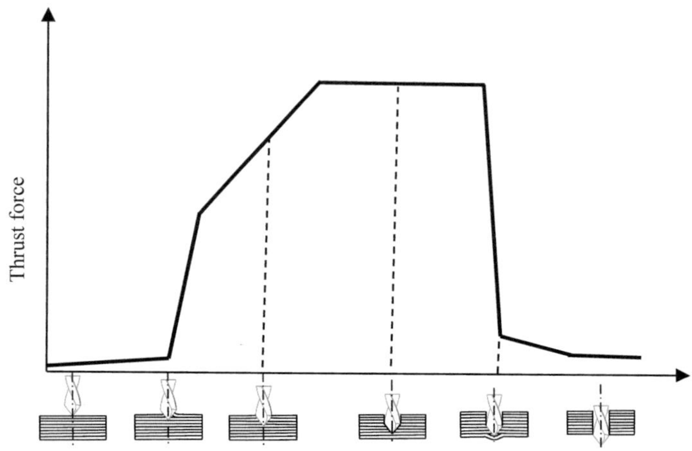

Fig. 2 Variation of thrust force during drilling

$$F_A = F_{CCA} + F_{LCA} + F_{SFA} + F_{GFA} \tag{1}$$

Figure 2 shows how typically thrust force varies during a drilling process. A steep increase in thrust force is observed during the initial phase of the drilling when mainly the chisel edge is engaged in the cutting process. In the second phase, the primary cutting edge gradually gets engaged. The rate of increase in the thrust force is reduced considerably when the primary cutting edge is fully engaged. The small slope in this section is due to the gradual increase in the frictional force. The thrust force remains almost constant afterwards till the point when chisel edge exits out from the other side of the workpiece. Sudden fall in the thrust force is experienced at this point. Nearly constant but relatively smaller value of thrust is measured for remaining part of the drilling [3]. The actual values and slopes of the thrust force variations depend upon the feed rate, spindle speed, workpiece material, tool geometry, tool material, coating, use of coolant, etc.

1.1.2 Torque

Torque on the drill is due to the component of drilling-induced force normal to the plane of axis of drill. Similar to the thrust force, torque is also constituted as sum of components of torque due to cutting force and frictional force normal to the plane of axis of drill. The cutting force varies along the primary cutting edge due to varying cutting angles and variation of cutting speed due to varying radius. For simplicity, the average force is assumed to be acting at the center of each of the cutting edges, i.e., at the half of the radius, $R/2$ as shown in Fig. 3.

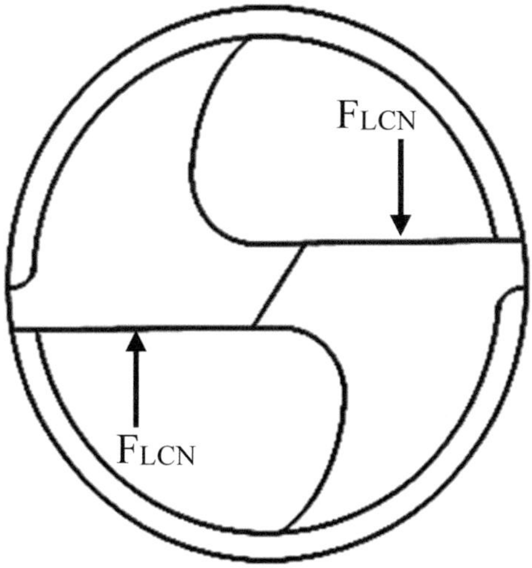

Fig. 3 Torque due to cutting action at cutting lips

Hence, the total torque due to the cutting force in normal direction at primary cutting edge T_{LCN} is given by [3]

$$T_{LCN} = 2 \times \frac{R}{2} \times F_{LCN} = RF_{LCN}, \quad (2)$$

where F_{LCN} is the average cutting force at each lip in the direction normal to axis.

The component due to friction at the chisel edge is given as, T_{CFN}, while components of torque due to the friction at the drilled surface and at the helical groove are T_{SFN} and T_{GFN}, respectively. Thus, the total torque, T, during drilling is given by [3]

$$T = T_{LCN} + T_{CFN} + T_{SFN} + T_{GFN} \quad (3)$$

The typical trend of variation of torque during drilling is shown in Fig. 4. Steep increase in torque is observed till the primary cutting edge gets fully engaged. Further increase in the torque is due to the frictional torque. The torque starts dropping once the chisel edge exits from the other end of the workpiece. Once the cutting edge exits completely then the torque remains constant for the remaining part of the drilling [3].

1.1.3 Radial Components of the Cutting Forces

Each lip has also a radial component of cutting force, F_{LCR} acting toward the axis of drill. The effect of the radial component cancels each other as they are equal and opposite as shown in Fig. 5.

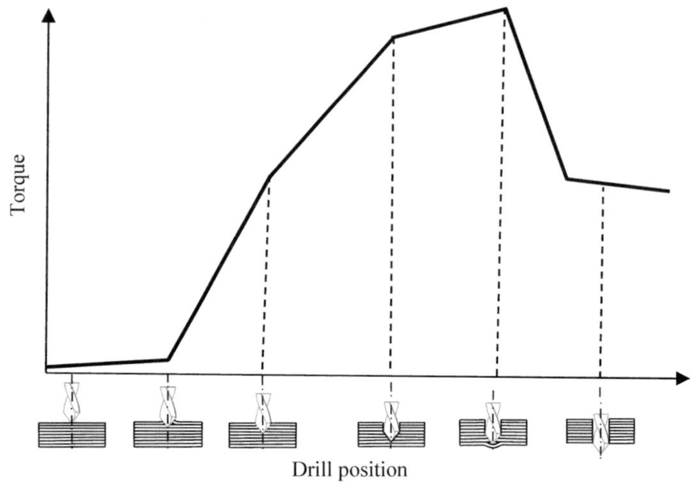

Fig. 4 Variation in torque during drilling

Fig. 5 Radial components of cutting forces on lips

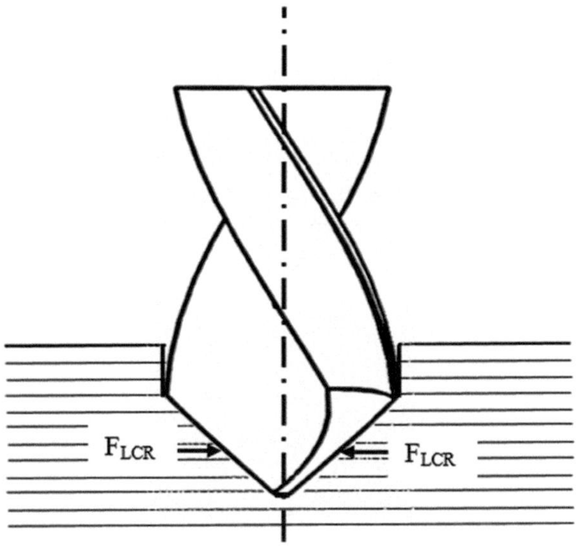

1.2 Problems Associated with Drilling of FRPs

FRP composites are manufactured to the near-net shape. Drilling is, however, required for the assembly purpose. Difficulties associated with drilling of FRP composites are quite different from those incurred during metal drilling. These are briefly discussed below.

1.2.1 Abrasive Tool Wear

A drill alternately comes in contact with the matrix and the fiber phase. The fibers are abrasive in nature. Abrasive nature of reinforcing fibers causes excessive wear of drill. This needs for frequent tool regrinding which causes loss of productive time.

1.2.2 Thermal Softening of Matrix

The rise in temperature and its effects are not significant while drilling with new or sharp drills. However, after drilling few holes, the drill wears out considerably. Drilling with such worn-out tool may cause excessive rise in temperature. Since the matrix as well as fiber are bad conductors of heat (only in case of FRP composites), the heat cannot be dissipated easily. This causes localized rise in temperature in a very close vicinity of the hole. This may result in thermal softening of the matrix material causing separation of fibers from the matrix. Matrix and fibers are brittle in nature. The machined hole surface may also be damaged due to this.

1.2.3 Matrix Cracking

The mechanism of material removal is due to the series of brittle fractures. This may cause brittle cracking in the matrix. Though these cracks do not propagate much deeper in the material, they reduce the strength of the laminate.

1.2.4 Fiber Pullout

The fibers sometimes stick to the cutting edge of the drill. If the matrix is softened due to thermal effects and the bonding between the fiber and matrix becomes weak, then it may cause the fibers to be carried away along with the cutting edge of the drill. This dislodging of the fiber causes damage to the hole surface as well as reduces the strength.

1.2.5 Fuzzing

Sometimes the fibers near the wall of the hole may bend under the action of drill. They remain uncut and get back to original position once the drill is removed. This causes poor surface finish of the drilled surface. This defect of uncut fibers around the drilled surface is called “fuzzing”.

1.2.6 Delamination

The heterogeneous nature of the work material imposes some problems in the workpiece. There is a possibility of breaking of the interlaminar bonds at entry and exit of the drill. This is called as "delamination". The delamination at the drill entry is called as "peel-up delamination" while that at the drill exit is called "pushout delamination".

Among the abovementioned defects, delamination has attracted major attention of the researchers as it reduces the strength of the workpiece significantly causing rejection of the components. For critical applications, such as aircrafts, almost 60% of the components are rejected due to delamination [4]. Drilling normally being the last operation just before assembly, all the costs associated with material and various processes required for manufacturing the components are wasted due to this.

1.3 Delamination Mechanism

Under the action of cutting forces, the layers of the laminate get separated due to their laminar nature resulting in delamination. Delamination is observed both at drill entry and at exit. But the mechanisms of both the types of delamination are different.

1.3.1 Peel-Up Delamination Mechanism

At the very first instant of drilling, the cutting edge tries to abrade the material coming in contact with it, thereby sweeping the material along with. The material flows over the flute surface. This abraded material is then removed in the form of chips. However, in case of FRP composites, a part of laminate is abraded. The flute exerts an upward force on the partially cut laminate while the drill thrust exerts a downward force on the uncut material. Unlike isotropic materials where the strength is same in all directions, the interlaminar bonds are weaker as compared to the in-plane strength of the laminate. The lamination away from the point of cutting also gets opened before the material is removed in the form of chip. This delamination may get extended beyond the hole diameter which results in peel-up delamination in the final drilled hole. The peripheral force exerted by the cutting lips is the main force responsible for the peel-up delamination. It is also a function of various geometrical parameters of the drill, especially the helix angle. The mechanism of peel-up delamination is demonstrated in Fig. 6a.

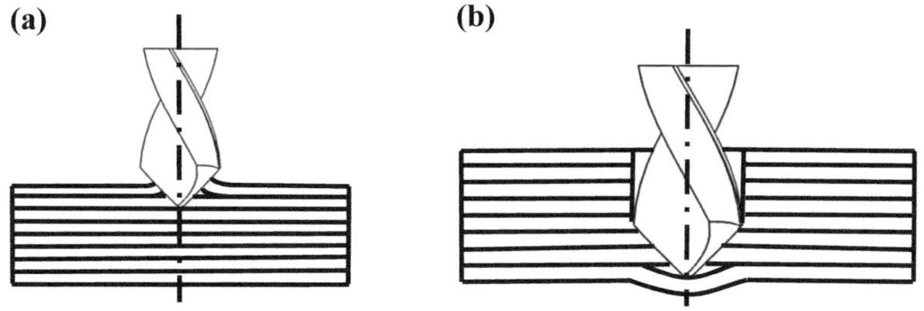

Fig. 6 Mechanism of **a** peel-up and **b** pushout delamination

1.3.2 Pushout Delamination

The axial thrust force always exerts a compressive force on the composite laminate under the drill tip. The laminate under the drill act as a simply supported plate with a central load acting on it. Initially, the bending strength of the laminate is sufficient to withstand the deformation due to thrust force and the deformation, if any, is experienced by the whole laminate under the drill. As the drill progresses, the uncut thickness of the laminate reduces. Especially at the exit side when very few layers are remaining, the resistance to the deformation is also reduced and more deformation is observed under the action of drilling thrust force. When this loading exceeds interlaminar bonding strength, the delamination occurs away from the point of drilling. If the delamination extends in the region beyond the maximum drill diameter, it remains present in the finally drilled hole as well. This reduces the strength of the parent material causing rejection of the component. The thrust force is the main reason behind the pushout delamination. The mechanism of pushout delamination is demonstrated in Fig. 6b.

2 Literature Review

Reduction in delamination during drilling of FRP composites has been a subject of interest for the researchers for long time. The approaches used by the researchers may broadly be classified under following three categories:

- Experimental Methods
- Analytical Methods
- Numerical Methods (Finite Element Method)

2.1 Experimental Methods

Experimental studies on drilling of FRP composites are very much popular. Most of the literature are available on experimental investigation of delamination of drilling of FRP composites. Experimental methods do not demand knowledge on micro-mechanism of the drilling process which is highly complex. They are well structured and help to arrive at optimum process parameters/conditions easily. Experimental studies are available on all three approaches, i.e., optimizing process parameters, tool geometries, and cutting conditions.

Bhatnagar et al. [5] studied the variation of thrust force and torque with a nondimensional parameter of ratio of cutting speed/feed rate for drilling of glass fiber reinforced plastic (GFRP) composites using four different drill bits, namely four-faceted, eight-faceted, parabolic, and Jodrill. Khashaba et al. [2] studied the effect of cutting speed, feed rate, and fiber volume fraction on delamination during drilling of GFRP composites. Linear regression relations were developed for torque and thrust force at each speed level. The thrust force and the torque were observed to be increasing with increase in fiber volume fraction and feed rate. However, with increase in cutting velocity, the thrust and torque were observed to be reduced. Taguchi experimental analysis technique and a multi-objective optimization were used by Enemuoh et al. [6] to study the effect of speed, feed rate, and point angle on delamination, damage width, surface roughness, and drilling thrust force. The thrust force was also found to be reduced with increase in speed [7]. Davim and Reis [8] have registered a contrary observation about the speed. They observed the thrust force increased with speed and feed rate. A regression correlation using multiple linear relation was tried by Davim and Reis [9] and Davim et al. [10]. Mohan et al. [11, 12] studied the influence of workpiece thickness, drill diameter, speed and feed rate on thrust force and torque using Taguchi's design of experiment. The response surface indicates that workpiece thickness has greater influence on both thrust force and torque, and both increase with increase in thickness. Thrust force was observed to be lower with lower drill size while smaller torque was observed with larger drill size [12]. A mathematical model using group method data handling (GMDH) was developed by Arul et al. [13]. Acoustic emission based analysis was studied by Ravishankar and Murthy [14, 15]. Four distinct zones, viz., contact friction zone, microchipping zone, piercing action zone, and peripheral friction zone, based upon acoustic emission, were identified during drilling process. Thrust force was observed to be increasing with increase in acoustic emission root mean square (AE-RMS). This correlation was suggested to be used for controlling the drilling-induced damage.

2.2 Analytical Methods

The linear elastic fracture mechanics (LEFM) based analytical model for critical thrust force at the onset of crack propagation was developed by Ho-Cheng and Dharan [16]. Classical plate bending theory was used. The boundary conditions were assumed to be clamped ends and the load was assumed to be concentrated at the center of the plate. The material was assumed to be homogeneous and isotropic. The material anisotropy was considered by Jain and Yang [17]. The critical thrust force was analytically estimated using the anisotropic elastic components making the model more general in nature. The shape of delamination zone was also predicted analytically. The delamination shape was found to be elliptical in case of unidirectional fiber-reinforced composites (UD-FRP) in agreement with the experimental observations. The authors suggested a strategy of variable feed rates to reduce the delamination. At the initiation of the drilling process when the uncut thickness of laminate below the drill is sufficient to bear higher load, a high feed rate was suggested to maximize the productivity. As the drill approaches the exit side of the laminate, the gradually reducing uncut layer thickness under the drill reduces the bending stiffness of the laminate. This results in excessive bending causing delamination. Lower feed rates have therefore been suggested in this region to minimize the thrust force and avoid delamination. In a latter study, Jain and Yang [18] used regression-based correlation for thrust force in terms of feed rate. Regression-based correlations help in capturing the implicit factors of the process which are not included in the analytical equations. The critical thrust equations were developed for both unidirectional (UD) and multidirectional (MD) FRP laminates.

An analytical model based on mechanistic approach and other well-established analytical models of metal drilling process have been developed by Chandrashekharan et al. [19]. The model is based on the geometric parameters of the process such as chip load, chip thickness and cutting angles. Thrust force and torque are predicted for the set of cutting parameters, i.e., feed rate and speed. Cutting mechanism at cutting lips and chisel edge is different. Different models have developed for the cutting lip and the chisel edge and were validated through experiments. In case of metals, an extrusion is observed at a small central portion at the chisel edge while orthogonal cutting with highly negative rake angle is observed at the remaining portion of chisel edge. FRP being brittle material, the extrusion may not occur. Orthogonal cutting over the whole chisel edge was assumed by the authors. This hypothesis could not be validated through experiments. A cutting force during drilling of FRP composites was also modeled analytically by Langella et al. [20]. The contribution over the cutting lips is modeled by integration over the cutting edge length. The cutting force at the chisel edge was calculated assuming a constant rake angle over the whole length. The contribution of chisel edge to the total cutting force was evaluated through developed analytical relations and experimental results. The effects of drill point angle and feed rate on delamination were studied. It was observed that the delamination increases proportionally with feed rate and point angle.

The location of delamination was identified through analytical model developed by Tsao and Chen [21]. Two different approaches were used for modeling: by considering the individual layers to be isotropic and then to be anisotropic. The location of delamination initiation was also found to be a function of ply thickness. For higher ply thickness, the location of delamination was at a higher height from bottom but the critical thrust force was higher.

Upadhyay and Lyons [22] considered a uniformly distributed load over the chisel edge. The critical thrust force was estimated considering the deformation to be small, i.e., without midplane stretching and also with large deformation, i.e., with midplane stretching. The comparison between the results with a localized concentrated point load and distributed force over the chisel edge and cutting lip has been presented by Frederic et al. [23]. The authors observed that the results of the analytical model considering the distributed load are closer to the experimental results compared to the predicted values from analytical model with concentrated point force.

An analytical model for estimating the critical thrust force resulting in delamination-free drilling of multidirectional FRP (MD-FRP) has been considered by Zhang et al. [24]. In case of multidirectional laminates, Zhang et al. [24] recorded that the delamination shape is elliptical while the orientation of the ellipse depends on the stacking sequence. The authors also noted that the critical thrust force is a function of ellipticity ratio. The models developed were applied to unidirectional laminate and unbalanced cross-ply laminate. Critical thrust force for MD angle-ply laminate was also derived by Pyo et al. [25]. In-plane stretching and transverse bending of the FRP lamina was considered. Thrust force was modeled as a concentrated point force and the delamination shape was assumed to be elliptical. Numerical results demonstrated that the critical thrust force increases with increasing ply thickness and fiber volume fraction. The critical thrust force was observed to be initially increasing with angle of fiber orientation between consecutive plies. This then reduces with further increase in this angle. Maximum value of critical thrust force was observed at inter-ply angle of $\pm 33.5^{\circ}$. The authors have also developed an optimization methodology for the design of asymmetric angle-ply laminate for maximum critical thrust force. The automated design procedure can derive design variables such as number of fibers per millimeter, fiber diameter, and interlaminar angle for maximum value of critical thrust force.

Vibration-assisted drilling of FRP composite was modeled analytically by Zhang et al. [26]. The main characteristic of vibration-assisted drilling is the dynamic variation in uncut chip thickness and cutting angles resulting in variations in cutting forces. Separate models were developed for vibratory cutting at the chisel edge and vibratory cutting at the lips. Experimental procedure was suggested for determination of the shear flow stress which depends upon cutting parameters and material. The reduction of 20–30% was observed in thrust force using vibratory-assisted drilling compared to the conventional drilling of FRP composites.

The effect of chisel edge and pilot hole during drilling of FRP composites was studied by Tsao and Hocheng [27]. A clamped plate with a central hole represents the laminate with pilot hole. No effect on thrust force was observed with change in feed rate while drilling the specimen with a pilot hole. Tsao and Hocheng [27] have also developed an analytical model for drilling with a pilot hole. The role of chisel edge is critical in case of delamination during drilling of FRP composites. A critical ratio of chisel edge to drill diameter has been derived.

The basic analytical model was modified by Hocheng and Tsao [28] for delamination analysis during drilling of FRP composites with various drill bits. Models were developed for twist drill, saw drill, core drill, step drill, and candlestick drill.

While most of the analytical models are based on LEFM, Kim and Lee [29] used energy balance equations to analytically model the critical thrust force for delamination-free drilling of FRP composites. The numerical results were compared with that obtained using the previous theories. It was observed that in the model presented, the thrust was over estimated.

Tsao and Hocheng [30] developed analytical model for understanding the effect of eccentricity for twist drill and candlestick drill. The results were compared with those obtained using drills having perfect geometry. For both twist and candlestick drills, the critical thrust force was observed to be reduced with increase in eccentricity. Similarly, the deviation may be observed in saw drill during its regrinding. The deviation obviously changes the location of cutting forces. An analytical model to study the effect of deviation in saw drill geometry due to regrinding was developed by Tsao [31]. The delamination damage was increasing with regrinding of special drill bits.

The effect of backup plate during drilling of FRP composites using saw drill and core drill was modeled analytically by Tsao and Hocheng [32]. The critical thrust force was observed higher after application of a backup plate during drilling with both saw and core drills. Hocheng and Tsao [33] presented analytical models for estimation of critical thrust force for a conventional twist drill and special geometry drills: saw drill, candlestick drill, core drill, core-center drill, core saw drill, and step drill. The effects of pilot hole and backup plate were also presented. Analytical model was developed by Tsao [34] for estimation of critical thrust force in drilling with core drill. Numerical and experimental study was conducted to understand the effect of ratio of pilot hole to drill diameter on critical thrust force. Drilling thrust force was observed to be reduced with increasing ratio of pilot hole to drill diameter.

Sedlacek and Humar [35] studied the fracture mechanism of delamination during drilling of FRP composite by considering the effect of chisel edge action and the cutting lips action separately. The uncut laminate was assumed to be isotropic. The effect of tool wear on delamination was also studied. Chisel edge was found to have maximum effect on delamination. Tsao and Hocheng [36] developed an analytical model to study the effect of tool wear on delamination. A concentrated point force was assumed in case of a fresh drill while a distributed force over a chisel edge was considered for a worn-out chisel edge. Numerical results from the analytical model with and without wear effect demonstrate reduction in critical thrust force with increase in tool wear.

2.3 Finite Element Method

The first finite element study on delamination during drilling of FRP composites, found in the literature, was by Sadat et al. [37]. A two-dimensional axisymmetric finite element model was developed to study the delamination. The tool was modeled as a rectangular block. The workpiece was clamped at the outer surface and a downward force was applied on the tool. Material anisotropy was accounted for using the elastic properties in various directions. Two different situations: with an aluminum backing plate and without any backing plate were analyzed. Finite element results confirmed that the delamination and deflection was reduced with the presence of backing plate. A three-dimensional finite element model was used by Bhattacharya and Horrigan [38]. Quarter portion of the plate and hole was modeled using 20-noded brick elements with symmetric boundary conditions accounting for the whole model. The drilled hole was assumed to be present up to the desired level where the initiation of delamination was intended to be studied. The drill thrust was modeled using uniformly distributed load. At the bottom of plate, two boundary conditions were analyzed: unsupported corresponding to without backing plate situation and supported with low elasticity corresponding to backing plate which is normally a soft material to avoid damage to drill. Through the stress distribution, efforts were made to calculate and compare the stress concentration factors in supported and unsupported condition. The stress concentration factor in unsupported condition was observed to be higher than that in case of supported. Effect of fiber volume fraction was analyzed through finite element analysis using commercial software Ansys by Budan et al. [39]. The two-dimensional axisymmetric model was developed similar to the model discussed by Sadat et al. [37]. The apparent properties of cut and uncut laminates were used. The tool was modeled as a rectangular block. Only thrust force was applied on the tool. The critical thrust force was observed to be reducing with increasing fiber volume fraction.

Finite element model for drilling of CFRP composites was developed by Durao et al. [40] by using ABAQUS software. The workpiece was modeled with a partial hole present. Tool was modeled as a rigid body. Two different drills: conventional twist drill and candlestick drill (called as C-shape drill) were modeled. The twist drill was modeled as a pointed indenter and the candlestick drill was modeled as a pointed indenter with peripheral cylindrical tube. The interface between the FRP layers was modeled using cohesive elements. Tool rotation was not considered in this case. The estimated value of the maximum thrust force during drilling was compared with the experimental values and was found in good agreement. A mixed mode failure was considered for damage modeling of cohesive zone in finite element study by Durao et al. [40]. The critical thrust values from various available analytical models and those predicted by finite element model were compared. The tool in this case is modeled similar to the earlier work by Durao et al. [40], as a rigid body and in the form of pointed indentor. The rotation of tool was not considered. The workpiece was modeled with an existing partial hole. Singh et al. [41] considered a twist drill, modeled using ProE software, used for finite element simulation of drilling of UD-FRP composites. Failure model was developed based on Tsai-Wu failure criterion. Effect of point angle on damage during drilling was observed. The damage was observed increasing with increase in point angle. The workpiece specimen was assumed to have a pilot hole. Drill rotation was not considered. Virtual crack extension (VCE) based finite element model was developed by Rahme et al. [42]. A preexisting circular crack was assumed to be present in the workpiece. The energy release rate was used to calculate the critical axial force. Critical thrust force values predicted by the model were compared with the values predicted by various load conditions in analytical model developed using classical plate theory.

A finite element model for critical thrust force during drilling of FRP composites was developed by Zitoune and Collombet [43] by using software code SAMCEF FE. The workpiece was modeled in the form of partially drilled hole with few layers remaining uncut. The bottom of the hole was modeled using various drill geometry parameters such as cone angle and chisel edge length. A crack was assumed at the bottom of the drilled surface. The total thrust force applied through drill was estimated using the analytical model. Since the total thrust exerted by the drill can be divided into component due to chisel edge and component due to cutting lips, accordingly the load was applied on the corresponding areas. Tsai-Wu criterion was used to quantify the delamination damage under the action of thrust force. The developed model was found suitable for UD laminates but could not be established successfully for woven fiber composites.

Finite element model using Ansys workbench was developed by Chakladar et al. [44]. The twist drill was modeled and used in the simulation. The workpiece was modeled as a rectangular plate. The variation of drill thrust force with respect to speed and feed was studied. The finite element predictions were found in good agreement with the experimental results. Drilling of CFRP composite using a twist drill was modeled by Isbilir and Ghassemieh [45] using finite element software ABAQUS/Explicit. The workpiece was modeled in the form of rectangular plate with fixed from bottom surface. The thermal effects were neglected. Damage modeling was developed based on Hashin's criteria. Failure in any of the four modes of failure as per the Hashin's criteria was used for sudden damage initiation and deletion of element for intra-laminar failure. The interface was modeled with cohesive elements and progressive damage was used for interface. Effect of feed rate and speed was studied for delamination damage through finite element simulations and experiments. Effect of drill geometry for step drill was studied by using finite element model in one another study by Isbilir and Ghassemieh [46]. The drill geometry was modeled using a CAD software SolidWorks and imported in the CAE software ABAQUS. The workpiece and boundary conditions were modeled similar to the earlier study by the same authors [45]. Effect of the step ratio on thrust force and peel-up and pushout delamination was studied. The both peel-up and pushout delamination were observed reducing with increase in step ratio. Finite element study on delamination during drilling of CFRP composites was conducted by Phadnis et al. [47]. The workpiece was modeled with mesoscale layered construction with equivalent homogenous laminate properties. The interface was modeled using cohesive elements. Hashin's criterion was used for damage modeling of fibers while the Puke's failure criterion was used for modeling damage of matrix. A VUMAT subroutine was used to incorporate the failure model and the elements fulfilling the failure criterion were deleted to simulate the material removal process. Effects of feed rate on thrust force, torque, and delamination during drilling were studied using finite element simulations. The results for peel-up delamination were fairly accurate but pushout delamination results were overestimated in shape and size.

3 Finite Element Formulation

Finite element analysis being a numerical tool and iterative process, there are many implicit and explicit parameters which are to be selected properly in order to get accurate solutions. In the following sections, the details about the finite element formulation for simulating drilling of FRP composites are discussed. The finite element analysis tool used for the analysis was the commercial FEM software ABAQUS 6.10.

3.1 Modeling of Twist Drill

Twist drill is a complex geometry. In general, FEA software is limited in modeling capabilities. Hence, it is advisable to model the geometry in a CAD software and then import the geometric CAD model in the FEA software. However, some losses are always associated with such transfer of models from one software to another. Such loss is predominantly more in case of complex geometries such as twist drill, where the number of CAD surfaces is large. Hence, it was decided to model the twist drill in ABAQUS. The development of the geometric CAD model of a twist drill is based on the concept of its manufacturing steps. The flute section of the twist drill is generated as a result of milling with drill blank translation and rotation about the drill axis simultaneously. The cross section is dependent on the cross section of the milling cutter and the grinding wheel. However, it is generated in such a way that after grinding of the flanks the secondary cutting edges are straight. Figure 7 shows the cross section of typical drill.

The coordinate axis is fixed in such a way that cross section of the flute is on the X-Y plane while the axis of the drill coincides with the Z axis. There are two distinct regions in the cross section of a twist drill. From point 1 to point 2, and from point 4 to point 5 are the portions of the drill blank and obviously are the arcs of a circle with diameter equal to the diameter of the twist drill. Point 2 to point 4 via point 3 and point 5 to point 1 via point 6 are the sections generated by milling as discussed above. This section actually makes the flute surface. They are symmetric, and hence

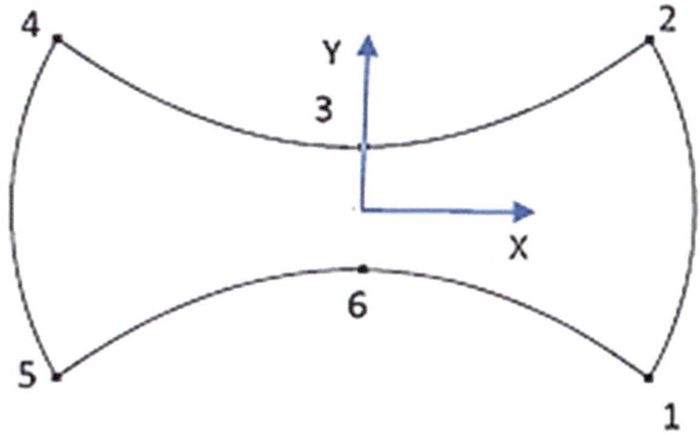

Fig. 7 Flute cross section

for simplicity only one of the sections, i.e., from point 3 to point 2 is defined by a polar equation, as follows [48]:

$$x = r\cos\xi \tag{4}$$

$$y = r\sin\xi, \tag{5}$$

where

$$\xi = \sin^{-1}\frac{w}{r} + \frac{\sqrt{r^2 - (w)^2}}{r}\tan\varphi\cot\left(\frac{\rho}{2}\right), \tag{6}$$

where w is the half-web thickness; φ is the helix angle and ρ is the point angle.

The Section 3 to 2 can be generated by varying the value of r from w to R, where R is the radius of the drill.

The cross section is extruded through the Z axis with a helical sweep. The pitch p is calculated as

$$p = 2\pi R\tan\varphi \tag{7}$$

Two conical sweep cuts are used for generating the flank surfaces. These are similar to the two grinding cones used for the flank grinding during manufacturing of the drill bits. The plane containing axes of the grinding cones is offset from the plane of axis of the drill by a distance $S = w$ on either side of the midplane as shown in Fig. 8. The axis of grinding cone makes an angle ϕ with the axis of drill while the half cone angle is θ. The angles ϕ and θ are chosen in such a way that $\theta + \phi = \frac{\varrho}{2}$, where ρ is the point angle [49, 50].

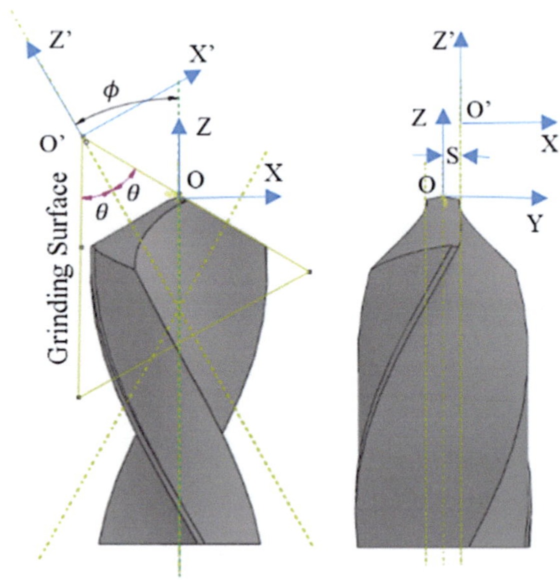

Fig. 8 Grinding cone for twist drill

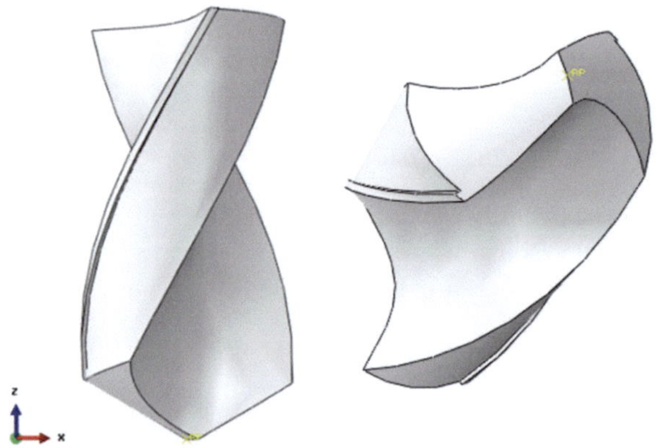

Fig. 9 Twist drill model in ABAQUS

This results into generation of the secondary cutting edge, called as the chisel edge at an angle ξ which is given by

$$\xi = \pi - \tan^{-1}\left(\frac{\sqrt{\tan^2\theta d^2 - S^2} - \tan^2\theta.d.\sin\phi}{S}\right), \tag{8}$$

where d is the offset of the apex of grinding cone from the drill tip along the grinding cone axis [48]. The geometric CAD model of the drill by using ABAQUS 6.10 is shown in Fig. 9.

3.2 Workpiece Modeling

The workpiece is modeled in the form of a circular disk with diameter 30 mm and thickness 3 mm. For simulation of the delamination process, the workpiece is to be modeled in the form of laminates. The individual lamina is modeled by partitioning. However, as the possibility of delamination is more in the bottom few layers and hence the workpiece was partitioned only for one-third of the thickness from the bottom side. An individual layer was assumed to be of 0.2 mm. An interfacing layer was modeled as a partition of 0.05 mm thickness in-between the two adjacent layers corresponding to the matrix rich region. The 3D model of workpiece, as modeled in ABAQUS 6.10, is shown in Fig. 10.

3.3 Material Modeling

Material modeling in finite element formulation refers to the modeling of the material constitutive relations, governing equations and defining the damage models. In this study, the work material is GFRP. This is inherently heterogeneous and anisotropic. The FEM tool used in the present work, i.e., ABAQUS 6.10 and a macroscopic modeling technique is used for modeling of the FRP workpiece. This will reduce the computational efforts considerably without compromising with the accuracy of the solution. A hybrid approach is also used for modeling the interface

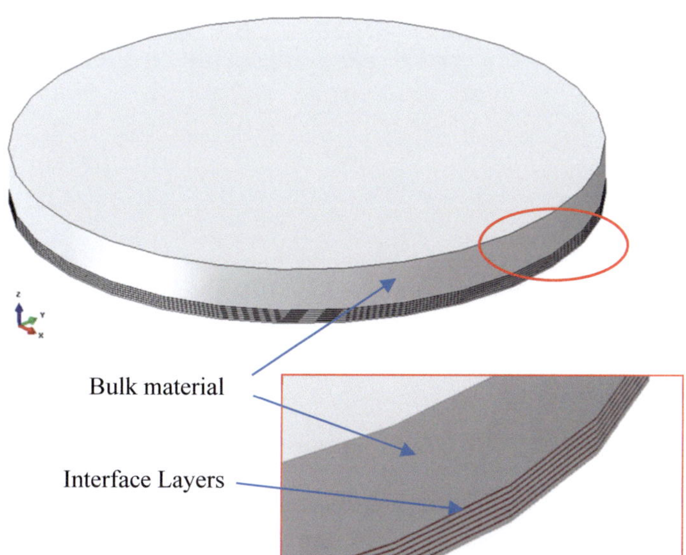

Fig. 10 Workpiece model in ABAQUS

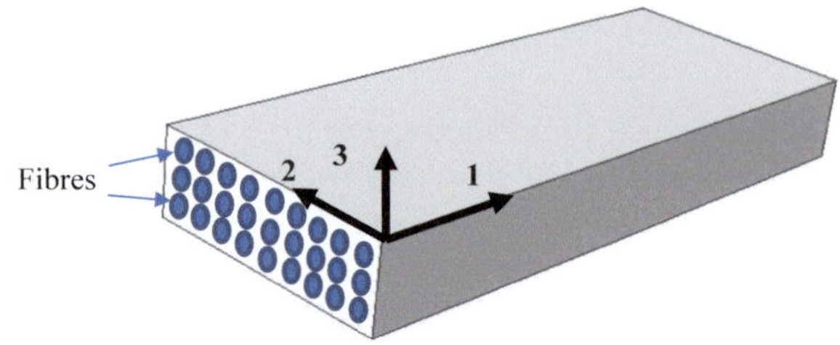

Fig. 11 Directions for laminate

between the consecutive layers especially near the exit side of the laminate where the possibility of delamination is the maximum. An equivalent homogeneous material cell is assumed with the overall material properties as that of the laminate. The laminate is assumed to be orthotropic with direction 1 as fiber direction and plane 1–2 parallel to the direction of layup, i.e., direction 3 normal to the layup as shown in Fig. 11. The interface in the region near the drill exit is modeled using a cohesive zone.

3.3.1 Material Constitutive Relation for the Bulk Material

The linear orthotropic elasticity of the material is defined using the nine elastic constants, viz., the three moduli of elasticity E_1, E_2, E_3, three Poison's ratios $\nu_{12}, \nu_{13}, \nu_{23}$, and three moduli of rigidity G_{12}, G_{13}, G_{23}. The stress–strain relations, in terms of these moduli for an orthotropic material, are given by Eq. 9.

$$\begin{Bmatrix} \varepsilon_{11} \\ \varepsilon_{22} \\ \varepsilon_{33} \\ \gamma_{12} \\ \gamma_{13} \\ \gamma_{23} \end{Bmatrix} = \begin{bmatrix} 1/E_1 & -\nu_{21}/E_2 & -\nu_{31}/E_3 & 0 & 0 & 0 \\ -\nu_{12}/E_1 & 1/E_2 & -\nu_{32}/E_3 & 0 & 0 & 0 \\ -\nu_{13}/E_1 & -\nu_{23}/E_2 & 1/E_3 & 0 & 0 & 0 \\ 0 & 0 & 0 & 1/G_{12} & 0 & 0 \\ 0 & 0 & 0 & 0 & 1/G_{13} & 0 \\ 0 & 0 & 0 & 0 & 0 & 1/G_{23} \end{bmatrix} \begin{Bmatrix} \sigma_{11} \\ \sigma_{22} \\ \sigma_{33} \\ \sigma_{12} \\ \sigma_{13} \\ \sigma_{23} \end{Bmatrix} \tag{9}$$

The Poison's ratio ν_{ij} physically means the transverse stress in the jth direction when the material is pulled in the ith direction. It is not necessary that $\nu_{ij} = \nu_{ji}$ but the relation $\nu_{ij}/E_i = \nu_{ji}/E_j$ has to be satisfied. The actual material properties used for modeling the workpiece material are listed in Table 1.

3.3.2 Damage Modeling for FRP Composites

Damage model refers to defining the mode of failure and the failure criteria for the material under the action of specified load conditions. There are mainly two types of failures observed in FRP composites: Intra-ply failures such as fiber-matrix debonding, fiber fracture, fiber pullout, and matrix cracking and inter-ply failure

Table 1 Material properties of FRP used for simulation

Stiffness parameters		
$E_1 = 48$ GPa	$G_{12} = 6.0$ GPa	$\nu_{12} = 0.25$
$E_2 = 12$ GPa	$G_{13} = 6.0$ GPa	$\nu_{13} = 0.25$
$E_3 = 12$ GPa	$G_{23} = 3.7$ GPa	$\nu_{23} = 0.42$
Strength parameters		
$X_t = 550$ MPa	$X_c = 325$ MPa	$S_{xy} = 76$ MPa
$Y_t = 34$ MPa	$Y_c = 128$ MPa	$S_{yz} = 52$ MPa
$z_t = 34$ MPa	$Z_c = 128$ MPa	$S_{xz} = 46$ MPa

such as delamination [51]. The intra-ply failure is modeled using the failure criteria for the laminate. The failure criteria are classified into two main categories, viz., microscopic criteria and macroscopic criteria. In microscopic analysis, the matrix and fiber phase are modeled separately. The failure criteria based on microscopic analysis is also same as homogeneous material failure criteria such as maximum principle stress criteria, maximum strain criteria, etc. Macroscopic criteria for modeling damage in FRP composites are further classified in two main categories, viz., energy-based criteria and phenomenological criteria. The criteria such as Tsai-Wu criteria and Tsai-Hill criteria are energy based while the criteria such as Hashin's criteria and Puck's criteria are phenomenological criteria. In the present study, Tsai-Hill criteria are used.

Tsai-Hill Criteria

This criterion is based on distortion energy failure theory given by von-Mises for isotropic materials. The total strain energy can be divided into two components: dilation energy responsible for bringing out the change in volume, and distortion energy responsible for change in shape. It is assumed that the failure of material takes place when distortion energy is higher than the failure distortion energy of the material. According to Tsai-Hill criteria, the material is assumed to be failed if the following condition is met [52]:

$$\left(\frac{\sigma_1}{X}\right)^2 - \left(\frac{\sigma_1 \sigma_2}{X^2}\right) + \left(\frac{\sigma_2}{Y}\right)^2 + \left(\frac{\tau_{12}}{S_{xy}}\right)^2 \geq 1, \tag{10}$$

where X and Y are the ultimate tensile/compressive strengths in longitudinal and transverse directions, respectively. In Eq. 10, S_{xy} is the ultimate shear strength. A typical safe/failure envelope for Tsai-Hill failure criteria under biaxial loading is shown in Fig. 12.

Rao et al. [53] used modified three-dimensional Tsai-Hill failure criteria for modeling the machining process of unidirectional FRP composite. The modified criteria were used through a user-defined subroutine in ABAQUS 6.10. The failure of material is based upon the value of failure index given by

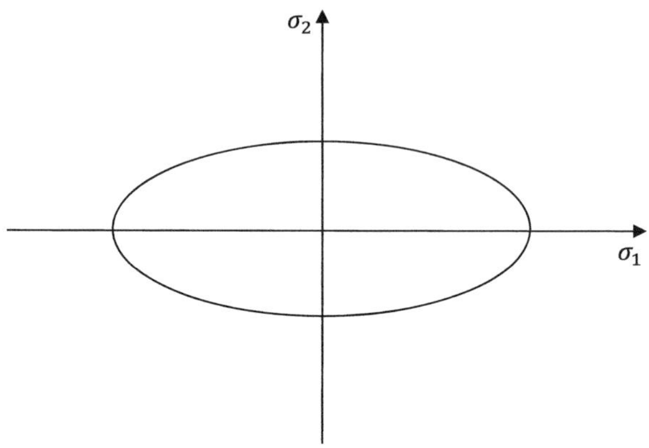

Fig. 12 Safe envelope for Tsai-Hill criteria

$$\begin{aligned} \mathrm{FI} = & \left(\frac{\sigma_{11}}{X}\right)^2 + \left(\frac{\sigma_{22}}{Y}\right)^2 + \left(\frac{\sigma_{33}}{Z}\right)^2 - \sigma_1\sigma_2\left(\frac{1}{X^2} + \frac{1}{Y^2} - \frac{1}{Z^2}\right) \\ & - \sigma_2\sigma_3\left(\frac{-1}{X^2} + \frac{1}{Y^2} + \frac{1}{Z^2}\right) - \sigma_1\sigma_3\left(\frac{1}{X^2} - \frac{1}{Y^2} + \frac{1}{Z^2}\right) \\ & + \left(\frac{\tau_{12}}{S_{xy}}\right)^2 + \left(\frac{\tau_{13}}{S_{xz}}\right)^2 + \left(\frac{\tau_{23}}{S_{yz}}\right)^2, \end{aligned} \quad (11)$$

where $\sigma_{11}, \sigma_{22}, \sigma_{33}$ are the induced direct stresses, $\tau_{12}, \tau_{13}, \tau_{23}$ are the induced shear stresses; X, Y, Z are the ultimate tensile/compressive strengths in the three principle directions, respectively, and S_{xy}, S_{yz}, S_{xz} are the shear ultimate strengths.

In the present work, the modified three-dimensional Tsai-Hill criteria is used for damage initiation in the FRP material to model the drilling process. However, in ABQUS 6.10 only one damage model, i.e., Hashin's damage model is available to model damage in FRP composites. A user subroutine was written by using Intel FORTRAN-10.1 for incorporating the user-defined damage model based on Tsai-Hill's failure theory. A solution-dependent state variable (SDV) is created for failure index (FI) as defined by Eq. 11. This is done using a user-defined subroutine VUSDFLD. The details about the subroutine are discussed in the following session.

Inter-Ply Failure: Delamination

Another major mode of failure in FRP composites is the inter-ply failure, i.e., delamination. Delamination is assumed to be one of the major defects during drilling of FRP composites and also is responsible for majority of rejections. Hence, the main focus of the present work is on the study of the delamination process. A special zone, called as cohesive zone, is modeled in between each consecutive plies. Based on the constitutive response of the cohesive zone, there are mainly three different modeling techniques used for modeling of cohesive zones [54].

- Continuum-based modeling
- Traction-separation based modeling
- Laterally unconstrained adhesive patches

Continuum-based modeling: It is used when the cohesive zone has a finite thickness such as in case of joining of two surfaces with the help of adhesive material like glue. The thickness, stiffness, and strength cohesive zone can be estimated using experimental methods. In case of continuum modeling, one direct stress along through thickness direction and two shear stresses mutually perpendicular and along the plane of adhesive are present.

Traction-separation based modeling: This modeling technique is preferred in applications where the cohesive zone has a negligible thickness. In such case, the macroscopic material properties such as stiffness and strengths of the adhesive material are not important. The damage analysis is based on the fracture mechanics, i.e., based on the energy required for creation of new surfaces. The behavior of the interface prior to initiation of damage is often described as linear elastic in terms of a penalty stiffness that degrades under tensile and/or shear loading but is unaffected by pure compression. One good part of this type of analysis is that it is not necessary to model any crack at the beginning of the analysis. Rather the precise locations where cracks will initiate are determined as a part of the solution.

Laterally unconstrained adhesive patches: This approach is mainly suitable for modeling joints with gaskets. The macroscopic properties of the gasket, such as strength and stiffness, are used for the analysis. Only unidirectional stress along the through thickness direction is considered in the analysis. The nonlinear and hyper elastic behavior of the materials used for gaskets such as rubber, foam, etc., can be captured in the constitutive relations used for the modeling techniques for laterally unconstrained adhesive patches.

Constitutive Behavior for Traction-Separation Modeling of Cohesive Zone

During modeling of the joints under traction-separation modeling technique, prior to damage initiation, linear elastic behavior is assumed. In case of three-dimensional problems, there are three components of stresses: normal stress along the through thickness direction, i.e., direction 3 and two shear stresses along direction 1 and 2. The linear elastic behavior prior to the initiation of damage can thus be governed by the constitutive relations as given by Eq. 12.

$$
\begin{gathered}
\sigma = D\varepsilon \\
\begin{Bmatrix} \sigma_{\mathrm{n}} \\ \sigma_{\mathrm{s}} \\ \sigma_{\mathrm{t}} \end{Bmatrix} = \begin{Bmatrix} D_{\mathrm{nn}} & D_{\mathrm{ns}} & D_{\mathrm{nt}} \\ D_{\mathrm{sn}} & D_{\mathrm{ss}} & D_{\mathrm{st}} \\ D_{\mathrm{tn}} & D_{\mathrm{ts}} & D_{\mathrm{tt}} \end{Bmatrix} \begin{Bmatrix} \varepsilon_{\mathrm{n}} \\ \varepsilon_{\mathrm{s}} \\ \varepsilon_{\mathrm{t}} \end{Bmatrix},
\end{gathered}
\tag{12}
$$

where σ_n is the normal stress along the local direction 3, i.e., the through thickness direction and σ_s and σ_t are the shear stress components along the local directions 1 and 2, respectively. K_{ij}s are the corresponding stiffness values. ε_n is the normal strain along the local direction 3, i.e., the through thickness direction, and ε_s and ε_t are the shear strain components along the local directions 1 and 2, respectively and

$$D = \begin{Bmatrix} D_{nn} & D_{ns} & D_{nt} \\ D_{sn} & D_{ss} & D_{st} \\ D_{tn} & D_{ts} & D_{tt} \end{Bmatrix}$$

The delamination damage is a gradual process rather than a sudden failure. The definition of failure mechanism consists of three parts: a damage initiation criterion, a damage evolution law, and a choice of element removal (or deletion) upon reaching a completely damaged state.

Damage initiation: A number of criteria are available to quantify the damage initiation. Few of the commonly used criteria are discussed here.

Maximum nominal stress criteria: Damage is assumed to initiate when the maximum nominal stress ratio is equal to or greater than one in any of the mode at any point of time. This criterion can be represented as [1]

$$\mathrm{Max}\left\{\frac{\sigma_n}{\sigma_n^0}, \frac{\sigma_s}{\sigma_s^0}, \frac{\sigma_t}{\sigma_t^0}\right\} \geq 1, \tag{13}$$

where σ_n^0, σ_s^0 and σ_t^0 are the maximum values of the normal stress and two shear stresses, respectively, without damage under pure mode of loading. It must be noted that no damage is initiated if σ_n is compressive in nature.

Maximum nominal strain criteria: Damage is assumed to initiate when the maximum nominal strain ratio is equal to or greater than one in any of the modes at any point of time. This criterion can be represented as [1]

$$\mathrm{Max}\left\{\frac{\varepsilon_n}{\varepsilon_n^0}, \frac{\varepsilon_s}{\varepsilon_s^0}, \frac{\varepsilon_t}{\varepsilon_t^0}\right\} \geq 1, \tag{14}$$

where $\varepsilon_n^0, \varepsilon_s^0$ and ε_t^0 are the maximum values of normal strain and two shear strains respectively, without damage under pure mode of loading.

Damage evolution law: Once the damage criteria are met in any one of the modes then the stiffness starts degrading causing a gradual failure. This is also called as softening. The evolution laws define the process of degrading the stiffness. The relationship between stress and relative displacement during this phase is given by

$$\sigma = (I - E)D\delta, \tag{15}$$

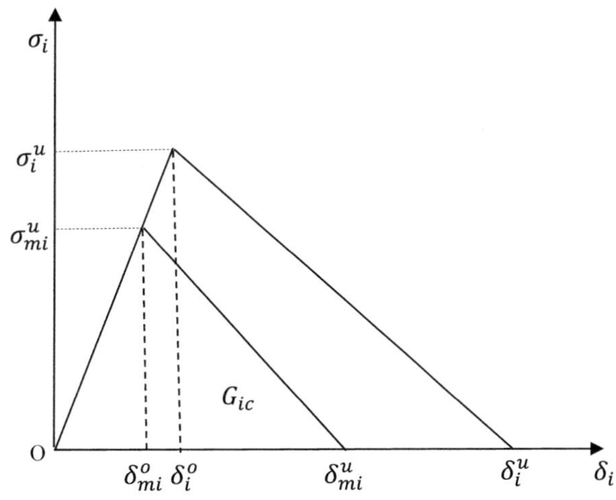

Fig. 13 Mixed mode of damage evolution

where I is an identity matrix and E is a diagonal matrix containing the damage variables given by [1]

$$e_i = \frac{\delta_i^u\left(\delta_i - \delta_i^0\right)}{\delta_i\left(\delta_i^u - \delta_i^0\right)}, \tag{16}$$

where δ_i^0 is the displacement corresponding to the onset of damage in pure mode of loading. δ_i^u is the maximum relative displacement when complete failure occurs and can be obtained by equating the area under the softening curve with the critical fracture energy in the respective mode given by Eq. 16. Figure 13 shows the critical fracture energy in pure mode and mixed mode of failure [1], which is calculated as

$$G_{ic} = \frac{1}{2}\sigma_i^u \delta_i^u \tag{17}$$

The above equation is valid when the pure mode of failure is observed. However, in most of the practical situations, like in the one under study, all the three modes of fracture are simultaneously contributing to the failure. A mixed mode of damage evolution using a quadratic stress based criteria is given as [1]

$$\begin{aligned} \left(\frac{\sigma_I}{\sigma_I^u}\right)^2 + \left(\frac{\sigma_{II}}{\sigma_{II}^u}\right)^2 + \left(\frac{\sigma_{III}}{\sigma_{III}^u}\right)^2 &= 1 \quad \text{if } \sigma_I \geq 0 \\ \left(\frac{\sigma_{II}}{\sigma_{II}^u}\right)^2 + \left(\frac{\sigma_{III}}{\sigma_{III}^u}\right)^2 &= 1 \quad \text{if } \sigma_I < 0 \end{aligned} \tag{18}$$

Two different equations are needed to prevent the contribution of compressive normal stress, σ_I, to the damage criteria.

The criteria based on maximum relative displacement is given by [1]

$$\left(\frac{\delta_I}{\delta_I^u}\right)^2 + \left(\frac{\delta_{II}}{\delta_{II}^u}\right)^2 + \left(\frac{\delta_{III}}{\delta_{III}^u}\right)^2 = 1 \tag{19}$$

Equivalent mixed mode displacement, δ_{m}, is given as

$$\delta_{\mathrm{m}} = \sqrt{\delta_I^2 + \delta_{II}^2 + \delta_{III}^2} \tag{20}$$

The mixed mode ratios, β_is, are given as

$$\beta_i = \frac{\delta_i}{\delta_I} \tag{21}$$

The criteria based on fracture energy for the damage propagation is given by [1]

$$\left(\frac{G_I}{G_{I\mathrm{c}}}\right) + \left(\frac{G_{II}}{G_{II\mathrm{c}}}\right) + \left(\frac{G_{III}}{G_{III\mathrm{c}}}\right) = 1 \tag{22}$$

where $G_{I\mathrm{c}}$ = Critical energy release rate per unit area in mode I; $G_{II\mathrm{c}}$ = Critical energy release rate per unit area in mode II and $G_{III\mathrm{c}}$ = Critical energy release rate per unit area in mode III.

The fracture energies in each mode can be calculated based on the area under the modified stress/relative displacement curve under mixed mode and is given by

$$G_{i\mathrm{c}} = \frac{1}{2}\sigma_{mi}^u \delta_{mi}^u, \tag{23}$$

where δ_{mi}^u is the ultimate relative displacement at the complete failure; and i varies from 1 to 3.

The relative displacements under the mixed mode at the onset of damage δ_{mi}^0 and ultimate relative displacement at the complete failure δ_{mi}^u are given by [1]

$$\begin{aligned} \delta_{mi}^0 &= \frac{\beta_i \delta_m^0}{\sqrt{1 + \beta_{II}^2 + \beta_{III}^2}} \\ \delta_{mi}^u &= \frac{\beta_i \delta_m^u}{\sqrt{1 + \beta_{II}^2 + \beta_{III}^2}}, \end{aligned} \tag{24}$$

where β_i represents the mode ratios and δ_m^0 and δ_m^u are the mixed mode relative displacements onset to the damage and at the ultimate failure, respectively, and are given by [1]

$$\delta_m^0 = \delta_I^0 \delta_{II}^0 \delta_{III}^0 \sqrt{\frac{1 + \beta_{II}^2 + \beta_{III}^2}{\left(\delta_{II}^0 \delta_{III}^0\right)^2 + \left(\beta_{II} \delta_I^0 \delta_{III}^0\right)^2 + \left(\beta_{III} \delta_I^0 \delta_{II}^0\right)^2}}$$

$$\delta_m^u = \frac{2\left(1 + \beta_{II}^2 + \beta_{III}^2\right)}{\delta_m^0} \left[\frac{1}{G_{Ic}} + \frac{\beta_{II}^2}{G_{IIc}} + \frac{\beta_{III}^2}{G_{IIIc}}\right]^{-1} \tag{25}$$

3.4 User Subroutine "VUSDFLD"

Whenever a material model, depicting the complex behavior of the material, is not readily available in the software, it is then to be defined in a material definition. The user needs to define the material behavior using user subroutine in the material definition module of the software. There are two popularly used user subroutines for defining the material behavior in ABAQUS, namely "UMAT" or "VUMAT" and "USDFLD" or "VUSDFLD". Subroutines UMAT and USDFLD are used in case of ABAQUS/Standard while VUMAT and VUSDFLD are used in case of ABAQUS/Explicit. The subroutines UMAT or VUMAT are used for defining the material constitutive relations if none of the readily available material models, plasticity models or failure models are suited for the material under study. Standard set of output variables is available in ABAQUS. Subroutines USDFLD or VUSDFLD provides the user an option to define new variables. The material properties may be defined as a function of user-defined field variable which is calculated based on the output values from previous increment. Additional subroutines may also be written to correlate the behavior of the material and define the material properties based on user-defined field variables. The user-defined field variables can be defined at every integration point within the element. However, the subroutines USDFLD and VUSDFLD are not available for all types of elements. It can only be used with the elements for which the output variables can be retrieved from the previous increment using functions such as GETVRM. During the element definition if "User-defined field" option is activated, then the subroutine USDFLD or VUSDFLD is called at the end of each increment. The values of user-defined field variables can then be calculated using the output of the field variables of the previous step. The user-defined field variables are also calculated as the nodal values similar to the inbuilt field variables. Unlike the field variable values, historical values of user-defined field variables are not stored. The user needs to write separate subroutine to store the values if they are needed for any further study. The values are overwritten at the end of every time increment [55].

In the present study, the VUSDFLD is written for defining the failure model for FRP composite material. A solution-dependent state variable is defined in material model for all continuum elements. A user-defined field variable is defined for the failure index, FI. The element deletion is linked with this SDV. In case of FRP composites, hardened matrix, as well as fibers which are already brittle, makes it brittle in nature. During the drilling of FRP composites, the material properties are assumed to be constant and are not linked with the SDV. The values of stress components at each material point are obtained at the end of each increment. The failure index, FI, is calculated based on three-dimensional Tsai-Hill's failure criteria for FRP composites. A user subroutine VUSDFLD is written for obtaining the material point data and to calculate FI. If FI for any of the element is higher than one, then the material is assumed to be failed and the corresponding element is deleted. In case of drilling of FRP composites, the chips are in the form of powder. Hence, the elements failed under the loads during drilling are deleted permanently and will not appear in any further increment or step corresponding to material removal in the actual drilling process. This results in the creation of hole in the composite plate. The process is better demonstrated with the help of flow diagram as shown in Fig. 14.

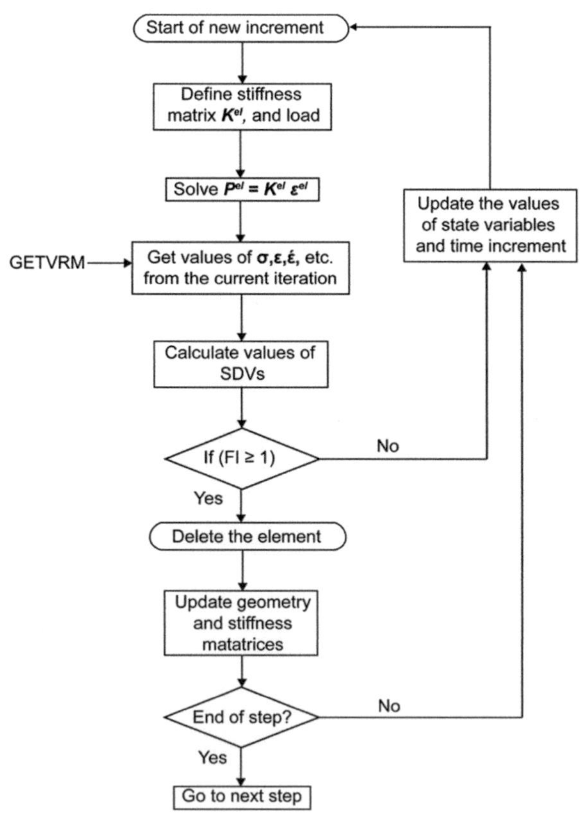

Fig. 14 Flow chart of the material erosion process using VUSDFLD

3.5 Mesh Generation

Proper selection of mesh parameters is a key step in FEA. The first step in meshing is the selection of appropriate element type. In the problem under study, there are two geometries which are to be meshed. For both the geometries, *top-down approach* was used for the meshing. However, for workpiece, due to its regular geometry, a swept mesh with hexahedral elements was used. Drill geometry being slightly complex with lot of curved surfaces, tetrahedral elements were preferred and a free mesh was used. Four-noded three-dimensional linear continuum elements, C3D4, were used for meshing of the twist drill. Average global size of 0.5 mm was used for generating mesh on 6 mm drill diameter. The mesh generation was accurate with geometric deviation factors and aspect ratios well within the permissible limit. Figure 15 shows the twist drill with mesh generated on it.

For meshing of the workpiece, two different types of elements were used. The workpiece was modeled with a total thickness of 3 mm. In the present study, the delamination analysis is one of the major objectives. A hybrid modeling consisting of macroscopic and mesoscale model is used for the workpiece. The workpiece is partitioned into two main regions, corresponding to the bulk composite material region modeled as equivalent homogeneous anisotropic cells and the interlaminar matrix rich region. However, since the study is mainly focused on the exit delamination, possibility of which is mainly towards the bottom side of the workpiece, the workpiece partitioning is done only up to a thickness of 1 mm from the bottom side. Obviously, the two different types of elements are used for

Fig. 15 Twist drill with mesh generated

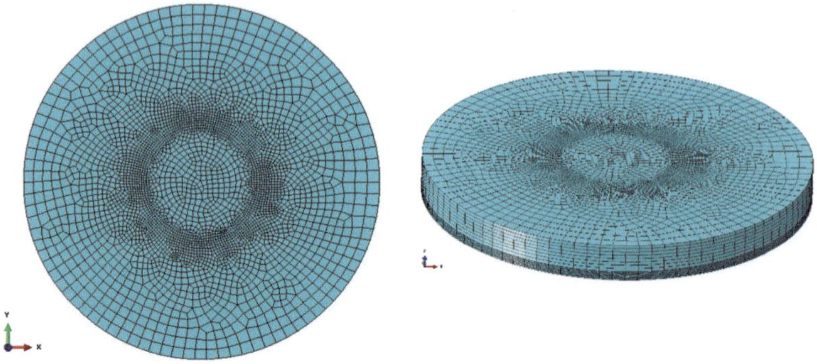

Fig. 16 Mesh on workpiece

meshing the two regions. Eight-noded three-dimensional linear continuum elements with reduced integration and hour glassing control C3D8R available in ABAQUS/ Explicit 6.10 were used for the meshing of the FRP region. For better understanding of the delamination phenomenon, a very fine mesh with approximate global size of 0.2 mm was used in the region near the periphery of the hole to be drilled, whereas, in order to get a computational advantage, graded mesh with increasing element size away from the peripheral zone of the hole is used. The interface region is meshed with cohesive elements. Three-dimensional eight-noded cohesive elements “COH3D8” are used for the meshing of the cohesive zone. The aspect ratio is not considered in mesh verification for the cohesive elements. The constitutive behavior for the cohesive elements is traction-separation. Cohesive elements with zero thickness can also be used under the traction-separation constitutive relation. Figure 16 shows the mesh generated on workpiece.

3.5.1 Mass Scaling Analysis

In this work, explicit dynamic analysis is carried out through large number of small time increments. An explicit central difference time integration rule is used. For the small time increment, this is relatively inexpensive compared to the implicit analysis where direct integration is used. However, many a time, direct solutions are not available for a set of simultaneous equations. This imposes the problems of non-convergence in case of implicit analysis. Fortunately, this problem is not faced in explicit analysis. In case of explicit analysis, the central difference operator satisfies the dynamic equilibrium equations at the beginning of the increment. This allows the explicit analysis to proceed without iterations through the small increments. This is why explicit analysis is not only popular for high-speed dynamic phenomena but also for the quasi-static processes even at slow speed involving large deformations, large energy gradients, etc. One of the drawbacks of the explicit analysis is that sometimes the results can be arbitrary as the accuracy of the results is not checked automatically unlike in implicit analysis. To overcome this drawback, it becomes very essential to select the time increments sufficiently small so that the solution is changed slightly in every increment. The limitation on the time increment comes from the conditional stability of the central difference operator. This smallest time increment is called as stable time increment and is estimated based on smallest transit time of a dilatational wave across the smallest side for any of the elements in the mesh.

The central time increment without considering the damping is given by [54]

$$\Delta t_{\min} \leq \frac{2}{\omega_{\max}} \tag{26}$$

while the same after considering the damping, is given by

$$\Delta t_{\min} \leq \frac{2}{\omega_{\max}}\left(\sqrt{1+\xi_{\max}^2}+\xi_{\max}\right), \tag{27}$$

where $\Delta t_{\min}$ is the minimum stable time increment; $\omega_{\max}$ is the highest eigenvalue in the model, and $\xi_{\max}$ is the critical damping in the highest frequency mode.

The minimum stable time increment can also be estimated based on the dilatational wave speed as given by Eq. 28.

$$\Delta t_{\min} \leq \frac{L_{\min}}{C_{\mathrm{d}}}, \tag{28}$$

where $L_{\min}$ is the critical element length, also known as characteristic length, which is equal to the smallest dimension of any of the element within the mesh and C_{d} is the dilatational wave speed given as [54]

$$C_{\mathrm{d}} = \sqrt{\frac{\hat{\lambda}+2\hat{\mu}}{\rho}}, \tag{29}$$

where ρ is the material density; $\hat{\lambda}$ and $\hat{\mu}$ are the Lame's constants calculated using Young's modulus, E and Poisson's ratio, v as

$$\hat{\lambda} = \frac{Ev}{(1+v)(1-2v)} \tag{30}$$

and

$$\hat{\mu} = \frac{E}{2(1+v)} \tag{31}$$

Thus, the minimum stable time increment for the whole model is given by [54]

$$\Delta t_{\min} \leq \text{Min}\left(L_{\min}\sqrt{\frac{\rho}{\hat{\lambda}+2\hat{\mu}}}\right) \tag{32}$$

The minimum stable time increment is calculated for each element within the model and the minimum of all these values is considered as the minimum stable time increment for the model.

The total simulation time is proportional to the total time of the process and inversely proportional to the minimum stable time increment. In case of linear anisotropic materials, the minimum stable time increment is reduced by a scaling factor equal to the square root of the ratio of effective material stiffness to maximum stiffness in a particular direction. This will ensure that the minimum time increment is sufficient for a wave to propagate across the element having critical characteristic length. During mesh refinement, the critical characteristic length may further be reduced. If the element size is made half, in case of two-dimensional problem, the number of elements will be increased by four times and the minimum time increment is reduced by half. This results in eight times increase in the solution time. Similarly, for three-dimensional mesh, if the element size is reduced by half, the number of elements is increased by eight times and the total solution time is increased by sixteen times. In the ABAQUS software, the stable time increment is calculated automatically. In this case, the time increment is not fixed and is calculated based on the critical length $L_{\min}$ for the deformed shape at the beginning of new increment. However, the user has a freedom to select a fixed time increment which can be defined at the beginning of the problem.

There are three methods to reduce the computational efforts without compromising with the element size.

- Reducing the time period
- Mass scaling
- Selective sub-scaling

Reducing time period: The solution time is directly proportional to total process time. The time can be scaled up by artificially reducing the time period for the process. This method is simple to implement but suffers from two main drawbacks. Firstly, increasing the simulation speed may increase the inertia forces. This may affect the solution accuracy. Secondly, if the process is rate dependent then the phenomena may not be captured properly. This may not allow the artificial reduction in time period.

Mass scaling: The solution time is dependent on the stable time increment. The stable time increment in ABAQUS/Explicit is given by

$$\Delta t = L_{\min}\sqrt{\frac{\rho}{E}}, \tag{33}$$

where Δt is the stable time increment; ρ is the bulk density, E is the bulk modulus of elasticity in case of isotropic material and the highest of E_{ii} in case of anisotropic materials, and $L_{\min}$ is the minimum element length.

One of the methods to reduce the total solution time keeping the element size unchanged is to increase the minimum time increment by increasing the density artificially. This method is called as mass scaling. By increasing the density, the minimum time increment can be increased by square root of the increase in density. The stable time increment is calculated for each element within the model and the stable time increment for the whole model is equal to the smallest of the stable increment from all the elements within the model.

There are two ways in which the mass scaling is used: fixed mass scaling and variable mass scaling. Also, sometimes in the model, there are only specific regions where mesh is very fine. But as the time increment is calculated from minimum of the time increment from the whole model, this quantity for the entire simulation is based on its value in such region. Hence, in mass scaling, options are available to scale the model as a whole or only a specified region called as selective mass scaling. This can address the issue of second limitations of reducing time period, i.e., of rate-dependent processes. But the problem of increased inertia forces still remains in this scale. Due to increase in density, the mass of element is also artificially increased which introduces additional inertia forces. This may affect the solution accuracy adversely.

Selective sub-cycling: Sometimes in the model, there are only specific regions where mesh is very fine. But as the time increment is calculated from minimum of the time increment from the whole model. In case of a nonuniform mesh, the minimum stable time increment is always dominated by the mesh regions where small time increment is required either due to fine mesh or material properties. Majority of regions may otherwise be solved using higher time increments. This can be achieved by selective sub-scaling. According to this method, the regions are created based on the minimum time increment. Different time increments are used for computation of the different regions. This may reduce the overall computational efforts.

In the present study as the mesh is nonuniform, the minimum stable time increment is always dominated by the mesh regions of fine mesh. Hence a selective mass scaling is used. Higher the mass scaling factor lower will be solution time. Higher mass scaling factor has to be selected properly. Otherwise it affects the solution accuracy due to increase in inertia forces as discussed earlier. A set of simulations was carried out for selecting the mass scaling factor. Only one set of cutting parameters, i.e., feed rate of 0.3 mm/rev and speed of 1500 rpm was used for the simulations. A graded mesh was used with 0.2 mm fine mesh at the region nearer to the periphery and of increased size at region away from it. Different mass scaling factors, viz., 10, 20, 30, 40 was used for the mass scaling analysis and the results were compared to the ones without any mass scaling. The assessment of the mass scaling factor was based on the simulation results for the thrust force arising due to the drilling action and the delamination. Figure 17 shows the computational advantage due to the use of mass scaling.

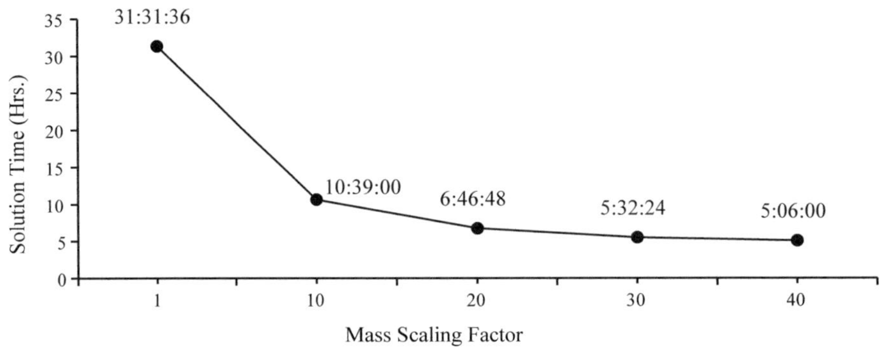

Fig. 17 Solution time with variation in mass scaling factor

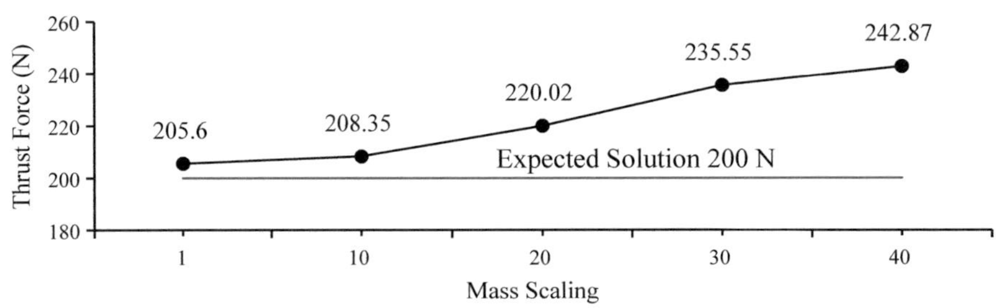

Fig. 18 Variation of thrust force with mass scaling factor

The solution time was observed reducing drastically with increase in mass scaling factor from 1 to 10. Further increase in mass scaling factor may reduce the stable time increment in the regions of fine mesh. However, as selective time mass scaling is used, the total solution time is dominated by the stable time increment for the other elements where mass scaling was not used. Hence, small or almost negligible advantage is observed by further increase in mass scaling factors. But the decision on selection of the mass scaling factor should also be based on the solution accuracy.

One of the drawbacks of increasing the stable time increment using mass scaling is the undue increase in inertia forces due to increase in density. As the mass scaling factor is small, the inertia force is also very small. Hence, the solution accuracy for thrust force estimation is not much affected in this region. However, as the mass scaling factor increases beyond a certain limit, the solution is majorly affected due to the inertia force resulting in increase in thrust force.

As shown in Fig. 18, the results varied drastically with introduction of higher values of mass scaling. Even though the solution time is reduced by more than six times, the solution value has observed to be deviated by about 20% from the

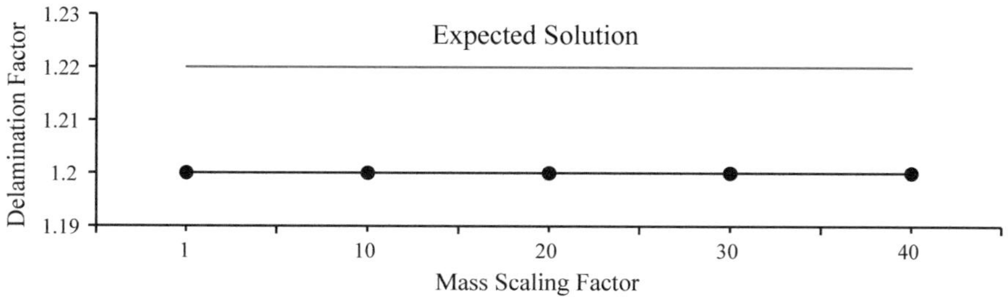

Fig. 19 Variation of delamination factor with mass scaling factor

expected value of 200 N. The results were also compared to study the effect of mass scaling on delamination factor. They are as shown in Fig. 19.

The delamination was not observed to be affected due to mass scaling. This is probably because the cohesive elements are responsible for delamination. Inertia forces are not participating in the damage mechanism of cohesive elements. Thus, based on the results for thrust force and delamination, the value of mass scaling factor equals to 10 was found to be safer for adopting for the simulation runs. The variation in the thrust force is less than 2% which is marginal compared to the large computational advantage. About three times less computational time is required in this case.

3.6 Interaction and Contact Definition

One of the most popular features of ABAQUS/Explicit is the efficient contact definitions. There are mainly two types of contact algorithms available in ABAQUS/Explicit:

- General contact
- Contact pair

The general contact algorithm is very simple which allows the definitions of contact imposing very few restrictions on the types of contact surfaces involved. Comparatively, the contact pair algorithm needs better understanding of the contact and imposes more restrictions on the types of contact surfaces involved. However, many of the interaction behaviors which are not currently available with general contact algorithm are available in contact pair algorithm. Although the general contact algorithm is more powerful and allows for simpler contact definitions, the contact pair algorithm must be used in certain cases where more specialized contact features are desired. In the present study, the drill is considered as a rigid body. This can be justified as the stiffness of the drill material is HSS, carbide, etc., which has quite high stiffness compared to that of work material. Second, the study is mainly focused only on the thrust force arising from the drilling process and the delamination defects during drilling of FRP composites rather than the study of tool wear, temperature distribution over tool, etc., are beyond the scope of the present study. Thus, we can comfortably assume the tool as rigid body. A reference point is created at the tip of the drill. All the boundary conditions on the tool are defined at this point and also all the reactions are calculated there at. The contact is defined as surface-to-surface contact with the drill as the master surface and the workpiece as the slave surface.

3.7 Boundary Conditions and Loads

The boundary conditions are imposed on the workpiece corresponding to the fixed boundary condition used in the development of the analytical models for drilling of GFRP composites [16]. The periphery of the workpiece was fixed in all the three directions, i.e., *X*, *Y*, and *Z* directions. The feed and rotation both were given on the drill through a linear speed in "mm/s" along the negative *Z* direction and a rotational speed in "rad/s" about the *Z* axis, defined on the reference point on the drill. Figures 20 and 21 show the boundary conditions applied on workpiece and the drill, respectively.

The feed is defined at the tip of the drill. In order to completely drill the hole, the drill should have over travel, which is at least equal to the cone height of the drilling cone. The cone height is dependent on the drill diameter and point angle. For example, the over travel for drill diameter 6 mm and 118° point angle is 2.178 mm. This means the total travel for the drill has to be 5.178 mm, i.e., 3 mm thickness of the plate and an over travel of 2.178 mm. As the feed on the drill is defined using velocity along negative *Z* axis, the travel distance can be controlled using the step

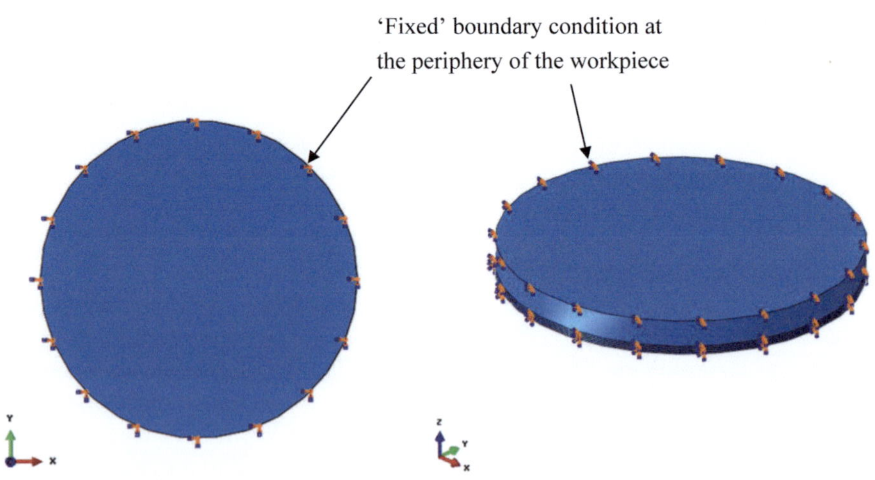

Fig. 20 Boundary conditions on workpiece

Fig. 21 Feed and rotation defined on drill

time. The step time was calculated using the linear feed rate and the abovementioned total travel time. Thus, the step time, T_s is given as

$$T_s = \frac{S_d}{f}, \tag{34}$$

where T_s is the step time in s; S_d is the total drill travel (plate thickness + minimum over travel for drill) in mm; and f is the feed rate in mm/s.

The feed rate f in mm/s is calculated using the feed rate in mm/rev, f_r and the rotational speed in rpm.

$$f = \frac{f_r \times N}{60} \tag{35}$$

where, f is the feed rate in mm/sec; f_r is the feed rate in mm/rev; and N is the rotational speed in rpm.

4 Results and Discussion

For the simulations, workpiece that was in the form of a circular disk with 30 mm diameter and 3 mm thickness made from unidirectional glass fiber reinforced plastic (UD-GFRP) was used. A twist drill with 6 mm diameter and three different point angles namely 90°, 104°, and 118° were used. The simulation plan was made to capture the effect of most influential drilling parameters such as feed rate, speed, and point angle. Since the feed rate is the most influential parameter, five different levels of feed rate and four levels of speed, and three different point angles were used for the simulations. Table 2 shows the different levels of various parameters used for simulations.

Table 2 Levels of various simulation parameters

S. No.	Parameter	Levels
01	Drill point angle	90°, 104°, 118°
02	Speed (rpm)	400, 800, 1200, 1600
03	Feed rate (mm/rev)	0.1, 0.15, 0.2, 0.25, 0.3

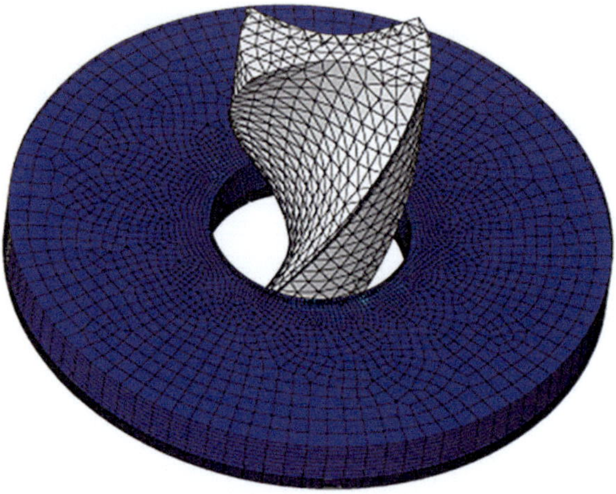

Fig. 22 Sample simulation result of drilled hole

A total of 60 simulations run were carried out to understand the effect of feed rate, speed, and point angle and drill diameter on thrust force and torque during drilling of UD-GFRP composites. Figure 22 shows a typical result of a simulation.

4.1 *Effect of Feed Rate on Thrust Force and Torque*

Feed rate being most influential parameter for thrust force, five different feed rate values were used to study its effect on thrust force variations. The results for thrust force variations with respect to the feed rate are shown in Fig. 23. The thrust force was observed to be increasing almost linearly with increase in feed rate for all speeds. A multiple linear regression was used for thrust force, and the correlation coefficient R^2, and adjusted R^2 values in all the cases are more than 0.89.

Mixed observations are reported in literature regarding torque. Some of the researchers have reported torque to be reducing with increasing thrust force [37] while few others have observed torque increasing linearly with increase in feed rate [38]. In the present study, the torque was observed increasing with increasing feed rate. Except in case of 6 mm drill diameter with 90° point angle, almost linear relationship was observed between torque and feed rate. In case of 6 mm drill diameter with 90° point angle, second order regression has given better correlation coefficient of 0.99.

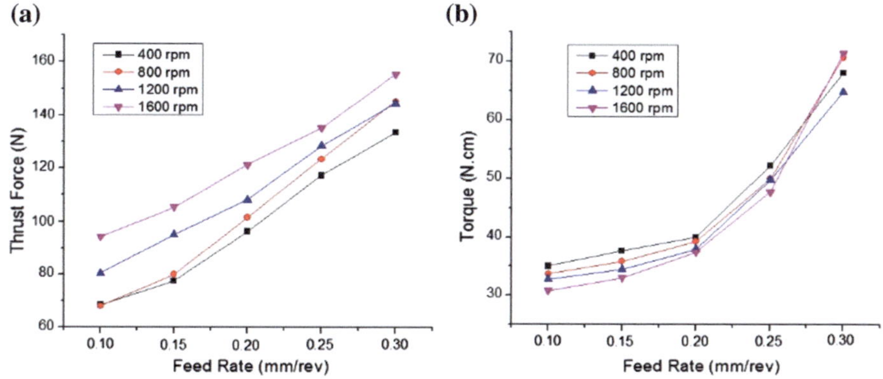

Fig. 23 Variation of **a** thrust force **b** torque with respect to feed rate (for 6 mm diameter drill with point angle 90°)

4.2 Effect of Speed on Thrust Force and Torque

The second important machining parameter in drilling is the speed. Four different levels of speed (in rpm) are considered for simulation and its effect on thrust force and the torque was studied through the graphs shown in Fig. 24. Thrust force was observed to be increasing almost linearly with increase in speed at all feed rates. This is because the linear feed rate in mm/sec increases with increase in speed. However, the variation in torque was very marginal. Also for lower point angle of 90°, the torque was observed decreasing slightly with increase in speed.

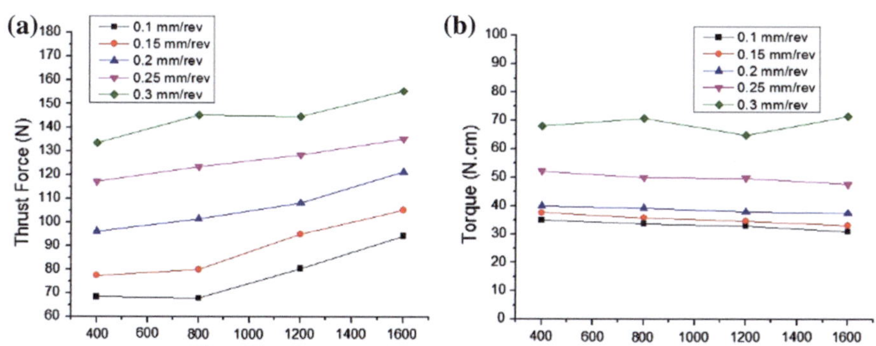

Fig. 24 Variation of **a** thrust force and **b** torque with respect to speed (for 6 mm diameter drill with point angle 90°)

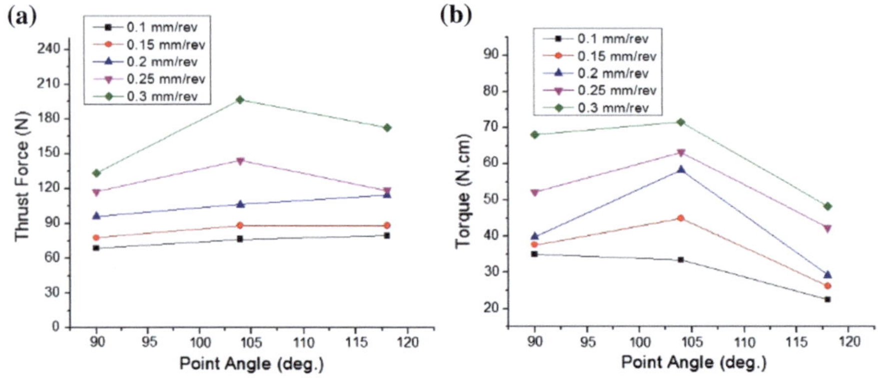

Fig. 25 Variation of **a** thrust force **b** torque with respect to point angle (for 6 mm diameter drill with speed 400 rpm)

4.3 Effect of Drill Point Angle on Thrust Force and Torque

Point angle is one of the important parameters of drill geometry. It has direct contribution on cutting angles and the secondary cutting edge length. The results obtained for variation of thrust force and torque are shown in Fig. 25. The thrust force was observed increasing marginally with increasing point angle from 90° to 118°. This is in line with the experimental observations reported by Feito et al. [56]. On the contrary, torque was observed reducing with increase in point angle. However, the trends of thrust force and torque are not very clear. Poor correlation was observed between the thrust force, torque and feed rate in case of all the four speeds under study.

5 Conclusions

The variation of thrust force and torque against speed and feed rate were studied through 60 simulations using finite element model. The Tsai-Hill's criteria based failure model developed using user subroutine was successfully used in the simulation. Based on the observations from the simulations, the following major conclusions can be drawn:

- Thrust force is most sensitive to the variation in feed rate and increases almost linearly with the increase in feed rate. The regression analysis showed that average correlation coefficient was 0.93.
- Thrust force was observed increasing with speed but very marginally.
- The variation of thrust force with respect to point angle was very small and relatively irregular.

- Torque was also observed increasing with increase in feed rate but it was not much affected by variation in speed.
- The correlations of torque with point angle are very poor.

References

1. Durao, L. M. P., de Moura, M. F. S. F., & Marques, A. T. (2008). Numerical prediction of delamination onset in carbon/epoxy composites drilling. *Engineering Fracture Mechanics, 75,* 2767–2778.
2. Khashaba, U., Seif, M. A., & Elhamid, M. A. (2007). Drilling analysis of chopped composites. *Composites Part A Applied Science and Manufacturing, 38,* 61–70.
3. Capello, E., Langella, A., Nele, L., Paoletti, A., Santo, L., & Tagliaferri, V. (2008). Drilling polymeric matrix composites. In J. P. Davim (Ed.), *Machining: Fundamentals and recent advances* (pp. 167–191). London: Springer.
4. Abrao, A. M., Rubio, J. C. C., Faria, P. E., & Davim, J. P. (2008). The effect of cutting tool geometry on thrust force and delamination when drilling glass fibre reinforced plastic composite. *Materials and Design, 29,* 508–513.
5. Bhatnagar, N., Singh, I., & Nayak, D. (2004). Damage investigation in drilling of glass fiber reinforced plastic composite laminates. *Materials and Manufacturing Processes, 19,* 995–1007.
6. Enemuoh, E. U., El-Gizawy, A. S., & Okafor, A. C. (2001). An approach for development of damage-free drilling of carbon fiber reinforced thermosets. *International Journal of Machine Tools and Manufacture, 41*, 1795–1814.
7. Gaitonde, V. N., Karnikb, S. R., & Davimc, J. P. (2008). Taguchi multiple-performance characteristics optimization in drilling of medium density fibreboard (MDF) to minimize delamination using utility concept. *Journal of Materials Processing Technology, 196*, 73–78.
8. Davim, J. P., & Reis, P. (2003). Study of delamination in drilling carbon fiber reinforced plastics (CFRP) using design experiments. *Composite Structures, 59*, 481–487.
9. Davim, J. P., & Reis, P. (2003). Drilling carbon fiber reinforced plastics manufactured by autoclave- experimental and statistical study. *Materials and Design, 24*, 315–324.
10. Davim, J. P., Reis, P., & Antonio, C. C. (2004). Experimental study of drilling glass fiber reinforced plastics (GFRP) manufactured by hand layup. *Composites Science and Technology, 64*, 289–297.
11. Mohan, N. S., Kulkarni, S. M., & Ramachandra, A. (2007). Delamination analysis in drilling process of glass fiber reinforced plastic (GFRP) composite materials. *Journal of Materials Processing Technology, 186*, 265–271.
12. Mohan, N. S., Ramachandra, A., & Kulkarni, S. M. (2005). Influence of process parameters on cutting force and torque during drilling of glass–fiber polyester reinforced composites. *Composite Structures, 71*, 407–4013.
13. Arul, S., Samuel Raj, D., Vijayaraghavan, L., Malhotra, S. K., & Krishnamurthy, R. (2006). Modeling and optimization of process parameters for defect toleranced drilling of GFRP composites. *Materials and Manufacturing Processes, 21*, 357–365.
14. Ravishankar, S. R., & Murthy, C. R. L. (2000). Characteristics of AE signals obtained during drilling composite laminates. *NDT&E International, 33*, 341–348.
15. Ravishankar, S. R., & Murthy, C. R. L. (2000). Application of acoustic emission in drilling of composite laminates. *NDT&E International, 33*, 429–435.
16. Ho-Cheng, H., & Dharan, C. K. H. (1990). Delamination during drilling in composite laminates. *Transactions of ASME, 112,* 236–239.

17. Jain, S., & Yang, D. C. H. (1993). Effect of federate and chisel edge on delamination in composites drilling. *Transactions of ASME, 115,* 398–405.
18. Jain, S., & Yang, D. C. H. (1994). Delamination-free drilling of composite laminates. *Journal of Engineering for Industry, 116,* 475–481.
19. Chandrasekharan, V., Kapoor, S. G., & DeVor, R. E. (1995). A mechanistic approach to predicting the cutting forces in drilling: With application to fiber-reinforced composite. *Journal of Engineering for Industry, 117,* 559–570.
20. Langella, A., Nele, L., & Maio, A. (2005). A torque and thrust prediction model for drilling of composite materials. *Composites: Part A, 36,* 83–93.
21. Tsao, C. C., & Chen, W. C. (1997). Prediction of the location of delamination in the drilling of composite laminates. *Journal of Materials Processing Technology, 70,* 185–189.
22. Upadhyay, P. C., & Lyons, J. S. (1999). On the evaluation of critical thrust for delamination-free drilling of composite laminates. *Journal of Reinforced Plastics and Composites, 18,* 1287–1303.
23. Frederic, L., Robert, P., Francis, C., & Laurent, S. (2001). Drilling of composite structures. *Composite Structures, 52,* 511–516.
24. Zhang, L. B., Wang, L. J., & Liu, X. Y. (2001). A mechanical model for predicting critical thrust forces in drilling composite laminates. *Proceedings of Institute of Mechanical Engineers Part B, 215,* 135–146.
25. Pyo, J. J., Kim, K. W., & Yong, L. K. (2005). Critical thrust force at delamination propagation during drilling of angle-ply laminates. *Composite Structures, 68,* 391–397.
26. Zhang, L., Wang, L. J., Liu, X. Y., Zhao, H. W., Wang, X., & Luo, H. Y. (2001). Mechanical model for predicting thrust and torque in vibration drilling fibre-reinforced composite materials. *International Journal of Machine Tools and Manufacture, 41,* 641–657.
27. Tsao, C. C., & Hocheng, H. (2003). The effect of chisel length and associated pilot hole on delamination when drilling composite materials. *International Journal of Machine Tools and Manufacture, 42,* 1087–1092.
28. Hocheng, H., & Tsao, C. C. (2003). Comprehensive analysis of delamination in drilling of composite materials with various drill bits. *Journal of Material Processing Technology, 140,* 335–339.
29. Kim, G. W., & Lee, K. Y. (2005). Critical thrust force at propagation of delamination zone due to drilling of FRP/metalic strips. *Composite Structures, 69,* 137–141.
30. Tsao, C. C., & Hocheng, H. (2005). Effect of eccentricity of twist drill and candle stick drill on delamination in drilling composite materials. *International Journal of Machine Tools and Manufacture, 45,* 125–130.
31. Tsao, C. C. (2007). Effect of deviation on delamination by saw drill. *International Journal of Machine Tools and Manufacture, 47,* 1132–1138.
32. Tsao, C. C., & Hocheng, H. (2005). Effect of exit back-up on delamination in drilling composite materials using a saw drill and a core drill. *International Journal of Machine Tools and Manufacture, 45,* 1261–1270.
33. Hocheng, H., & Tsao, C. C. (2005). The path towards delamination-free drilling of composite materials. *Journal of Material Processing Technology, 167,* 251–264.
34. Tsao, C. C. (2006). The effect of pilot hole on delamination when core drill during composite materials. *International Journal of Machine Tools & Manufacture, 46,* 1653–1661.
35. Sedlacek, J., & Humar, A. (2008). Analysis of fracture mechanism and surface quality in drilling of composite materials. *Strength of Materials, 40,* 40–43.
36. Tsao, C. C., & Hocheng, H. (2007). Effect of tool wear on delamination in drilling composite materials. *International Journal of Mechanical Sciences, 49,* 983–988.
37. Sadat, A. B., Chang, W. S., & Wang, B. P. (1992). Delamination of graphite/epoxy laminate during drilling operation. *Journal of Energy Resources Technology, 114,* 139–141.
38. Bhattacharya, D., & Horrigan, D. P. W. (1998). A study of hole drilling in kevlar composites. *Composite Science and Technology, 58,* 267–283.
39. Budan, A. D., & Vijayarangan, S. (2002). Quality assessment and delamination force evaluation in drilling glass fiber-reinforced plastic laminates—A finite element analysis and

linear fracture mechanics approach. *Proceedings of Institute of Mechanical Engineers, 216,* 173–182.
40. Durao, L. M. P., de Moura, M. F. S. F., & Marques, A. T. (2006). Numerical simulation of the drilling process on carbon/epoxy composite laminates. *Composites: Part A, 37,* 1325–1333.
41. Singh, I., Bhatnagar, N., & Vishwanath, P. (2008). Drilling of uni-directional glass fiber reinforced plastics: Experimental and finite element study. *Materials and Design, 29,* 546–553.
42. Rahme, P., Landon, L. Y., Piquet, R., & Lagarrige, P. (2011). Analytical models of composite material drilling. *International Journal of Advance Manufacturing Technology, 52,* 609–617.
43. Zitoune, R., & Collombet, F. (2007). Numerical prediction of the thrust force responsible of delamination during the drilling of the long-fibre composite structures. *Composites: Part A, 38,* 858–866.
44. Chakladar, N. D., Pal, S. K., & Mandal, P. (2012). Drilling of woven glass fiber-reinforced plastic—An experimental and finite element study. *The International Journal of Advanced Manufacturing Technology, 58,* 267–278.
45. Isbilir, O., & Ghassemieh, E. (2012). Finite element analysis of drilling of carbon fibre reinforced composites. *Applied Composite Materials, 19,* 637–656.
46. Isbilir, O., & Ghassemieh, E. (2013). Numerical investigation of the effects of drill geometry on drilling induced delamination of carbon fiber reinforced composites. *Composite Structures, 105,* 126–133.
47. Phadnis, V. A., Makhdum, F., Roy, A., & Siberschmidt, V. (2013). Drilling in carbon/epoxy composites: Experimental investigations and finite element implementation. *Composites: Part A, 47,* 41–51.
48. Vijayaraghavan, A., & Dornfeld, D. A. (2007). Automated drill modeling for drilling process simulation. *Journal of Computing and Information Science in Engineering, 7,* 276–282.
49. Astakhov, V. P. (2010). *Geometry of single-point turning tools and drills: Fundamentals and practical applications*. London: Springer.
50. Paul, A., Kapoor, S. G., & DeVor, R. E. (2005). A chisel edge model for arbitrary drill point geometry. *Journal of Manufacturing Science and Engineering, 127,* 23–32.
51. Ochoa, O. O., & Reddy, J. N. (1992). Finite element analysis of composite laminates. In *Solid mechanics and its applications* (pp. 37–109). Dordrecht: Springer Science + Business Media.
52. Kaw, A. K. (2006). *Mechanics of composite materials* (2nd ed.). Boka Raton: Taylor & Francis.
53. Rao, G. V. G., Mahajan, P., & Bhatnagar, N. (2008). Three dimensional macro-mechanical finite element model for machining of unidirectional-fiber reinforced polymer composites. *Materials Science and Engineering A, 498,* 142–149.
54. ABAQUS analysis users' manual ABAQUS 6.10.
55. ABAQUS theory manual, ABAQUS 6.10.
56. Feito, N., Jose, N. D. A. A., Jose, L. C. & Maria, H. M. (2014). Experimental analysis of the influence of drill point angle and wear on drilling of woven CFRPs. *Materials, 7*, 4258–4271.

Computational Fluid Dynamics Analysis of MQL Spray Parameters and Its Influence on MQL Milling of SS304

J. N. Rohit, K. Surendra Kumar, N. Sura Reddy, P. Kuppan and A. S. S. Balan

1 Introduction

Austenitic stainless steels have a widespread application in the nuclear industry due to their corrosion resistance, and high strength properties. Mechanical properties such as low thermal conductivity and work hardening make stainless steels difficult-to-machine materials. Machining of such steels is associated with built-up edge formation, higher order temperature and irregular wear on the respective rake and flank faces of the tool [1]. High strength, high fatigue, and fracture toughness are the main causes for such undesirable tool behavior. In order to deal with such issues, effective lubrication is necessary. Conventional flood cooling application fails to lubricate the tool–chip interface efficiently [2]. In recent years, the cost of conventional flood cooling supply has been raised due to the strict enforcement of new health regulations and standards. Due to flood cooling, the machine operator can get affected by skin and breathing problems on the shop floor [3]. Several types of research stated that the cutting fluids cost is higher than the cost of machining tools [4]. This trend can be certainly observed in machining materials which are usually termed as difficult-to-machine materials. Therefore, cooling lubricant utilization during the machining operations is very vital. Studies show that over 2 billion gallons of cooling fluid are wasted every year. Many manufacturers moved toward dry cutting to eliminate the cost of cutting fluids and also to make the process safe in terms of pollution [5]. As cutting fluid plays a predominant role in enhancing the quality of machining, it is impossible to completely eliminate the

J. N. Rohit · K. Surendra Kumar · N. Sura Reddy · P. Kuppan
School of Mechanical Engineering, Vellore Institute of Technology, Vellore 632014, India

A. S. S. Balan (✉)
Centre for Innovative Manufacturing Research, Vellore Institute of Technology, Vellore 632014, India
e-mail: balan.sheytra@vit.ac.in

U. S. Dixit and R. Kant (eds.), *Simulations for Design and Manufacturing*, Lecture Notes on Multidisciplinary Industrial Engineering,
https://doi.org/10.1007/978-981-10-8518-5_2

consumption of cutting fluids. There are several advantages of dry machining but manufacturers and industrialists will be willing to accept this method only if researchers find a solution for the drawbacks of dry machining. Dry milling without any amount of coolant being supplied is not viable. In dry machining processes, the cutting edges and workpiece are subject to high mechanical, thermal, and chemical loads. So, appropriate machinability, high wear resistance, and good performance can be expected from dry machining only if, the cutting materials meet certain requirements in terms of their physical properties. The cutting material for dry machining is expected to have high resistance to pressure, high hardness, high toughness, high thermal fatigue limit, critical stress intensity factor, high chemical resistance, and high hot hardness [6]. These demands resulted in the evolution of more eco and an environmentally friendly technique called Minimum Quantity Lubrication (MQL). MQL machining involves the consumption of cutting fluids generally in order of 50–500 ml/h [7]. MQL refers to the use of a precision dispenser to supply a minuscule amount of cutting fluid to the tool–workpiece interface which is around 3–4 orders of magnitude lower than the amount consumed in a flood cooling condition. Balan et al. concluded that MQL is advantageous over flood cooling as it facilitates a reduction in fluid consumption, cost, environmental pollution, and lower infrastructural demands [8]. A good quality oil mist could minimize cutting zone temperature and tool wear to an appreciable extent on the application at the tool–chip interface [2]. Dhar et al. investigated the effect of MQL in machining of AISI 1040 steel and revealed that MQL enhances considerable reduction of temperature in the cutting zone and has an appreciable impact on work–tool interactions, thereby reducing friction, thermal distortion of the work, and built-up edge formation [9]. Rahman et al. studied the impact of MQL in the milling of ASSAB 718 HH steel. They concluded that MQL is effective at low speed, depth of cut, and feed in comparison to flood cooling in terms of reduction in cutting force [10]. Alper Uysal et al. concluded that parameters like droplet size, flow rate, and pressure will have a major influence on MQL spray quality. MQL technique gave minimum surface roughness when compared with dry milling. Reduction in tool wear and surface roughness can be observed by an increase in the MQL flow rate [11]. Arunachalam et al. developed an analytical model to predict Sauter Mean Diameter (SMD) of coolant spray. They revealed that the model can be used to determine the optimum range of input spray parameters in MQL [12]. Lutao et al. found that surface finish improved mainly due to the reduction of tool wear and damage at the tool tip by the application of MQL [13]. Tawakoli et al. experimentally investigated the effect of air atomizing pressure and flow rate on the quality of grinding and concluded that positioning of the nozzle is an important factor for the effective application of oil mist [14]. Rahman et al. have modeled a standard MQL nozzle and studied the flow pattern through a nozzle using CFD. Their study revealed the flow distribution at the tip of the nozzle for a particular pressure which helps in designing the nozzle [15].

1.1 Classification of MQL

Classifying the MQL setup is based on the feed mechanism. A differentiation is made between external and internal feed of the lubrication medium, which makes a noticeable difference to the cost of the device technology. The lubricant is applied along the circumference of the tool with the help of spray nozzles in the external feed. This type of system is suitable for entry-level standard process implementation such as turning, milling, drilling, and grinding. Figure 1 shows the external feed MQL setup used in the present investigation. In internal feed, transportation of the lubricant is done through the spindle system of the machine and through the channels in the tool to the machining point. This system is used primarily when flexible processing centers with high-speed cutting (HSC). The internal feed can further be divided into subcategories based on the number of infeed channels. The aerosol is generated before entering the spindle in the tank and hence called 1-channel system. Separate feeding of the media through the spindle before and after the exit, producing aerosol directly at the tool in two-channel system [16].

1.2 Mist Formation in MQL

The process of mist formation is called atomization. Atomization involves forcing of liquid through a nozzle. The energy along with the geometry of the nozzle causes the liquid to emerge as small ligaments. Due to the application of pressure in MQL, the ligaments fragment into a number of droplets. Each spray provides a range of droplet sizes and is referred as drop size distribution. Impingement of nanoscale lubricant droplets of high pressure into the tool–work interface is MQL. Figure 2 shows the mist formation in MQL.

Generally, the oil-based types of lubricants possess lower heat capacity when compared with water-based coolants. Reduction of friction between tool and work material is the main criteria of the lubricants used in MQL. The adequate covering of the contact area, i.e., the tool–work and tool–chip interfaces by the lubricant is

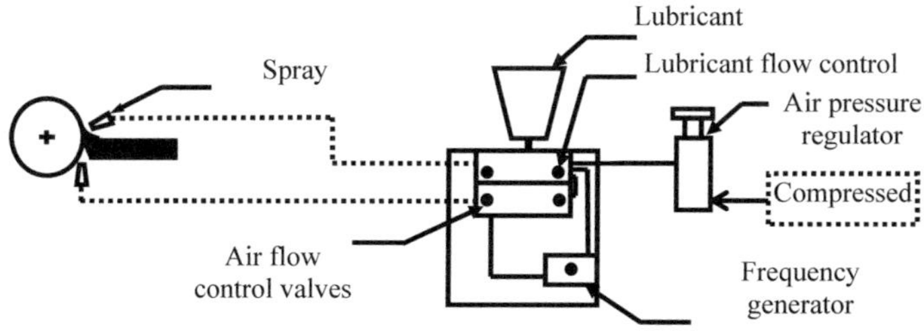

Fig. 1 External feed MQL setup

Fig. 2 Mist formation in MQL

the most important function in MQL. The mass flow rate of the oil, nozzle distance, and air pressure are the considerable parameters which influence the droplet size and distribution. Smaller droplet means more deposition distance but the low wetting area. On the other hand, larger droplets mean larger wetting area but small deposition distance. The analytical models to explain better performance by MQL were not studied until 2010. Tawakoli et al. investigated the performance of MQL by considering various MQL parameters such as air pressure, mass flow rate, and distance of the nozzle from the workpiece. They concluded that these parameters affect the performance of MQL translated through the wetting area and droplet deposition distance [14, 17].

1.3 Effect of Pressure and Flow Rate Over Droplet Size

The size of the droplet has a significant effect on the lubrication in MQL. The droplet size is greatly influenced by the MQL spray parameters such as mass flow rate and air pressure. The wetting area of the mist spray depends both on the volume of a single droplet as well as the number of droplets coming through mist per second. According to Radoslaw et al., droplet diameter tends to become smaller with an increase in MQL flow rate, thereby increasing the number of droplets [18]. The increase in the wetting area can be noticed with increase in air flow. Therefore, to achieve best results higher lubrication at the cutting zone is necessary which lowers frictional shear stress, consequently lowers the cutting forces in milling process. So, the MQL droplets should completely cover the tool–work interface for a good lubrication effect.

1.4 Effect of Droplet Deposition Distance

To enhance the penetration of oil mist into the contact zone MQL nozzle distance plays a critical role. A horizontal oil mist jet with a spherical oil droplet of diameter d_{drop} with velocity u_{drop}, on neglecting the effect of gravity, has the deposition distance S_{drop} as

$$S_{\text{drop}} = \frac{\rho_{\text{drop}} \Delta u_{\text{drop}} d_{\text{drop}}^2}{18 \mu_g C_R} \left(1 - \frac{\rho_{\text{air}}}{\rho_{\text{drop}}} \right), \tag{1}$$

where ρ_{drop} is the density of the droplet, μ_g is the viscosity of the oil.

The moving droplet has a drag coefficient in the air jet as follows [19, 20]:

$$C_R = \frac{\pi \left[24 + 24 Re_g + 8 Re_g^2 + Re_g^3 + 4 e^{Re_g} \left(Re_g^2 - 6 \right) \right]}{6 \left[2 \left(Re_g + 1 \right) + e^{Re_g} \left(Re_g^2 - 2 \right) \right]}, \tag{2}$$

where

$$Re_g = \frac{\rho_{\text{drop}} \Delta u_{\text{drop}} d_{\text{drop}}}{\mu_g} \tag{3}$$

Larger droplets with large inertia may fall early and not travel in the air stream whereas very small droplets may not enter the tool–work interface thereby resulting in inefficient lubrication [21].

1.5 Effect of Nozzle Position

MQL milling can involve positioning of one or multiple nozzles. The ideal horizontal angle for end milling should be 135° from the feed direction or 45° from the spindle direction. This allows lubrication of the milling cutting edges before the cut and avoids chip injection area. In milling operation where cutting tool changes its direction with respect to the workpiece, as a result, the tool itself blocks the entry of the lubrication into the cutting zone in such case multiple nozzles has to be used. This optimum angle has to be maintained throughout the operation. As the cutting tool edge enters and exits the workpiece several times per second, milling is considered as an interrupted operation. Formation of hydrodynamic boundary layer around the rotating tool results in a back (reversed) flow when there is close proximity between the milling cutter and the workpiece. The flow of cutting fluid is opposed by this reversed air flow direction and thereby possesses adequate momentum which holds the cutting fluid back. This affects the MQL spray transported towards the workpiece and hence incomplete lubrication of the workpiece surface in the cutting zone can be

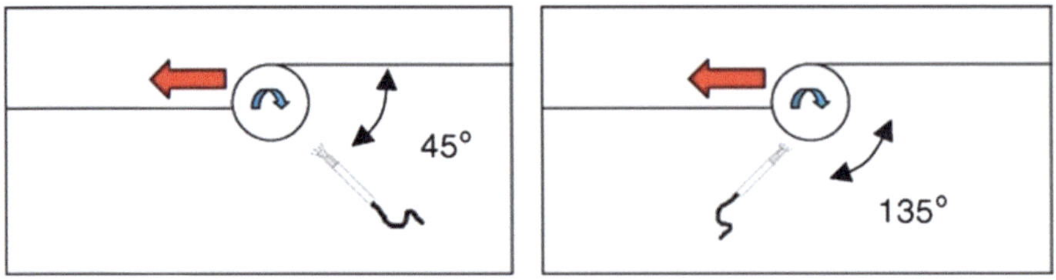

Fig. 3 Nozzle positioning in MQL [22]

observed. Figure 3 shows the different locations of the nozzle in MQL setup. Lopez et al. concluded that nozzle position in relation to feeding direction plays a vital role in terms of achieving the optimum effect of MQL [22].

1.6 Experimental Techniques to Determine Droplet Size

In the previous section, it is seen that estimation of droplet size in MQL is very vital in determining the optimum MQL conditions to achieve best results. As droplet size plays a major role in minimum quantity lubrication, in the present research attempt has been made in developing an appropriate measuring procedure to determine the size of the droplet with respect to spray conditions. There are numerous techniques available to experimentally find the size of the droplet.

1.6.1 Optical Imaging Analyzers

Optical imaging analyzers include the spatial technique of measurement. The basic components of these analyzers are a video camera, a light source (generally a strobe light or laser), and computer. Illumination of spray is done by the light and the video camera is used for recording. Separation of the droplets into different classes based on the size is carried out after scanning the image. A typical arrangement of optical imaging analyzer is shown in Fig. 4.

Improper sample size, blurring, and depth of field variations are the sources of error in this device. These error sources can be recognized and corrected. The system optics determines the upper range where the range of these systems is generally from one to unlimited micrometers.

1.6.2 Laser Diffraction Analyzers

Spatial sampling devices which fall into the category of non-imaging (ensemble) are called laser diffraction analyzers. The chief components are a receiver, computer, and transmitter. The principle of measurement is based on scattered light

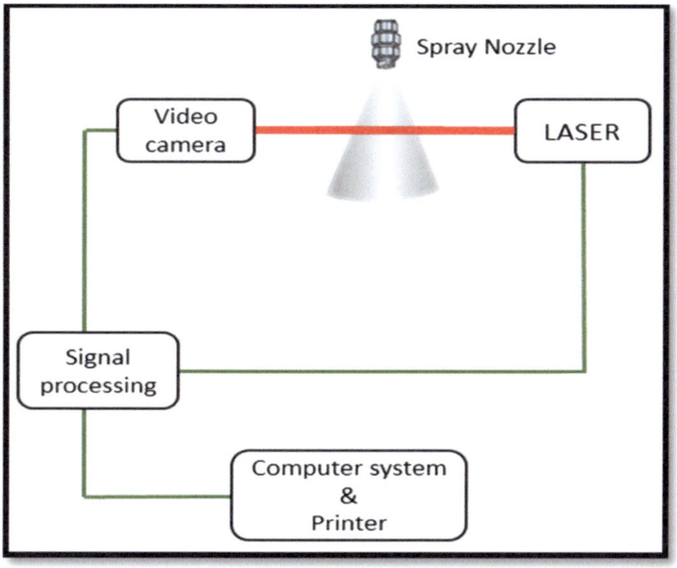

Fig. 4 Optical image analyzer

intensity. Intensity is measured by a continuously housed series of semicircular photodiodes in the receiver unit. To convert the intensity of light distribution into several drops size, empirical distribution functions based curve fitting program is used and shown in Fig. 5. The range lies between 1.2 and 1,800 μm. This instrument is ideally suited for measuring two fluids of smaller capacity, flat spray nozzles, and hydraulic nozzles. This is also used for quick comparison and evaluation of prototype nozzles. Multiple scattering is the major limitation of this instrument. Higher spray densities resulted due to the scattering of the light by multiple drops before reaching the detector are the main reason for occurring of multiple scattering. As a result, errors occur while computing the distribution of droplet size [17].

1.6.3 Phase-Doppler Particle Analyzers (PDPA)

PDPA are generally termed as flux-sampling devices which fall into the category of non-imaging (single particle counter) devices. This analyzer consists of a signal processor, transmitter, computer, and receiver. The lower intensity laser used is separated into two or four beams in case of a two-dimensional system in PDPA. At the probe volume, the laser beams are made to intersect by using a beam splitter and frequency module. Formation of the interference fringe pattern by the scattered light occurs at the probe volume while drop passes through the probe volume. At Doppler difference frequency, the scattered interferences sweep past the receiver unit, which is proportional to the droplet velocity. Inverse proportionality is

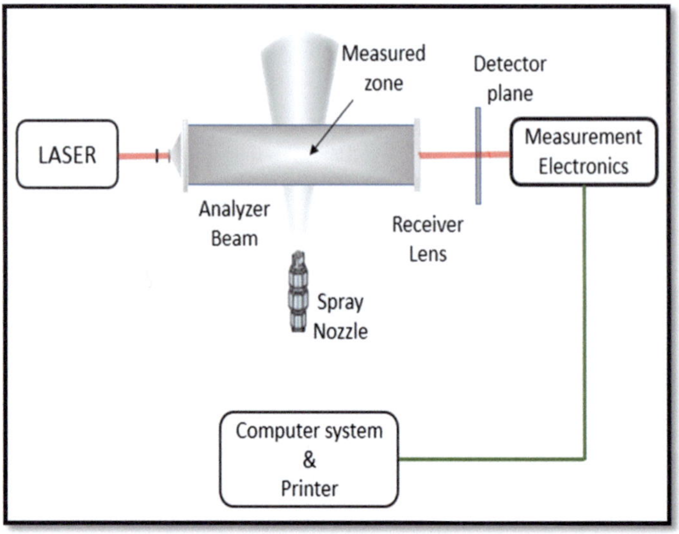

Fig. 5 Laser diffraction analyzer

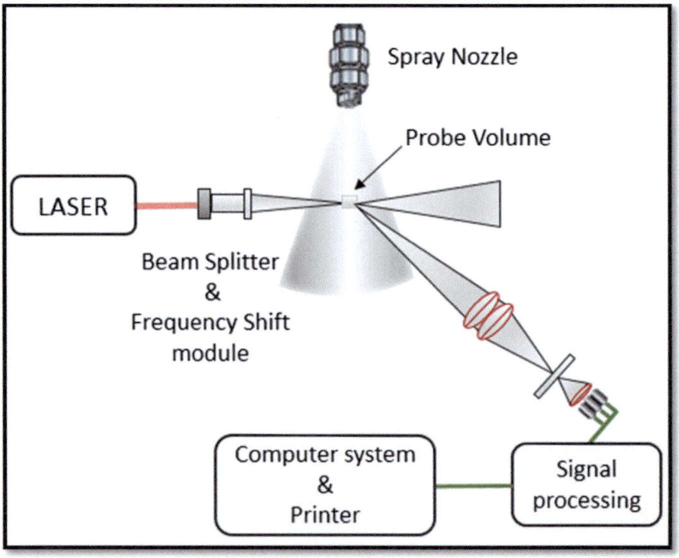

Fig. 6 Schematic of PDPA

maintained between the spatial frequency of the fringe pattern and the droplet diameter. Sizes ranging between 0.5 and 10,000 μm can be measured with the help of variable optical configurations in PDPA [17]. Figure 6 shows a schematic diagram for PDPA.

This section begins with a comparison between different methods of lubrication and cooling namely dry machining (milling), wet milling, and MQL. Different cutting parameters were compared to judge the performance of these methods of lubrication. This section also mentions the lack of simulated model in MQL and discusses the analytical models which have been discovered very recently.

2 CFD Simulations for MQL Milling

A literature survey reveals that there are very few simulation models available for MQL in Milling. CFD simulation is necessary because it helps in determining the effect of nozzle position on the penetration of MQL spray and also to study the effect of a change in air pressure and mass flow rate on the wetting area thus affecting the performance of MQL. Toshiyuki et al. performed CFD analysis for better spraying of oil mist in finish-turning of superalloys. They concluded that oil droplet flight determines the distance traveled by the droplet and low-sized droplets are unfavorable for MQL machining. Both cooling and lubrication are predominately improved by oblique spraying in the cutting point vicinity. Lopez et al. assessed through CFD the efficiency of MQL and the position of the injection nozzle for high-speed milling. They simulated the conventional flood cooling process and MQL and found out that MQL is very efficient in lubrication because of the high jet velocity which penetrated through the backflow of milling [22]. Park et al. concluded that the combination of image processing techniques and confocal laser scanning microscopy (CLSM) including of wavelet transforms are used to analyze the droplet distribution and size of the droplet has been spraying the oil onto a polished silicon wafer. They also studied the distribution of droplets to determine the MQL optimum distance between the nozzle and the workpiece and the nozzle discharge pressure [23]. Optimum conditions of MQL process are determined in this research work.

From the above literature survey, it can be observed that MQL has better performance as compared to dry machining and flood cooling. One of the most significant factors which determine the performance of MQL is the size of the droplet. Only a very few types of research have discussed the quality of MQL, defined by parameters like droplet velocity, pressure, and diameter. Hence the main objective of this work is to simulate the oil mist atomization using computational fluid dynamics (CFD) and to investigate the significance of spray parameters over milling performance.

2.1 *Numerical Modeling and Simulation of MQL*

Machining with chip removal is largely influenced by the condition of tool–work interface, which influences the friction and tool wear. A major constraint is a relative inaccessibility of cutting fluid (lubrication) to the tool–work interface. The problem is aggravated in milling by the presence of a pressure belt around the tool. So, the film lubrication through pressurized oil has been necessitated. One such technique is MQL, in which the lubricant is atomized as finer droplets to enable easier accessibility. In the previous section, the influence of the position of the nozzle, the effect of air pressure, mass flow rate, and nozzle standoff distance on the performance of MQL was analyzed. This section deals with the analytical models for the effect of these underlying parameters on the performance of MQL. The analytical model suggests that the only parameter which is unknown for a given experimental setup is the size of the droplet. Experimental ways to find droplet size are expensive and time-consuming and the literature survey essentially highlighted the lack of any simulation model for MQL in milling and the necessity to develop one. The present work is a simulated study of the atomization of oil mist using Computational Fluid Dynamics (CFD) model using FLUENT 6.3.26 flow solver. The parameter varied is the mass flow rate and air pressure. Simulation of the oil mist atomization was carried out considering turbulent flow under discrete phase model (DPM) owing to the fact that oil is considered as a discrete phase medium in air. Optimum values of mass flow rate and air pressure are determined by the obtained jet velocity and droplet diameter values of oil to achieve the optimum results in MQL milling.

2.2 CAD Model

The nozzle part of the experimental setup was modeled in AutoCAD using the details provided by the manufacturer (Spraying Systems Co., USA) for the better visualization. The 2D views of this modeled were used to create a mesh in Gambit 2.3.16. Figure 7 shows AutoCAD model of the nozzle. The air orifice for this nozzle is 0.6 mm and the liquid orifice is 0.7 mm. The ratio of the orifices, i.e., 1.16 matches the standard range of ratio 1.1–1.3 for air atomizing nozzles.

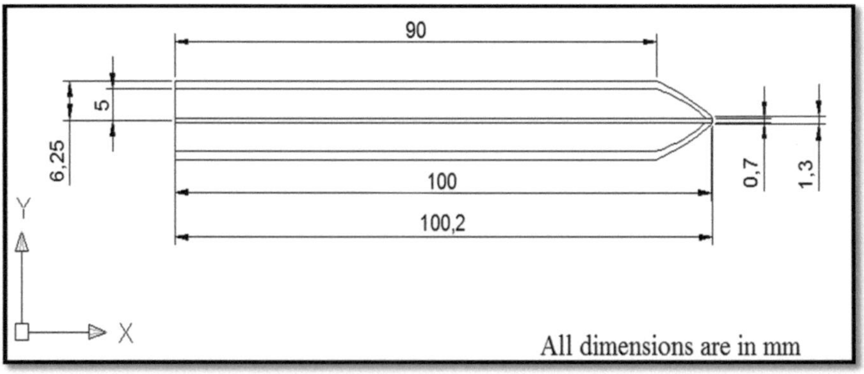

Fig. 7 AutoCAD 2D model of the air atomizing nozzle

2.3 *Grid Partitioning and Meshing*

Modeling and grid partitioning were completed using Gambit 2.3.16. A virtual box of dimension 100 mm × 100 mm was created because the actual stand of distance between the tool and nozzle is 100 mm and assigned all the properties of atmosphere to observe the dynamics of droplets in the atmosphere. All the dimensions provided by the nozzle manufacturer were used to the utmost accuracy so that the simulation replicates the real process as closely as possible.

2.3.1 Grid Partitioning

In order to achieve better accuracy in the simulation results and to reduce the computational time, the whole 2D model has to be divided into three parts. These three parts namely the air and oil stream, the mixing area, and the atmosphere need to be meshed appropriately depending upon the activity of fluids in that region. Figure 8 shows the 2D model in Gambit 2.3.16 with grid partitioning and boundary conditions.

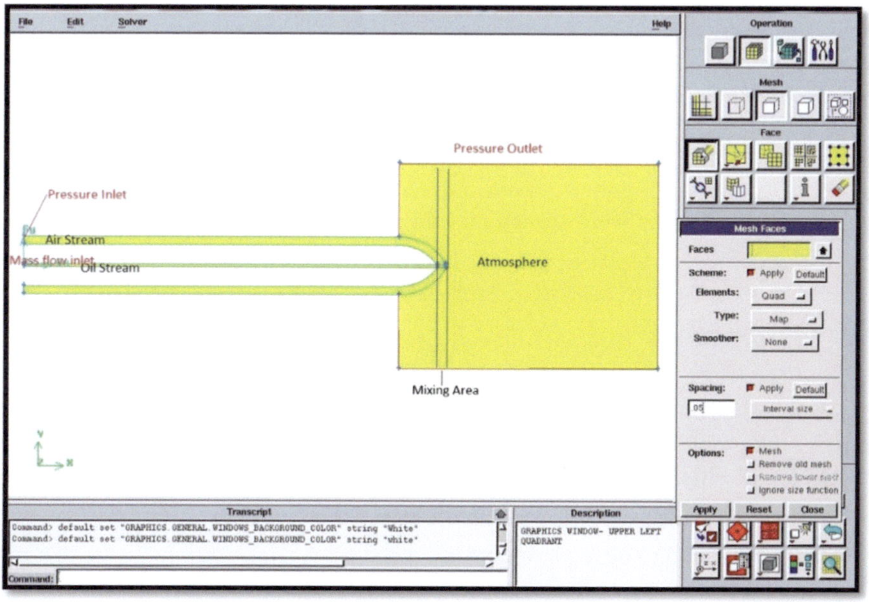

Fig. 8 Grid partitioning and boundary conditions

Meshing
In all the regions, the grid shape used is quadrilateral because of its less complexity and less computational time.

Air and Oil Stream
Air and oil stream was the least densely gridded pertaining to the fact that mixing is not happening in this region. The mesh size used for this region is 0.5 mm.

Mixing Area
It is known that a lot of turbulence and mass transfer takes place in this region. So, this region is densely meshed as compared to other regions to achieve better accuracy. The mesh size used for this region is 0.05 mm.

Atmosphere
Fluid activity in this region can be considered moderate and thus this region was meshed moderately dense. The mesh size used for this region is 0.1 mm.

2.3.2 Boundary Conditions

The boundary conditions for different regions were decided based on the constant inputs that were given for that particular region. The air stream input boundary was selected as "pressure inlet" owing to the fact that air is supplied at constant pressure from that boundary. Similarly, oil stream input and boundaries of the wall were selected as mass flow inlet and pressure outlet, respectively. Different domains such as air/oil streams, mixing area, and nozzle outlet were separated using "interior" boundary condition. Figure 9 shows different boundary conditions for the model and nozzle dimensions were shown in Table 1.

2.4 Assumptions

The structure of nozzle, spray characteristic, and the boundary conditions enable to consider the problem as a 2D thereby simplifying the overall simulation. Therefore, the model of simulation was built in 2D for computational simplification. The assumptions made for fluid properties are given below:

1. Air is considered to be a continuous fluid and oil is considered to be the discrete phase owing to the fact that volume fraction of oil occupies lower volume fraction (less than 10%) that perfectly fits the model of MQL.

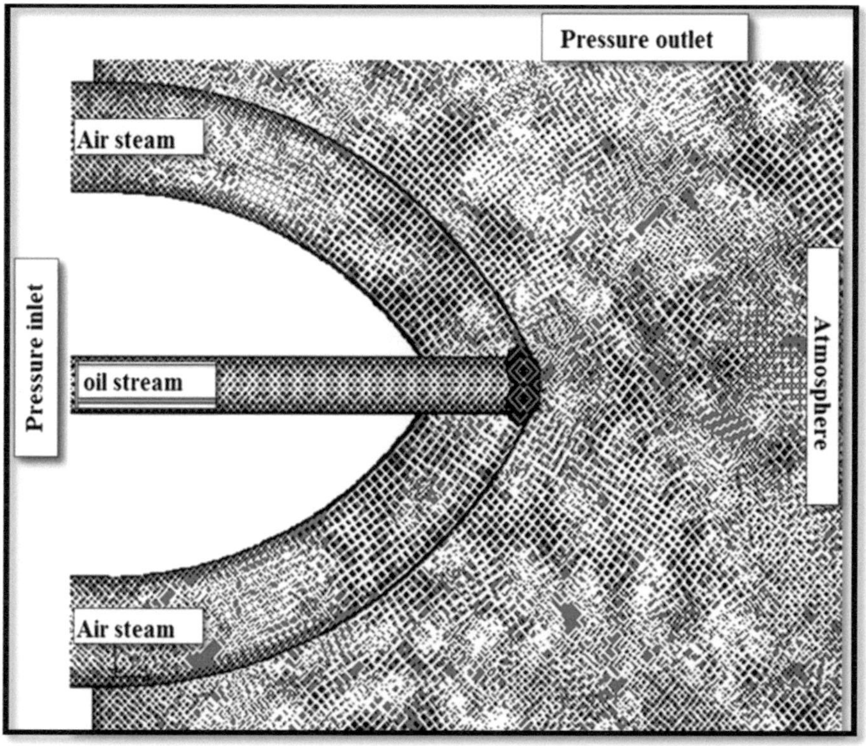

Fig. 9 Meshed model with boundary conditions

Table 1 Dimensions of the nozzle

S. No.	Parameters	Value
1	Air orifice diameter	0.6 mm
2	Oil orifice diameter	0.7 mm
3	Ratio	1.16
Coolant properties		
4	Density	900 kg/m^3
5	Viscosity	24 mm^2/s

2. There is no heat exchanged between air and oil and temperature of both the fluids is constant.
3. Particles are injected at the interval of every 1 μs which can be considered as a continuous flow.
4. Oil droplets are assumed to be perfectly spherical.

2.5 CFD Simulation

The three set of values for the two input boundary conditions were tried to create simulation results for nine cases which exhaust all the possible combinations with three sets of air pressure values and three sets of mass flow rates. The standoff distance is considered to be constant for all the case. Table 2 shows the values for two different variable parameters.

The computing models for atomization of mist were selected after observing similar types of simulations in different fields like IC engines, medical application, and fire extinguishers. Among the three different multiphase models available namely the Mixture model, the Volume of Fluid (VOF) model, and the Discrete Phase Model. Normally Eulerian approach is used in VOF model for studying the two-phase flow in fluids, but it has its own limitation during simulation due to the presence of finer droplet sizes in MQL process. The computational time and cost will be increased for simulating such micron-sized droplet breakup. Hence, simulation of the atomization under turbulent conditions was done in discrete phase model (DPM). The Euler–Lagrange approach is followed by DPM model in which air is considered as a continuum and cutting fluid particles as discrete particles. Mass, momentum, and energy equations are used for obtaining continuous or primary phase in this model. Solving transportation and force balance equations gives properties of DPM. Coalescence and particle breakup were determined by predefined sub-models in the CFD software. The discrete secondary phase was simulated using Lagrangian approach [24]. Lower volume fraction, i.e., less than 10% is occupied by secondary discrete phase. Table 3 shows the different models used and conditions applied. The continuous phase interaction was included during the simulation in order to minimize the approximations and bring the simulated observations close to the experimental, because of the higher velocity difference between the two phases. This allows the discrete phase, which is oil droplets, in this case, to exchange momentum, mass, and energy with air. As the spray was observed for 0.2 s, unsteady particle tracking was kept on. Among the four different drag law available in FLUENT namely Stokes–Cunningham, spherical, nonspherical and high-Mach-number laws. The spherical law best fits this model. Considering particle as a sphere is the acceptable simplification for the oil mist formation in MQL in this spherical drag law. Droplet breakup simulation uses WAV breakup model [24].

Simulations conditions were solved until it achieves residual convergence. The limit for mass and energy imbalance is 0.1% input. The details about minimum and

Table 2 Parameters for different cases

Parameters	Value
Pressures	2, 3, 4 bars
Mass flow rate	60, 80, 100 ml/h
Standoff distance	100 mm

Table 3 Models and parameters used in FLUENT 6.3.26

Model/parameter	Conditions
Multiphase flow model	Discrete Phase Model (DPM)
No of elements	1, 66, 368
Mesh type	Uniform quad
Solver	DP, SIMPLE, second order
Turbulence model	Realizable $k - \varepsilon$
Interaction with continuous phase	On
Unsteady particle tracking	On
Particle time step size (s)	0.000001
Droplet breakup	Wave type
Drag law	Spherical

Table 4 Parameters for the droplet distribution

Model/parameter	Value
Diameter distribution	Rosin-Rammler
Minimum diameter (μm)	10
Maximum diameter (μm)	100
Mean diameter (μm)	55

the maximum diameter of the droplets produced by the nozzle are provided by the manufacturer (Spraying Systems Co., USA). Table 4 shows the parameters used for droplet distribution. Rosin-Rammler distribution model was used. This model used laser diffraction particle sizer Malvern 2600 which always shows better experimental fit to the data. This distribution provides data sizes based on the exponential relationship between diameter and mass fraction.

2.5.1 Turbulence Model

There are a number of turbulent models available. A single turbulence model is never accepted universally as being superior for various classes of problems. Appropriate models are being used based on application definition. The turbulent models used for computing generally consist of zero equations model, one or two-equation models. Among these models, the realizable $k - \varepsilon$ model is the most preferred turbulence model in engineering computations and more accurate results can be attained in case of high Reynolds number. It is a two-equation RANS (Reynolds averaged Navier–Stokes) based model which describes the turbulence by means of two transport equations (PDEs). For this high-speed, high-pressure simulation, realizable $k - \varepsilon$ model was selected, which made use of realizable $k - \varepsilon$ model equations.

Continuity equation is given by

$$\frac{\partial u}{\partial x} + \frac{\partial u}{\partial y} = 0 \tag{4}$$

X-axis momentum equation is

$$\begin{aligned}\frac{\partial(\rho u^2)}{\partial x} + \frac{\partial(\rho uv)}{\partial y} &= \frac{\partial}{\partial x}\left(\mu_e \frac{\partial v}{\partial x}\right) + \frac{\partial}{\partial y}\left(\mu_e \frac{\partial u}{\partial y}\right) \\ &\quad - \frac{\partial P}{\partial x} + \frac{\partial}{\partial x}\left(\mu_e \frac{\partial v}{\partial x}\right) + \frac{\partial}{\partial y}\left(\mu_e \frac{\partial u}{\partial x}\right)\end{aligned} \tag{5}$$

Y-axis momentum equation is

$$\begin{aligned}\frac{\partial(\rho uv)}{\partial x} + \frac{\partial(\rho v^2)}{\partial y} &= \frac{\partial}{\partial x}\left(\mu_e \frac{\partial v}{\partial x}\right) + \frac{\partial}{\partial y}\left(\mu_e \frac{\partial v}{\partial y}\right) \\ &\quad - \frac{\partial P}{\partial y} + \frac{\partial}{\partial x}\left(\mu_e \frac{\partial u}{\partial x}\right) + \frac{\partial}{\partial y}\left(\mu_e \frac{\partial v}{\partial y}\right)\end{aligned} \tag{6}$$

Equation of turbulent kinetic energy is

$$\frac{\partial(\rho uk)}{\partial x} + \frac{\partial(\rho vk)}{\partial y} = \frac{\partial}{\partial x}\left(\left(\mu + \frac{\mu_t}{\delta_k}\right)\frac{\partial k}{\partial x}\right) + \frac{\partial}{\partial y}\left(\left(\mu + \frac{\mu_t}{\delta_k}\right)\frac{\partial k}{\partial y}\right) + PV - \rho\varepsilon \tag{7}$$

Dissipation rate equation ε of turbulent kinetic energy is

$$\begin{aligned}\frac{\partial(\rho u\varepsilon)}{\partial x} + \frac{\partial(\rho v\varepsilon)}{\partial y} &= \frac{\partial}{\partial x}\left(\left(\mu + \frac{\mu_t}{\delta_k}\right)\frac{\partial k}{\partial x}\right) + \frac{\partial}{\partial y}\left(\left(\mu + \frac{\mu_t}{\delta_k}\right)\frac{\partial k}{\partial x}\right) \\ &\quad + (C_1 P_P - C_2 \rho\varepsilon)\frac{\varepsilon}{k},\end{aligned} \tag{8}$$

where u, v are the axial and radial velocity of air, respectively. ρ is the density of air. P is the static pressure of air, μ is the dynamic coefficient,

$\mu_e = \mu + \mu_t$ is the effective viscosity coefficient,

$\mu_t = C_\mu \rho k^2/\varepsilon$ is the vortex viscosity coefficient, C_μ, C_1, C_2 are the constants and gained through experiments.

More accurate prediction of the rate of spreading of both spherical and planar jets is realizable $k - \varepsilon$ model's benefit. Flows involving rotation, separation, boundary layers under strong adverse pressure gradient, and recirculation superior performance can be likely provided by this model. The equations of this turbulence model are inbuilt within FLUENT and thus input was given so that FLUENT could solve these equations to generate a simulation.

2.5.2 Discrete Phase Model (DPM)

DPM follows Euler–Lagrange approach in FLUENT. Discrete second-phase simulation in a Lagrangian frame of reference can be achieved through FLUENT. The spherical droplets of oil dispersed in the continuous phase constitute the secondary phase. FLUENT performs the trajectory computation of the droplets by solving the time-averaged Navier–Stokes equations tracking a large number of droplets through flow field calculation for air and oil. The DPM takes into account the exchange of mass, energy, and momentum of droplets along with the air. Low-volume fraction occupied by the dispersed phase is the fundamental assumption of this model, which perfectly fits the model of MQL in which the mass flow rates are of the order of microliters. FLUENT already has the governing equations of DPM incorporated in the software. Inputs need to be provided to the software so that it can solve these equations to generate a simulated result.

Governing Equations for DPM

Integration of the force balance on the particle is used for predicting the trajectory of the droplet, stated in a Lagrangian reference frame.

The equation of force balance in the *x*-direction is

$$\frac{\partial u_{\mathrm{drop}}}{\partial t} = F_{\mathrm{D}}\left(u - u_{\mathrm{drop}}\right) + \frac{gx\left(\rho_{\mathrm{drop}} - \rho\right)}{\rho_{\mathrm{drop}}} + F_{\mathrm{x}}, \tag{9}$$

where F_{x} is called as additional acceleration (force per unit particle mass) term, $F_{\mathrm{D}}\left(u - u_{\mathrm{drop}}\right)$ is the drag force/unit particle mass and

$$F_{\mathrm{D}} = \frac{18\mu}{\rho_{\mathrm{drop}} d_{\mathrm{drop}}^2} \frac{C_{\mathrm{D}} Re}{24}, \tag{10}$$

where u is the fluid phase velocity, u_{drop} is termed as droplet velocity, μ is called molecular viscosity of the fluid, ρ is air density, ρ_{drop} is the density of the oil, and d_{drop} is the particle diameter. *Re* is the relative Reynolds number, defined as

$$Re = \frac{\rho d_{\mathrm{drop}} \left| u_{\mathrm{drop}} - u \right|}{\mu} \tag{11}$$

and drag coefficient C_{D} is given by,

$$C_{\mathrm{D}} = a_1 + \frac{a_2}{Re} + \frac{a_3}{Re^2}, \tag{12}$$

where a_1, a_2, a_3 are smooth spherical particle's drag constants.

FLUENT 6.3.26 incorporates other forces like the force of "virtual mass" which is defined as the force required accelerating the fluid surrounding the particle. The additional force includes forces on particles that are exhibited due to reference frame rotation. Turbulence in the fluid phase causes dispersion and this is predicted by a tracking model called stochastic model which is included in FLUENT 6.3.26 software.

2.6 MQL Droplet Size Prediction

In the previous section, computational models and conditions used and the input values of the parameters were discussed. The simulation was run for all the nine conditions as inputs were fed to the boundary conditions. The iterations were run for unsteady state till the simulation for 0.2 s for each condition was obtained. The iterations were run till the residual error on the mesh was less than 10^{-4}. Summary after post-processing for DPM was obtained for all the simulations and SMD was taken as the droplet diameter. SMD is most widely used in these types of models where aerosol sprays are simulated. SMD is defined as the diameter of a spherical droplet which has the same ratio of surface area/volume as that of the whole spray. Mean droplet velocity was also recorded for further reference. Grid independence tests were performed to achieve an optimum element size. The standoff distance is considered to be constant for all the cases. Summary after post-processing for DPM was obtained for all the simulations. The experiments were repeated to check the repeatability. Simulated SMD and mean velocity are given in Table 5. In a two-phase (gas–liquid) jet of MQL, droplet characteristics are mostly affected by the primary phase (compressed air). The increase in the air supply pressure results in a reduction of droplet size and the droplet size decreases with increased flow rate. Particles velocity increases, with an increased rate of pressure and mass flow rate. The maximum droplet size (24.9 μm) was found to in the case of 2 bar air pressure

Table 5 SMD and mean velocity for different simulation condition

Pressure (in bars)	Flow rate (ml/h)	Sauter mean diameter SMD (μm)	Velocity of droplet (m/s)
2	60	24.9	21.9
2	80	14.2	27.5
2	100	8.1	30.1
3	60	22.4	25.2
3	80	11.7	33.1
3	100	7.3	36.2
4	60	20	29.4
4	80	9.3	38.7
4	100	6.6	42.4

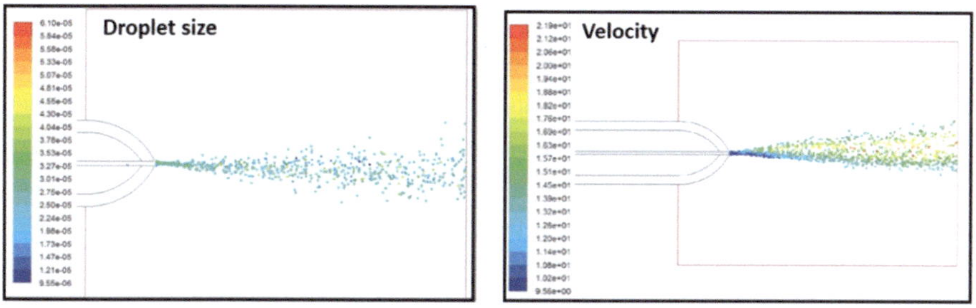

(a) Maximum droplet (24.9μm) size with respective mean velocity (21.9m/s)

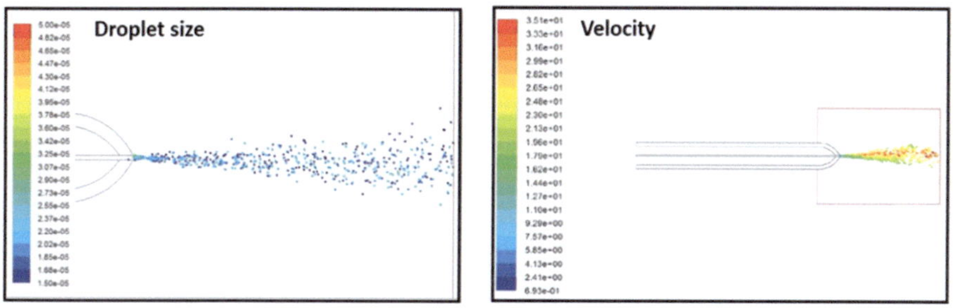

(b) Medium droplet (20μm) size with respective mean velocity (29.4m/s)

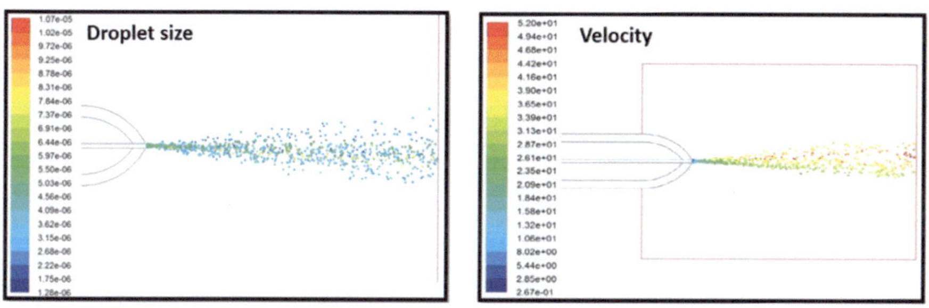

(c) Minimum droplet (6.6 μm) size with respective mean velocity (42.4 m/s)

Fig. 10 Contours of **a** higher **b** medium and **c** lower droplet size with its velocity

and 60 ml/h mass flow rate whereas the minimum droplet size (6.6 μm) was found in the case of 4 bar air pressure and 100 ml/h are shown in Fig. 10.

The effect of MQL depends on the accessibility of the oil droplet and consequent wetting of the tool–work interface. As discussed in earlier section, larger droplets with large inertia may fall early and not travel in the air stream whereas very small droplets are not able to carry up to the cutting zone due to their low mass and deviate from the flow path thereby resulting in inefficient lubrication. So, in the simulation results, a range of droplet size is observed to obtain a favorable distribution of droplet.

2.7 Experimental Validation

In order to validate the simulation results, a droplet size analyzer Malvern Spraytec (manufactured by Malvern Instruments, UK) was used. Laser diffraction techniques were used to find the distribution of droplet size. The setup of the Malvern Analyzer is illustrated in Fig. 11. It is rapid, nondestructive and requires no external calibration. The extremely fast, 10 kHz, data acquisition rate of the system produces real-time particle size distributions with a 100-μs resolution.

Figure 12b shows the mist formation in MQL. The droplet velocity was measured using a high-speed camera (*N*-type) of Panatech Asia and shown in Fig. 12a. The droplet tracking was done using proanalyst software as shown in Fig. 13. The comparisons between numerical and experimental data of droplets size were carried out and it is shown in Fig. 14. The comparisons between numerical and experimental data of droplets size were carried out. The numerically simulated values are matching with experimental values and these deviations are due to the limitations in submodels and assumptions in discrete phase modeling. The numerically simulated values are similar with experimental values at lower and medium mass flow rate. At higher mass flow rate and air pressure of (4 and 6 bar), the predicted SMD is higher and values are tabulated in Table 6. The observation on droplet size shows for all mass flow rate and pressure are overpredicted. The comparisons between numerical and experimental data of droplet size and velocity were compared and it is shown in

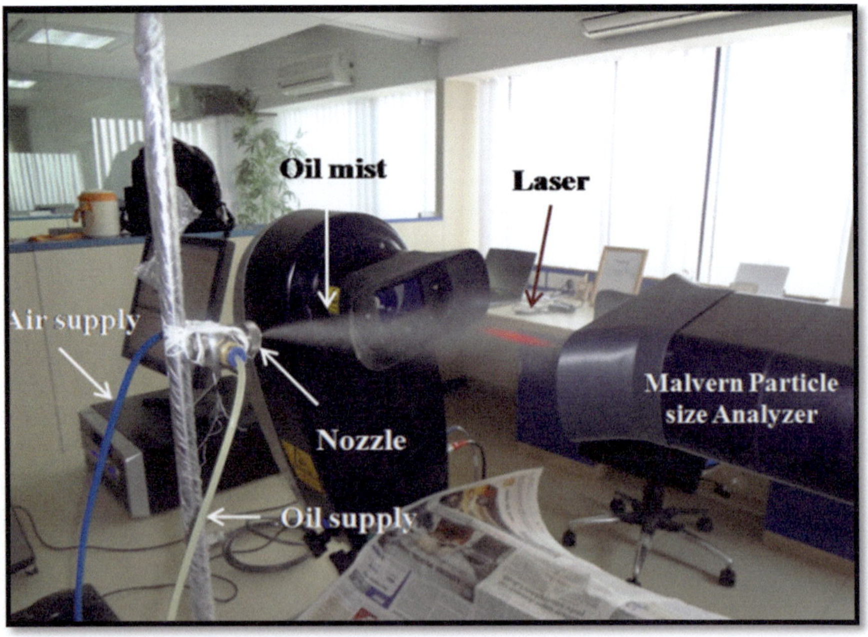

Fig. 11 Setup for measuring droplet size (Malvern Instruments, UK)

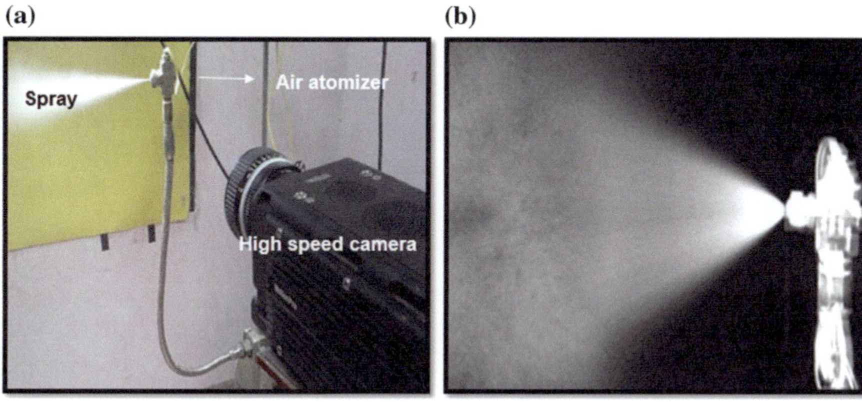

Fig. 12 **a** High-speed camera. **b** Mist formation

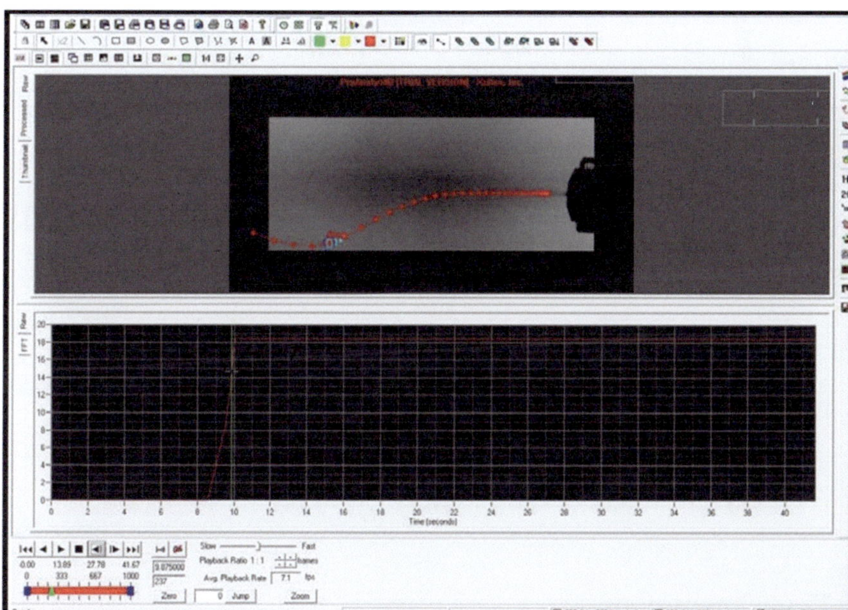

Fig. 13 Droplet velocity tracking

Figs. 14 and 15. Simulated results were closely matching with the experimental results. For velocity of 60 ml/h, the experimental velocity was lower than the predicted velocity. For 100 ml/h, experimental values were higher. The simulated results for 80 ml/h, with 4 and 6 bar are over predicted when compared with experimental values.

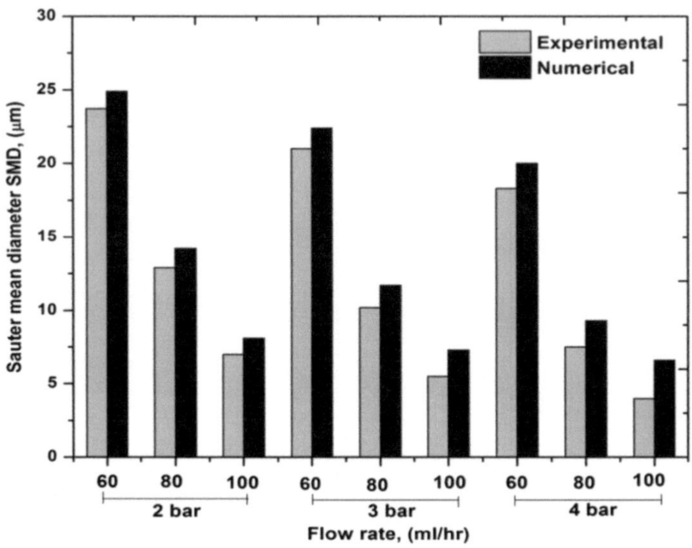

Fig. 14 Comparison of numerical versus experimental data

Table 6 Experimental and numerical SMD and velocity values

Pressure (bar)	Flow rate (ml/h)	Experimental SMD (μm)	Numerical SMD (μm)	Experimental velocity (m/s)	Numerical velocity (m/s)
2	60	23.7	24.9	19.7	21.9
2	80	12.9	14.2	25	27.5
2	100	7	8.1	32	30.1
3	60	21	22.4	23.3	25.2
3	80	10.2	11.7	33.5	33.1
3	100	5.5	7.3	38.8	36.2
4	60	18.3	20	27	29.4
4	80	7.5	9.3	42	38.7
4	100	4	6.6	45.7	42.4

3 Experimentation

Stainless steels are desirable for their excellent creep-rupture strength, high corrosion and oxidation resistance in high-temperature environments. The difficulties encountered when machining such material can be attributed to their thermo-physical properties. During milling, the heat generated does not dissipate quickly, and as a result, most of the heat remains concentrated on the workpiece surface and produces a negative effect on its surface integrity and dimensional accuracy. Thereby, material work hardens rapidly and makes it tougher for the cutting edge to remove material efficiently, resulting in rapid tool wear.

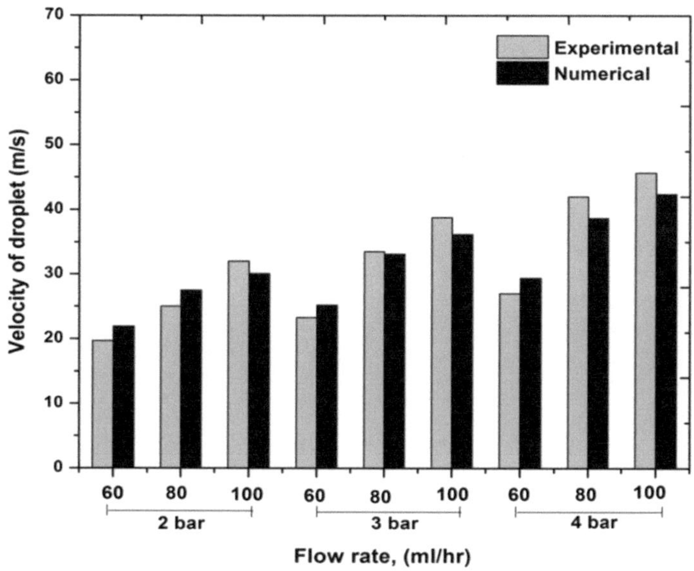

Fig. 15 Comparison of numerical versus experimental data

3.1 Experimental Setup

End milling experiments were carried on Austenitic stainless steel 304 grade. Milling conditions are listed in Table 7.

The performance indicators namely cutting and normal forces and surface roughness were determined. The workpiece was placed on a three-component piezoelectric dynamometer (Maker: Kistler, Model: 9257B) in order to measure

Table 7 Milling conditions

Parameters	Conditions
Cutting tool insert	TiAlN PVD carbide inserts
Cutting velocity (V_s)	125.6 m/min
Feed rate (f)	0.2 m/min
Depth of cut	0.5 mm
Flow rate of MQL oil	60, 80, 100 ml/h
Pressure	2, 3, 4 bar
Nozzle angle from feed direction/ spindle	135°/45°
Standoff distance	100 mm
MQL oil used	LRT 30 (Viscosity 40 °C:24cSt), Specific weight kg/l at 150C:0.900

Fig. 16 Machining setup

both cutting force (F_x) and thrust force (F_y) components of milling force. The MQL nozzle setup was attached to the spindle column so that it moves linearly along with it in order to maintain the nozzle position constant throughout the machining. The machining setup is shown in Fig. 16.

3.2 Experimental Results

The experimental values of cutting, normal force, and the surface roughness are shown in Table 8.

3.2.1 Effect of Air Pressure

Parameters like feed, cutting speed, and depth of cut being constant and their influence on MQL parameters such as varying flow rate and air pressure have been studied in the present work with a constant standoff distance of 100 mm and 135° nozzle angle with respect to the feed direction. Nozzle angle of 135° allows the MQL reach the tool–chip interface more effectively [22]. Typically monitored parametric (MQL) influence on surface roughness and cutting force components for varying air inlet pressure are shown in Fig. 17a–c. Varying air pressure has a noticeable effect on surface roughness and the cutting forces. As the air pressure increases, reduction in cutting forces and surface roughness can be observed.

Table 8 Experimental data for cutting forces and surface roughness under dry and MQL milling environments

Milling environments	Pressure (bar)	Flow rate (ml/h)	Cutting force F_x (N)	Thrust force F_y (N)	Surface roughness R_a (μm)
DRY	–	–	345	258	0.4827
MQL	2	60	194	181	0.4232
	2	80	132	143	0.3434
	2	100	167	161	0.3294
	3	60	181	172	0.4175
	3	80	133	139	0.3238
	3	100	158	149	0.3097
	4	60	169	163	0.4111
	4	80	134	134	0.3042
	4	100	150	150	0.2900

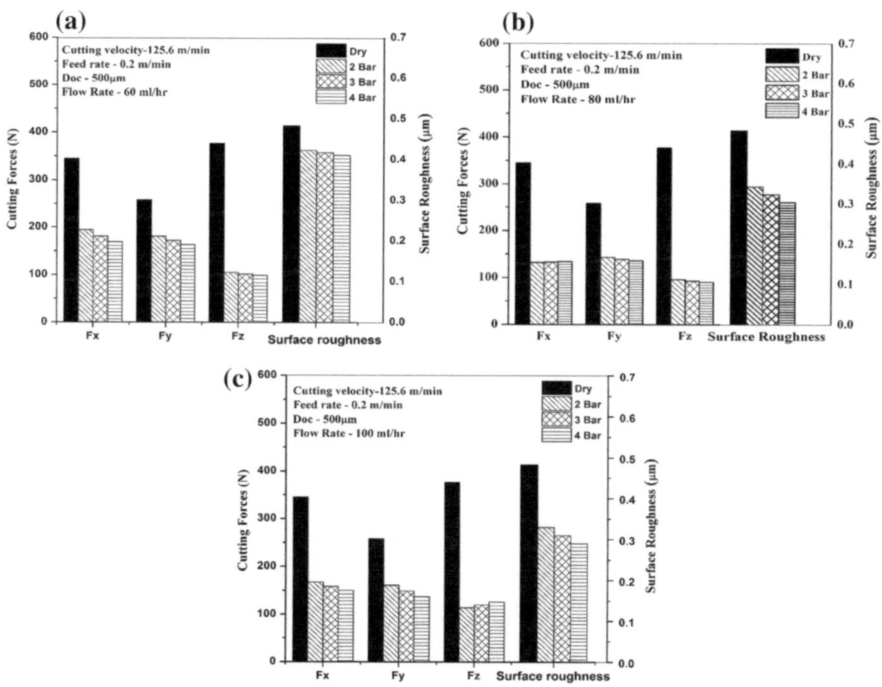

Fig. 17 **a** Variation of cutting forces and surface roughness with various air pressure. **b** Variation of cutting forces and surface roughness with various air pressure. **c** Variation of cutting forces and surface roughness with various air pressures

The reason is that higher order pressure (4 bar) causes an effective penetration of droplet into the pressure belt facilitating enhanced wetting (boundary lubricating environment) of the tool–chip interface. Pressure variation also has an influence on oil droplet diameter. Increasing air pressure will further divide oil mist droplets into tiny droplets and increases the exit velocity of the fluid which carries the droplets to the chip–tool interface [13]. This is replicated in the cutting forces and surface roughness values as shown in Fig. 17a–c. This fact can also be justified by the empirical relation given below. The following equation gives the relationship between oil droplet diameter, oil flow velocity and air pressure [25].

$$v = c_1 d^{c_2} \tag{13}$$

$$d = c_3 \ln(p) + c_4, \tag{14}$$

where c_1 c_2 c_3 c_4 are four coefficients

c_1, $c_4 > 0$ and c_2, $c_3 < 0$

The effective penetration of the oil mist will reduce the friction at the tool–chip interface which results in a reduction of cutting forces. Machining of SS304 incorporates formation of built-up edge, due to its lower thermal conductivity and higher ductility. Increasing air pressure allows the droplets of oil mist to carry away the heat from the machine tool interface more effectively thereby avoiding weld of chips to the tool surface and built-up edge formation which enhances the surface finish. It is clearly seen that there is a reduction by 47.3% in cutting forces when compared with dry milling as shown in Fig. 17b due to effective penetration of droplet in the machining zone.

3.2.2 Effect of Flow Rate

Typically monitored parametric (MQL) influence on surface roughness and cutting force components for varying oil flow rates is shown in Fig. 18a–c. As the oil flow rate increases, the cutting forces decreases up to certain optimum limit and then increases. It is clearly seen that there is a reduction in cutting forces by 36.8% compared to dry milling as shown in Fig. 18c. With increasing flow rate from 60 to 80 ml/h at various pressures a decreasing trend in the cutting forces were observed. The reason is an increase in mass flow rate that leads to thin film formation which will reduce the friction at the chip–tool interface thereby decreasing the cutting forces. The energy formation of material drops down (Rehbinder effect) as the film-forming tendency in the MQL increases thereby lubrication will be more effective which enhances surface finish.

Further, increase in oil flow rate from 80 to 100 ml/h at various pressure, results in reflection of droplets from the surface due to higher exit velocity leads to insufficient lubrication. Therefore, increasing trend in cutting forces is observed. Medium size droplets with higher pressure can penetrate easily into the machining zone and thereby provides effective lubrication.

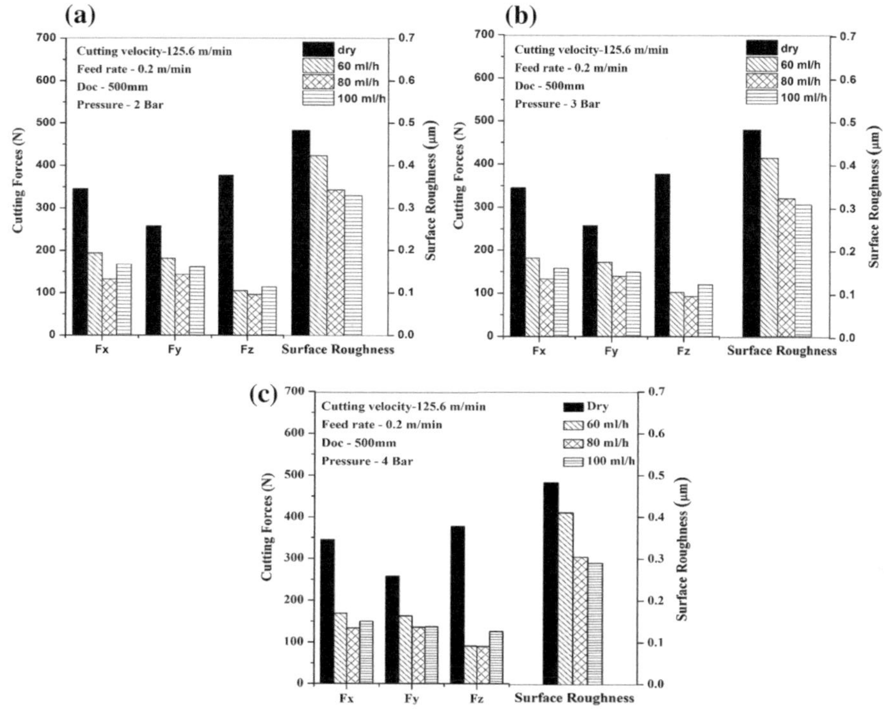

Fig. 18 **a** Variation of cutting forces and surface roughness with various mass flow rates. **b** Variation of cutting forces and surface roughness with various mass flow rates. **c** Variation of cutting forces and surface roughness with various mass flow rates

3.2.3 Effect of Droplet Size

Figure 19 illustrates the significance of spray parameter (droplet size) on milling performance. The regime of influence can be seen as region I (lower droplet size up to 7.5 μm), region II (medium size droplet size 7.5–12.9 μm), and region III (Larger droplet size > 12.9 μm). The normal force component tends to increase up to 7.5 μm droplet size then sudden decrease up to 12.9 μm droplet size and then increases drastically. The same trend was observed in surface roughness also. The droplet size has the most significant effect on the flight distance and is proportional to the square of the diameter [21]. Smaller droplets up to 6 μm (region I) find it difficult to move in the free air. So, it can easily change its track due to high cutting velocity and finally leads to insufficient lubrication at the tool–chip interface.

Larger droplets (>12.9 μm, region III) cannot be carried to machining zone due to their higher mass and leads to lubricant starvation in the milling zone. Hence,

Fig. 19 Effect of droplet size on milling force components and surface roughness

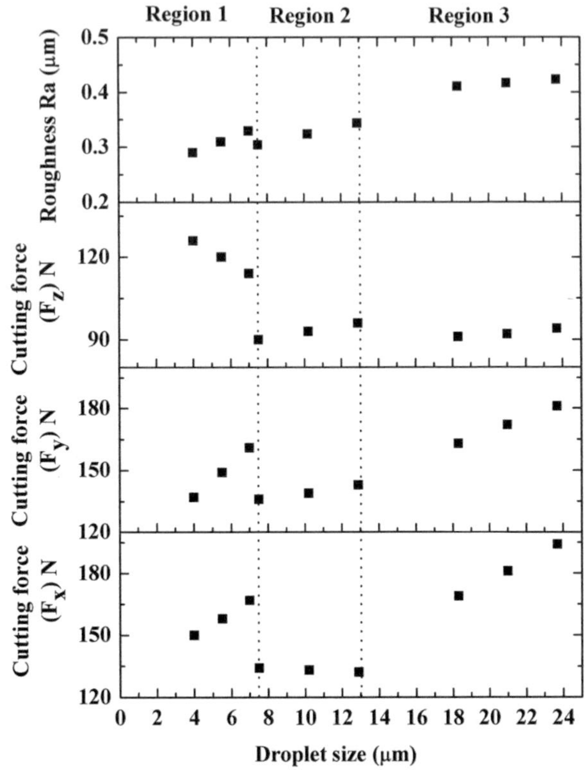

increased droplet size results in increased surface roughness and force components. Effective lubrication environment can be provided by medium-sized droplet as they can easily penetrate into cutting zone and increase surface quality. Flushing of the chips by high pressurized air (flushing effect) helps in maintaining better surface finish. The decrease in feed force can be noticed with MQL droplet sizes of (7.5–10.2 μm) due to better lubrication at the tool–chip interface. Hence, it can be clearly seen from the figure that medium-sized droplets (7.5–10.2 μm) with higher pressure (4 bar) gives effective lubrication in milling of SS304

3.2.4 Effect of Wetting Area

Tawakoli et al. suggested the following model to observe the effect of droplet size on the lubrication. The droplet radius in the MQL oil mist was assigned r_{drop}, thus the droplet volume in the outlet of the nozzle is calculated as follows:

$$V_{\mathrm{Drop}} = \frac{4}{3}\pi r_{\mathrm{drop}}, \quad (15)$$

where V_{Drop} is the volume of each droplet in the MQL oil mist. The number of oil droplets n_d along with volume V_{drop} in the spray jet per unit time can be expressed as

$$n_d = \frac{Q}{V_{\mathrm{drop}}}, \quad (16)$$

where Q is the MQL flow rate in $\mu m^3/s$ and V_{drop} is the droplet volume in μm^3. Deformation of the droplet is caused when it meets the surface of the workpiece and the contact diameter is $2r$. As shown in Fig. 20, the volume of droplet before and after impingement should be equal. Hence, wetting area r, with oil droplet after impingement on the surface, is

$$r = \sqrt[3]{\frac{V_{\mathrm{drop}}}{\frac{\pi}{2}\frac{1}{\cos\alpha}\left(\frac{1}{\cos\alpha} - 1\right)\left[1 + \frac{1}{3}\frac{1}{(\cos\alpha)2}\left(\frac{1}{\cos\alpha} - 1\right)2\right.}}, \quad (17)$$

where α is the droplet contact angle on the surface. Hence, the wetting area of droplet deformed on the surface can be defined as

$$A_{\mathrm{drop,surf}} = \pi r^2 \quad (18)$$

The total wetting area obtained by all spray droplets per unit time on the surface will be as follows:

$$A_{\mathrm{total,drop,surf}} = \mathrm{n_d}\, \pi r^2 \quad (19)$$

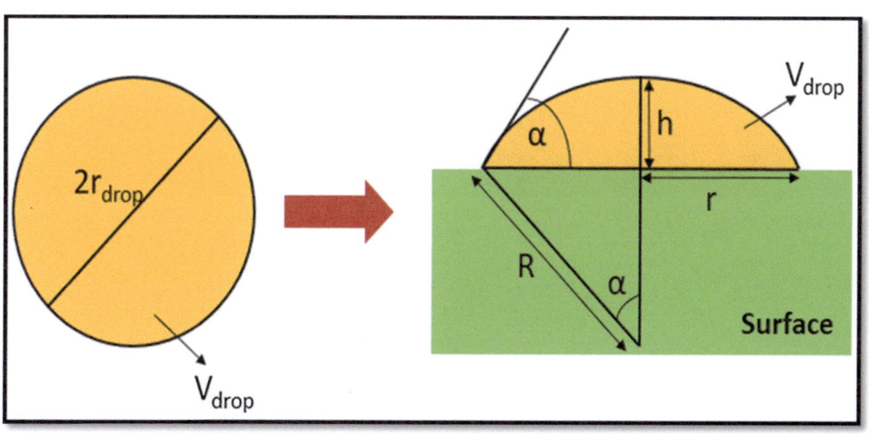

Fig. 20 Shape of the deformed oil droplet

Thus, it can be concluded that the only unknown parameter is the radius of the droplet. The objective of this simulation is to determine this radius to save time and cost of experimentation.

The wetting area is calculated based on the experimental and numerical data using equations and tabulated in Table 9. Figure 21 illustrates the significance of droplet size on the wetting area. Third-order polynomial fit was fitted and the wetting area was calculated based on the above analytical model. Droplet size plays a key role in wetting area calculation. Larger wetting area decreases machining temperature. The combined effect of lubrication and cooling results in such temperature decrement.

Lubrication helps in reducing the cutting forces and consumption of energy for cutting, whereas convectional transfer of heat and/or boiling carries away most of the cutting zone heat. From Table 9, it was found that finer droplet covers maximum tool–workpiece interface because smaller sized droplets result in higher n_d which lubricates more area, with the higher wetting area. From Fig. 21, it can be seen that increasing droplet size results in a reduction of wetting area, attributable to reduced accessibility. Increased droplet size with associated lower order velocity cannot penetrate the pressure belt. It is seen that up to around 12.9 μm of droplet size can lead to desirable MQL environment.

3.2.5 Micrograph Analysis

Tool wear is one of the important criteria for determining the machining quality and has an influence on quality and integrity of the machined surface. It is an

Table 9 Calculated values of deposited radius, number of droplets and the wetting area from experimental data

Air pressure (bar)	Mass flow rate (ml/h)	Droplet diameter (mm)	Droplet radius (r_{drop}) (mm)	V_{drop} (μm^3)	No of droplet n_d	Deposited radius (wetting area) (mm)	Wetting area (m^2/s)
2	60	23.7	11.85	6970	2,390,244	66	0.0325
2	80	12.9	6.45	1124	19,770,660	36	0.0802
2	100	7	3.50	180	154,320,988	19	0.1743
3	60	21	10.5	4847	3,437,178	58	0.0362
3	80	10.2	5.1	555	40,040,040	28	0.0984
3	100	5.5	2.75	87	319,284,802	15	0.2256
4	60	18.3	9.15	3209	5,191,648	23	0.0133
4	80	7.5	3.75	221	100,553,042	21	0.1393
4	100	4	2.00	33.5	829,187,396	11	0.3152

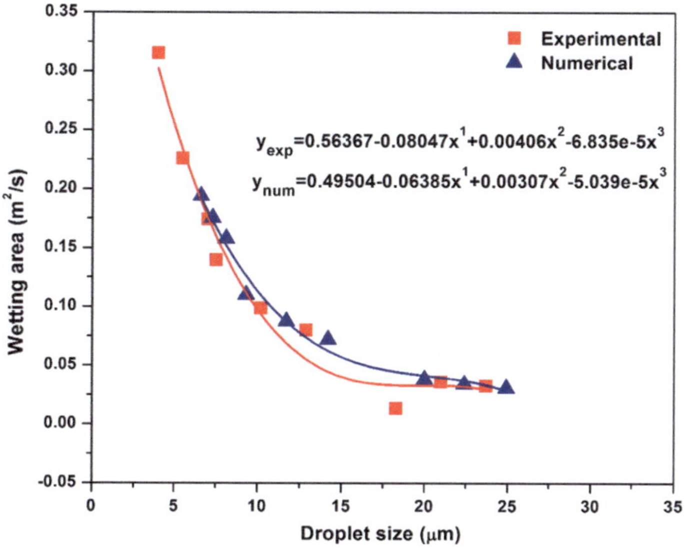

Fig. 21 Effect of droplet size on wetting area

unavoidable phenomenon which gradually increases by adhesion, diffusion, abrasion, chemical erosion, galvanic action, etc. [26]. In the present investigation, coated carbide inserts were used. Tool wear was examined at different MQL conditions and compared with that of dry milling. The micrograph images of the tool wear for different MQL conditions along with dry milling is as shown in Fig. 22.

Microchipping was visible in dry milling due to lack of lubrication Fig. 22a, whereas relatively less wear was observed in milling under MQL condition. The fragmented medium size droplet region 2 (7.5–12.9 μm) discussed in Fig. 19 associated with medium flow rate at various pressures of the cutting oil was able to penetrate into the tool–chip and tool–work interfaces more effectively and as a result, reduces the chip–tool contact area. Also, the high pressure will minimize the welding of chips to the tool rack face and overcome the problem of built-up edge thereby enhancing effective lubrication and reduction in friction and temperature which leads to minimum tool wear. Whereas, droplets size which falls in region 1 (lower droplet size up to 7.5 μm) and region 3 (Larger droplet size >12.9 μm) do not penetrate the cutting zone more effectively as discussed earlier which is reflected on the tool wear in the micrograph images. The flow rate has a major influence on the spray parameters. Therefore, minimum tool wear was observed for 80 ml/h flow rate and 4 bar pressure corresponding to the droplet size of 7.5 μm as shown in Fig. 22c and relative higher wear was observed for 60 ml/h flow rate and 2 bar pressure corresponding to the droplet size of 23.7 μm as shown in Fig. 22b.

Milling parameters	Magnification (200 X)	Magnification (1000 X)
A **Dry milling** **V_s = 125.6 m/min** **Feed rate = 0.2 m/min** **Doc = 0.5 mm**		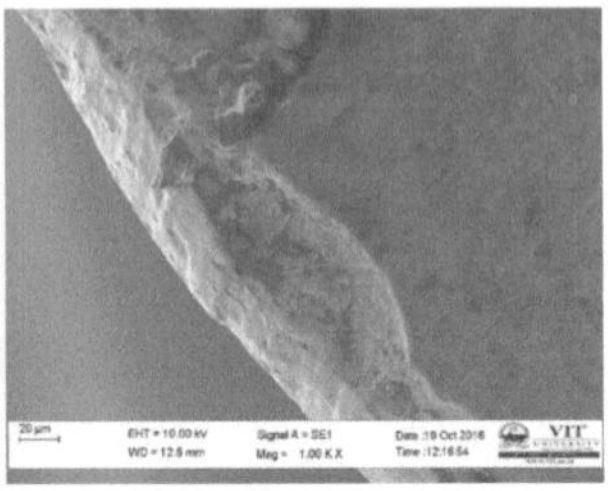
B **MQL milling** **2 bar 60 ml/hr** **Droplet dia = 23.7μm** **V_s = 125.6 m/min** **Feed rate = 0.2 m/min** **Doc = 0.5 mm**		
C **MQL milling** **4 bar 80 ml/hr** **Droplet dia = 7.5μm** **V_s = 125.6 m/min** **Feed rate = 0.2 m/min** **Doc = 0.5 mm**	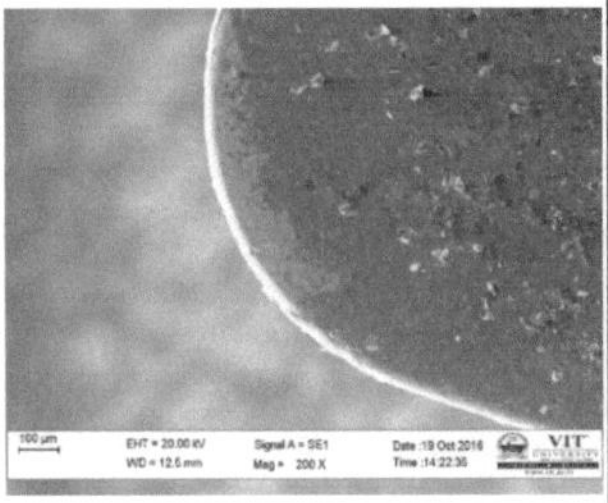	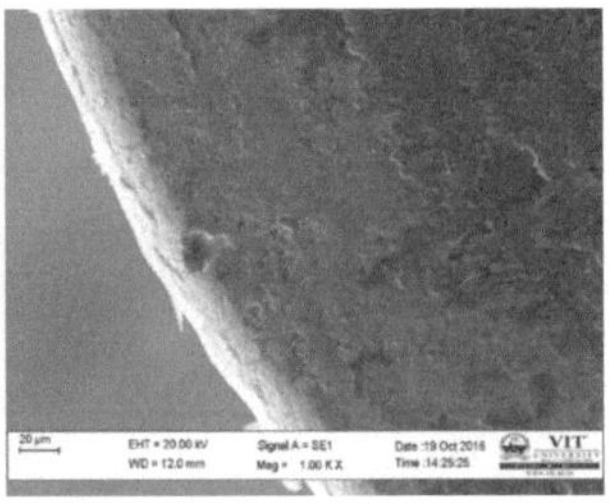

Fig. 22 Tool wear under different MQL condition and dry milling

4 Conclusions

A suitable environment can be chosen to mitigate the problems associated with milling. The present study is concerned with milling of SS304 in MQL environment to evaluate the significance of spray parameters like droplet size and velocity. The potential of the experimental results combined with the numerical modeling is used as a powerful tool for better understanding of the process observed during the MQL milling of SS304 and to analyze the influence of droplet size on milling performance.

The main conclusions obtained from the present investigations are as follows:

- The effectiveness of the MQL lies on the nature of film formation and wetting area on the milling zone, this is largely influenced by droplet size and velocity.

- The mass flow rate and air pressure influence the droplet size and velocity.
- The effectiveness of the MQL condition depends on the effective penetration of pressure belt around the tool and consequent access to the tool–work interface.
- The numerical simulation clearly indicates that the droplet size decreases with increasing flow rate and atomization pressure. The simulated results are in accordance with existing empirical model and experimentally measured values.
- Droplet size plays a key role in MQL milling. The medium-sized droplet with higher pressure can effectively lubricate the tool–chip interface more effectively.
- The droplets ranging from (7.5–12.9 μm) give effective lubrication and reduce cutting forces and enhance surface finish.
- It was observed that among the MQL parameters, varying oil mass flow rates have more influence than the air pressure on cutting forces and surface roughness.

References

1. Abou-El-Hossein, Z. Y. K. A. (2005). High speed end milling of AISI 304 stainless steel using new geometrically developed carbide inserts (pp. 1–6). In *Proceedings of the 13th International Science Conference*.
2. Kedare, S. B., Borse, D. R., & Shahane, P. T. (2014). Effect of minimum quantity lubrication (MQL) on surface roughness of mild steel of 15HRC on universal milling machine. *Procedia Materials Science*, *6*, 150–153.
3. Park, K. H., Olortegui-Yume, J., Yoon, M. C., & Kwon, P. (2010). A study on droplets and their distribution for minimum quantity lubrication (MQL). *International Journal of Machine Tools and Manufacture, 50*(9), 824–833.
4. Sharif, S., Yusof, N. M., Idris, M. H., Ahmad, Z. B., Sudin, I., Ripin, A., et al. (2009). *Feasibility study of using vegetable oil as a cutting lubricant through the use of minimum quantity lubrication during machining* (pp. 1–37). Malaysia: University of Technology.
5. Weinert, K., Inasaki, I., Sutherland, J. W., & Wakabayashi, T. (2004). Dry machining and minimum quantity lubrication. *CIRP Annals—Manufacturing Technology, 53*(2), 511–537.
6. Dudzinski, D., Devillez, A., Moufki, A., Larrouquère, D., Zerrouki, V., & Vigneau, J. (2004). A review of developments towards dry and high speed machining of Inconel 718 alloy. *International Journal of Machine Tools and Manufacture, 44*(4), 439–456.
7. Tiwana, J. S., & Pal, A. (2014). A review study on minimum quantity lubrication in machining. *IJRDET, 2*(5), 2012–2015.
8. Balan, A. S. S., Vijayaraghavan, L., & Krishnamurthy, R. (2013). Minimum quantity lubricated grinding of Inconel 751 alloy. *Materials and Manufacturing Processes, 28*(04), 430–435.
9. Dhar, N. R. A., Ahmed, M. T., & Islam, S. (2007). An experimental investigation on effect of minimum quantity lubrication in machining AISI 1040 steel. *International Journal of Machine Tools and Manufacture, 47,* 748–753.
10. Rahman, M., Senthil Kumar, A., & Salam, M. U. (2002). Experimental evaluation on the effect of minimal quantities of lubricant in milling. *International Journal of Machine Tools and Manufacture, 42*(5), 539–547.
11. Uysal, A., Demiren, F., & Altan, E. (2015). Applying minimum quantity lubrication (MQL) method on milling of martensitic stainless steel by using nano Mos 2 reinforced vegetable cutting fluid. *Procedia-Social and Behavioral Sciences, 195,* 2742–2747.

12. El, C., Arunachalam, N., & Vijayaraghavan, L. (2015). Analytical model to predict Sauter mean diameter in air assisted atomizers for MQL in machining application. *Procedia CIRP, 37,* 117–121.
13. Lutao, Y. A. N., Songmei, Y., & Qiang, L. I. U. (2012). Influence of minimum quantity lubrication parameters on tool wear and surface roughness in milling of forged steel. *Chinese Journal of Mechanical Engineering, 25*(3), 419–429.
14. Tawakoli, T., Hadad, M. J., & Sadeghi, M. H. (2010). Influence of oil mist parameters on minimum quantity lubrication—MQL grinding process. *International Journal of Machine Tools and Manufacture, 50*(6), 521–531.
15. Rahman, M. (2012). A computational fluid dynamics analysis of single and three nozzles minimum quantity lubricant flow for milling. *International Journal of Automotive and Mechanical Engineering*, *6*, 768–776.
16. Heisel, U., Lutz, M., Spath, D., Wassmer, R. A., & Walter, U. (1994). Application of minimum quantity cooling lubrication technology in cutting processes. *Production Engineering, 2*(1), 49–54.
17. Schick, R. J. (1997). Spray technology reference guide: Understanding drop size preface (p. 6). In *47th Chemical Processing Industry Expo*.
18. Maruda, R. W., Krolczyk, G. M., Feldshtein, E., Pusavec, F., Szydlowski, M., Legutko, S., et al. (2016). A study on droplets sizes, their distribution and heat exchange for minimum quantity cooling lubrication (MQCL). *International Journal of Machine Tools and Manufacture*, *100*, 81–92.
19. Trela, M., Zembik, J., & Durkiewicz, B. (1982). Droplet deposition on a flat plate from an air/water turbulent mist flow. *International Journal of Multiphase Flow, 8*(3), 227–238.
20. Veysey, J., & Goldenfeld, N. (2007). Simple viscous flows: From boundary layers to the renormalization group. *Reviews of Modern Physics, 79*(3), 883–927.
21. Obikawa, T., Asano, Y., & Kamata, Y. (2009). Computer fluid dynamics analysis for efficient spraying of oil mist in finish-turning of Inconel 718. *International Journal of Machine Tools and Manufacture, 49*(12–13), 971–978.
22. Lopez de Lacalle, L. N., Angulo, C., Lamikiz, A., & Sanchez, J. A. (2006). Experimental and numerical investigation of the effect of spray cutting fluids in high speed milling. *Journal of Materials Processing Technology, 172,* 11–15.
23. Park, K. H., Olortegui-Yume, J. A., Joshi, S., Kwon, P., Yoon, M. C., Lee, G. B., et al. (2008). Measurement of droplet size and distribution for minimum quantity lubrication (MQL) (pp. 447–454). In *International Conference on Smart Manufacturing Application*.
24. ANSYS (2013). ANSYS FLUENT User's Guide", vol. 15317, no. November, p. 2498.
25. Liu, Z. Q., Cai, X. J., Chen, M., & An, Q. L. (2011). Investigation of cutting force and temperature of end-milling Ti-6Al-4 V with different minimum quantity lubrication (MQL) parameters. *Proceedings of the Institution of Mechanical Engineers, Part B: Journal of Engineering Manufacture, 225*(8), 1273–1279.
26. Khan, M. M. A., Mithu, M. A. H., & Dhar, N. R. (2009). Effects of minimum quantity lubrication on turning AISI 9310 alloy steel using vegetable oil-based cutting fluid. *Journal of Materials Processing Technology, 209,* 5573–5583.

Finite Element Method and Experimental Study of Self-reacting Friction Stir Welding of Aluminium Alloy AA6061-T6

Piyush Singh, Pankaj Biswas and Sachin D. Kore

1 Introduction

Friction Stir Welding (FSW) was invented by Wayne Thomas at The Welding Institute (TWI) in 1991. The invention of FSW was an important breakthrough as it could overcome many issues related to fusion welding techniques. It produces high-quality welds in difficult-to-weld materials like aluminium alloys which are prone to solidification cracking, shrinkage, and porosity when fusion welded. It is a solid state welding process in which a non-consumable rotating tool as illustrated in Fig. 1 traverses along the abutting edges of the plates to be joined and produces a joint without melting the materials. Due to the low heat input, the joints produced have lower distortion and residual stresses. The tool has a pin or probe and a shoulder. The tool primarily has three functions: (a) heating the workpiece, (b) movement of the heated plasticized material to form the joint and (c) containing the hot plasticized material beneath the shoulder. The heating is achieved by the friction between the tool and workpiece surface and also the plastic deformation of the workpiece material. The welding cycle consists of four main stages: (1) Tool plunge, (2) Dwell, (3) Tool advance/traverse and (4) Tool withdrawal. There are a few variants of FSW as shown in Fig. 2 based on variations in tool design and type of weld produced.

Bobbin Tool Friction Stir Welding (BTFSW) or Self-Reacting Friction Stir Welding (SRFSW) is a variant of FSW in which there is a modification in the tool

P. Singh (✉) · P. Biswas · S. D. Kore
Department of Mechanical Engineering, IIT Guwahati, Guwahati, Assam, India
e-mail: s.piyush@iitg.ernet.in; mailstopiyush@gmail.com

P. Biswas
e-mail: pankaj.biswas@iitg.ernet.in

S. D. Kore
e-mail: sdk@iitg.ernet.in

U. S. Dixit and R. Kant (eds.), *Simulations for Design and Manufacturing*,
Lecture Notes on Multidisciplinary Industrial Engineering,
https://doi.org/10.1007/978-981-10-8518-5_3

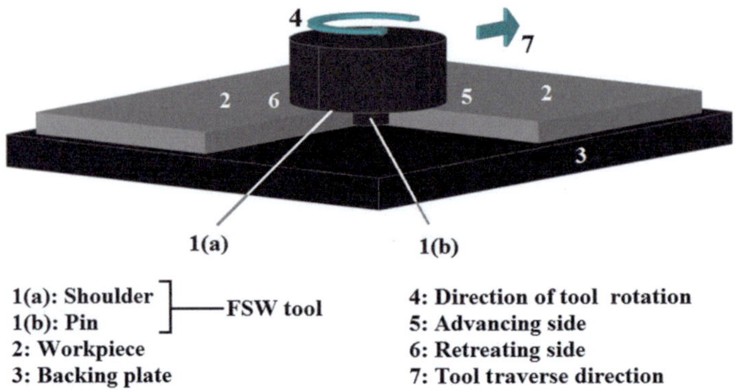

Fig. 1 Schematic illustration of the friction stir welding (FSW) process

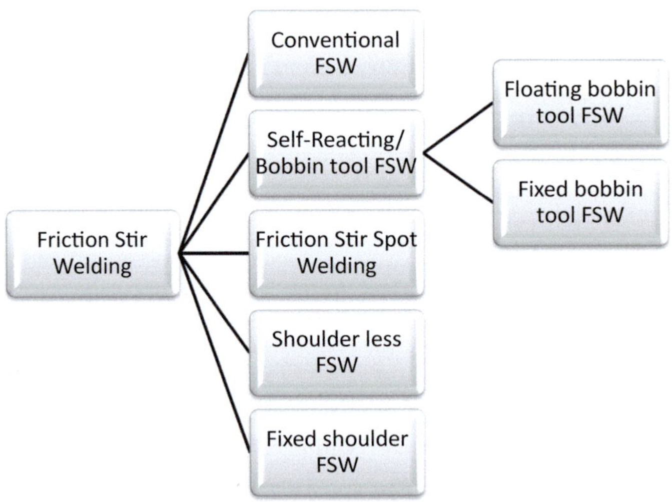

Fig. 2 Variants of friction stir welding

design. In this variant, the tool has two shoulders as shown in Fig. 3 connected by a pin which eliminates the requirement of a backing plate which is used in Conventional FSW (CFSW). The two plates to be welded are clamped in the fixture without any support below the plates along the joint line. The bottom shoulder of the rotating tool supports the hot plasticized material as it moves along the joint line. This simple modification brings about significant changes in the process and the results of various studies that have been carried out on conventional FSW may not necessarily be applicable in this variant. In this variant, there is heat input from both shoulders on either side of the joint. The plunge stage of the welding cycle is absent and therefore it consists of three stages only: (1) Tool approach and dwell, (2) Tool advance/traverse and (3) Tool withdrawal. An important advantage of this process is that there is no

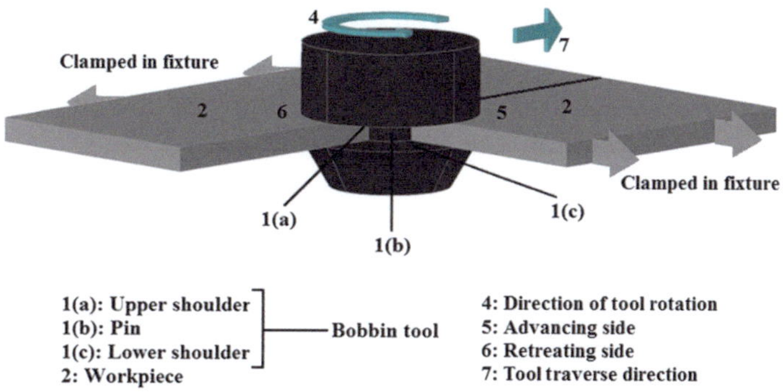

Fig. 3 Schematic illustration of the bobbin tool friction stir welding (BTFSW) process

requirement of backing plate and the normal load is significantly lower than that in conventional FSW process. However, since the tool cannot be withdrawn anywhere along the weld line, this method can weld only open-ended joints which limits its applications. Table 1 summarizes salient points of the comparison between CFSW and SRFSW. Some studies have been carried out on BTFSW but the volume of work is very small in comparison with conventional FSW.

Thomas et al. [1] successfully demonstrated the feasibility of successful defect free joints by BTFSW in 8-mm-thick 12% Cr steel. The tool used was a composite tool with the shoulder and probe made from different materials (Refractory alloys with different levels of Tungsten). Lafly et al. [2] carried out a comparative study between CFSW and BTFSW of AA6056 sheets. Microstructural study indicated that the two techniques produce similar grain structure which can be related to the thermo-mechanical conditions induced by the process but optimum process parameters for both the techniques are different. Threadgill et al. [3] studied BTFSW in 25-mm-thick AA6082-T6 and found a deformation which was more probe dominated with very less influence of shoulders on both sides unlike CFSW. Sued et al. [4] carried out experiments in AA6082-T6 and reported that the tool features do not necessarily have the same effect as in CFSW and hence are not directly transferable. Most of the studies were limited to feasibility and general understanding of the process. Some studies tried to understand the effect of process parameters like traverse speed [5–7], rotational speed [6, 8, 9] on weld quality. Liu et al. [5] reported an increase in the joint strength of AA6061-T6 with increase in the traverse speed (50–200 mm/min) with the maximum joint strength of 69% of the base material. Also the grain size in the stir nugget zone was found to increase with the traverse speed. Li et al. [6] also reported an increase in ultimate tensile strength (UTS) of Mg AZ31 alloy with increase in traverse speed. However, the traverse speed and rotational speed had slight impact on the yield strength and elongation. Zhang et al. [7] reported similarity in the hardness distribution across the thickness in the weld nugget of BTFSW 2A14-T6 alloy at different traverse speeds indicating homogeneity in mechanical properties. The joint strength first

Table 1 Comparison of SRFSW with conventional FSW

Bobbin tool FSW	Conventional FSW
• Full penetration welds	• Welds are keyhole type but not full penetration
• Backing plate is not required	• Backing plate is required
• The temperature profile is rectangular	• The temperature profile is triangular
• Two shoulders	• Only one shoulder
• Eliminates lack of penetration defects	• Lack of penetration defects is possible
• Less distortion due uniform heat input	• Chances of distortion are higher
• Lower forces on fixtures and machine	• Higher forces experienced by fixtures and machine
• Very low force in the *z*-direction in case of floating bobbin FSW	• Comparatively very high forces experienced in the *z*-direction
• Can weld open-ended joints only	• Does not suffer from this limitation

increased with increase in traverse speed and then abruptly decreased due to the formation of defects at 150 mm/min. The maximum joint strength was observed to be 75% of the base material and the failure always occurred in the weld region. Hou et al. [8] used a tool with smaller lower shoulder diameter for BTFSW of AA 6061-T6. It was observed that the visual quality of the weld increased with lowering bottom shoulder diameter till an optimum value, below which groove defects were observed on both sides of the weld. The grain size and hardness in the weld nugget zone was found to increase with the increase in the tool rotational speed. The joint strength was found to increase with the increase in tool rotational speed (400–600 rpm) and then abruptly decrease due to the formation of defects. The fracture location was located in the Thermo-Mechanically Affected Zone (TMAZ) at lower rotational speed which shifted to the Heat Affected Zone (HAZ) at higher traverse speed. Wang et al. [9] welded Al–Li 2198 alloy using BTFSW. Joint strength was found to first increase with the rotational speed and then decrease. The maximum joint efficiency was found to be 80%. Joint line remnants were observed in all welded joints and were compressed at higher rotational speeds. The hardness profile changed from U-shaped at lower rotational speed to W-shaped at higher rotational speed. The presence of triangular shaped, ellipse-shaped regions or band patterns [5, 7] were reported but a detailed explanation for the formation of these zones and the effect on weld quality was absent. This necessitates the separate study into BTFSW as the existing literature on CFSW though huge in volume does not directly apply to BTFSW. In the present study, a Computational Fluid Dynamics (CFD) based FEM model has been developed to explain the formations of different features observed in experimental results. CFD-based models have been used for both CFSW [10, 11] and BTFSW [12–14]. In the CFD-based approach, a small region along the joint line or the entire workpiece is assumed to behave like a fluid whose viscosity is calculated as a function of the strain rate and temperature. In addition to this, the effect of traverse speed on mechanical properties of welded joints has been studied.

2 Experimental Procedure

The experiments in the present study have been carried out on 4-mm-thick aluminium alloy AA6061-T6 plates. The tool used had a simple cylindrical geometry as shown in Fig. 4 and was made from AISI H13 tool steel. The tool pin and shoulder diameters were 8 and 18 mm, respectively, with the shoulder gap 4 mm. The compositions of the workpiece and tool materials are as given in Table 2.

The experiments were carried out on a vertical milling machine at a fixed rotational speed of 600 rpm and three different traverse speeds (36, 63 and 98 mm/min). The joints were produced in butt configuration. The dwell time was fixed at 30 s for all experiments. The tensile test specimens were cut according to the standard ASTM E8 along the perpendicular to the traverse direction such that the entire weld nugget lies within the gage length. The tensile tests were carried out on INSTRON, Model 8801 Dynamic Testing Machine with a constant crosshead speed of 1 mm/min. Three-point flexural tests were carried out for joints produced at traverse speed of 63 mm/min. The top and bottom surfaces were subjected to tension to understand the difference on the two sides of the joint. The samples for the microstructural study were cut from the parent material and from the welded joints perpendicular to the welding direction. The samples were polished using Silicon

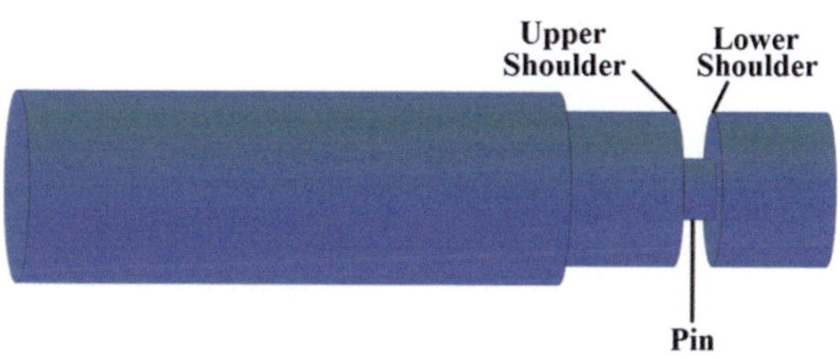

Fig. 4 Design of the bobbin tool used in the study

Table 2 Composition of AA6061-T6 and AISI H13 tool steel [15, 16]

AA6061-T6		AISI H13 tool steel	
Element	Weight %	Element	Weight %
Al	95.85–98.56	C	0.32–0.40
Cu	0.15–0.4	Cr	5.13–5.25
Si	0.4–0.8	Fe	$\geq$ 90.95
Fe	0.7 (max)	Mo	1.33–1.4
Mg	0.8–1.2	Si	1
Mn	0.15 (max)	V	1
Cr	0.04–0.35		
Zn	0.25 (max)		
Ti	0.15 (max)		
Others	0.05 each (max), total 0.15		

Carbide emery papers up to #2000 grit followed by a liquid abrasive suspension till mirror finish was achieved. The samples were then etched at room temperature to reveal the microstructure.

3 Numerical Model

A three-dimensional fully coupled thermo-mechanical FEM model has been developed based on CFD approach in COMSOL Multiphysics software to understand the observations of the experiments. Following subsections present the detailed methodology of the model.

3.1 *Model Description*

The dimensions of the workpiece and tool have been taken exactly as those in the experiments. However, the fixtures used in the experiments have not been considered. A small region along the weld line with width 1 mm greater than the tool shoulder radius on both sides has been considered to behave as fluid as shown in Fig. 5. The remaining region has been considered as solid.

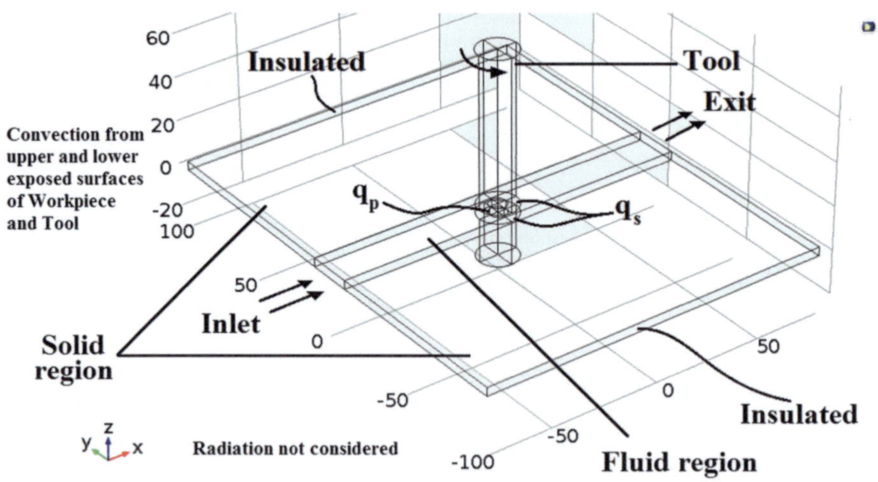

Fig. 5 Geometry and boundary conditions of the FEM model

3.2 Material Properties

Temperature-dependent properties for both workpiece and tool material have been used in the model. The variations in thermal conductivities of the workpiece material and the tool material are shown by the graphs in Figs. 6 and 7. The thermal conductivity of aluminium changes significantly with temperature as compared with H13 tool steel. The variation of specific heat of workpiece material has been plotted in Fig. 8. Beyond the available data range, the values for the properties have been assumed to be constant. The density of the materials has been assumed to remain constant.

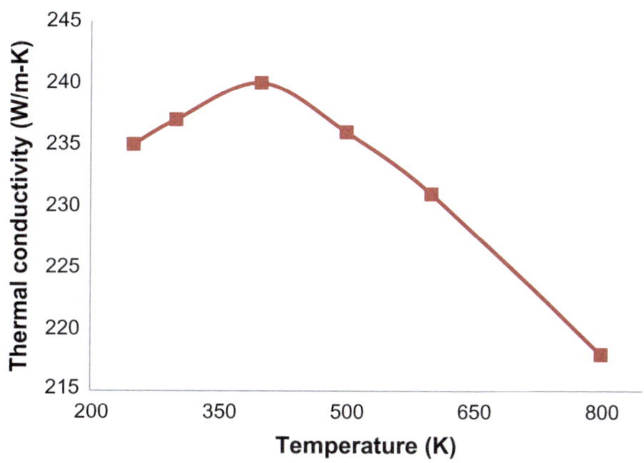

Fig. 6 Variation of thermal conductivity of the workpiece material with temperature [17]

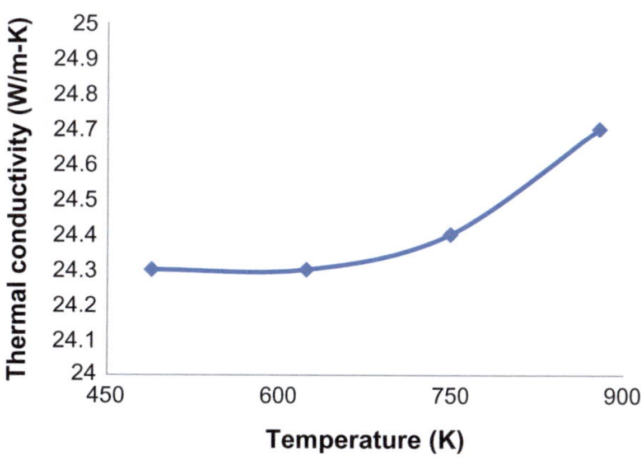

Fig. 7 Variation of thermal conductivity of AISI H13 tool steel with temperature [16]

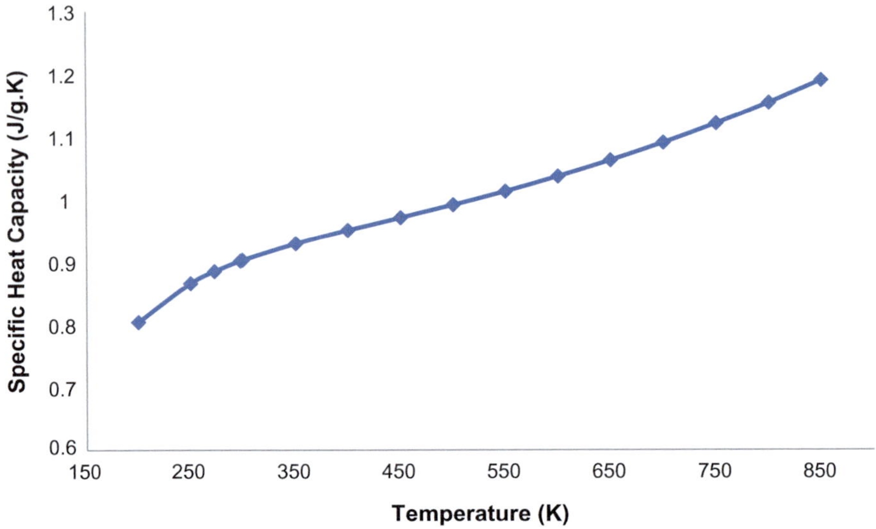

Fig. 8 Variation of specific heat of the workpiece material with temperature [18]

3.3 Boundary Conditions

Partial slip and stick has been considered as the contact condition which requires a slip factor (δ) as given by Eq. 1, where x, y are the coordinates from the origin at the centre of the tool and R_s is the shoulder radius.

$$\delta = \frac{\sqrt{x^2 + y^2}}{R_s} \quad (1)$$

The relative velocity between the workpiece and tool surfaces determines the actual coefficient of friction at each position which is given by the relation in Eq. 2 [19], where μ and $\mu_0 = 0.4$ are the actual and nominal coefficient of friction at various locations respectively, ω the angular velocity and $\lambda = 1$ s/m a constant to make the expression dimensionless.

$$\mu = \mu_0 e^{-\lambda\delta\omega\sqrt{x^2 + y^2}} \quad (2)$$

The strain rate has been calculated at all locations using the relation in Eq. 3 [10], where u, v and w are the velocity components along x, y and z respectively.

$$\dot{\varepsilon} = \left[\left(\frac{2}{3}\right) \times \left\{ \left(\frac{\mathrm{d}u}{\mathrm{d}x}\right)^2 + \left(\frac{\mathrm{d}v}{\mathrm{d}y}\right)^2 + \left(\frac{\mathrm{d}w}{\mathrm{d}z}\right)^2 + 0.5 \times \left(\frac{\mathrm{d}u}{\mathrm{d}z} + \frac{\mathrm{d}w}{\mathrm{d}x}\right)^2 + 0.5 \times \left(\frac{\mathrm{d}w}{\mathrm{d}y} + z\right)^2 + 0.5 \times \left(\frac{\mathrm{d}u}{\mathrm{d}y} + \frac{\mathrm{d}v}{\mathrm{d}x}\right)^2 \right\} \right]^{\frac{1}{2}} \tag{3}$$

The flow stress at the different locations is calculated using the Eq. 4 where Z is the Zener–Holloman parameter calculated using Eq. 5 [20, 21]. The values of the various constant are $Q = 134158.4\ \mathrm{J\ mol^{-1}}$, $A = 1.26 \times 10^{8}\ \mathrm{s^{-1}}$, $\alpha = 0.03055\ \mathrm{MPa^{-1}}$ and $n = 3.24644$.

$$\sigma = \frac{1}{\alpha} \sinh^{-1} \left\{ \left(\frac{Z}{A}\right)^{\left(\frac{1}{n}\right)} \right\} \tag{4}$$

$$Z = \dot{\varepsilon}\, \mathrm{e}^{\left(\frac{Q}{RT}\right)} \tag{5}$$

The dynamic viscosity in the fluid region has been calculated using Eq. 6 [10, 11].

$$\eta = \frac{\sigma}{(3\dot{\varepsilon})} \tag{6}$$

The heat input due to the shoulder and the pin are calculated using Eqs. 7 and 8 respectively [20]. No external load has been applied and the pressure (P) developed in the process has been used to calculate the heat developed. A step function has been used in the relations to prevent heat input if the temperature in the workpiece at a location reaches solidus temperature ($T_{\mathrm{m}} = 855.15$ K). r_{p} is the pin radius and T is the actual temperature at any location.

$$q_s = \left[\left\{ (1 - \delta) \times \frac{\sigma}{\sqrt{3}} \right\} + \delta \times \mu \times P \right] \times \sqrt{u^2 + v^2 + w^2} \times \mathrm{step}(T_{\mathrm{m}} - T) \tag{7}$$

$$q_p = \left[\left\{ (1 - \delta) \times \sqrt{u^2 + v^2 + w^2} \times \frac{\sigma}{\sqrt{3}} \right\} + \delta \times \frac{\mu}{\sqrt{3 \times (1 + \mu^2)}} \times r_p \times \omega \times \sigma \right] \times \mathrm{step}(T_{\mathrm{m}} - T) \tag{8}$$

The inlet surface of the workpiece as shown in Fig. 3 has been assumed to remain at the ambient temperature of 300 K. Heat flow due to radiation has been neglected. However, heat dissipation due to convection has been considered from the upper and lower exposed surfaces.

4 Results and Discussion

Figure 9 shows one of the welded samples at traverse speed of 36 mm/min. The top and bottom surfaces of the joints had almost similar appearance unlike joints produced by CFSW. The joints had zero distortion and good surface finish. The typical entry and exit of the tool from the joints are shown in Fig. 10. A weld initiation length was observed at the starting of the joints which was dependent on the process parameters. Colligan et al. [22] attempted to address the difficulty in weld initiation in BTFSW of below 6 mm thickness (approx.) materials. The reason was attributed to the plugging of the bobbin tool with the workpiece material similar to the one as shown in Fig. 11.

Figure 12 shows stress–strain curves for the base material and joints produced at different traverse speeds. The Ultimate Tensile Stress (UTS) was found to increase with traverse speed. However, the highest UTS observed at traverse speed of 98 mm/min was only 66.04% of that of the base material. The yield stress which is a very important design criteria for structural applications, also increased with traverse speed as shown in Fig. 13 but the increase was small in comparison with the increase in UTS. The maximum yield stress observed at 98 mm/min was considerably smaller in comparison with that of the base material. Figure 14 shows the fracture surface of a tensile specimen at 98 mm/min which developed a dimpled surface with micro-voids due to ductile fracture. Three-point flexural test (up to 135°) revealed a similar stress–strain curve when the top and bottom surfaces were subjected to tension for a joint produced at traverse speed of 63 mm/min as shown in Fig. 15. However, the maximum stress was slightly higher for the case in which the top surface was in extension but it developed a surface crack as shown in Figs. 15 and 16 respectively.

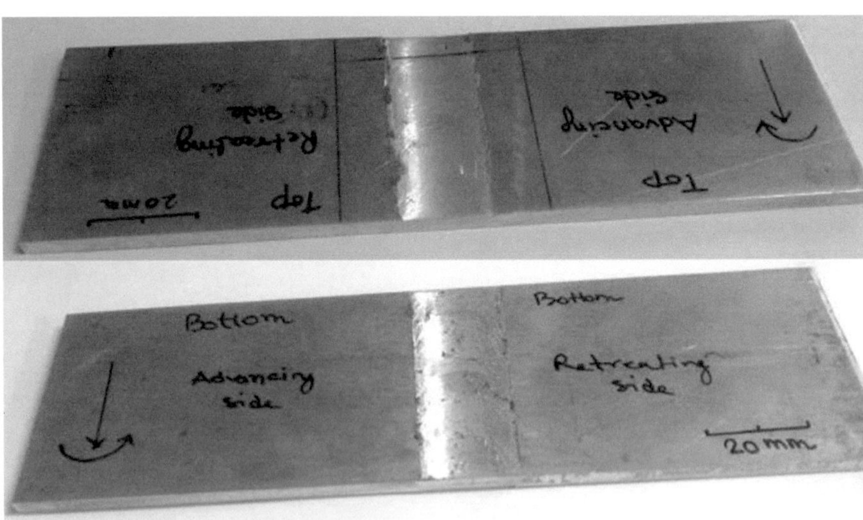

Fig. 9 A BTFSW joint at traverse speed of 36 mm/min

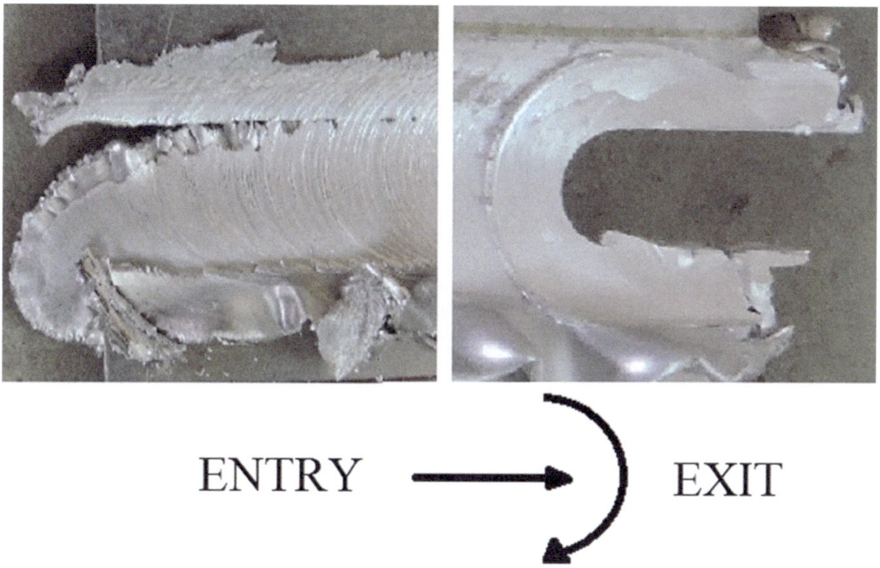

Fig. 10 Typical entry and exit locations of the tool

Fig. 11 A bobbin tool plugged with workpiece material

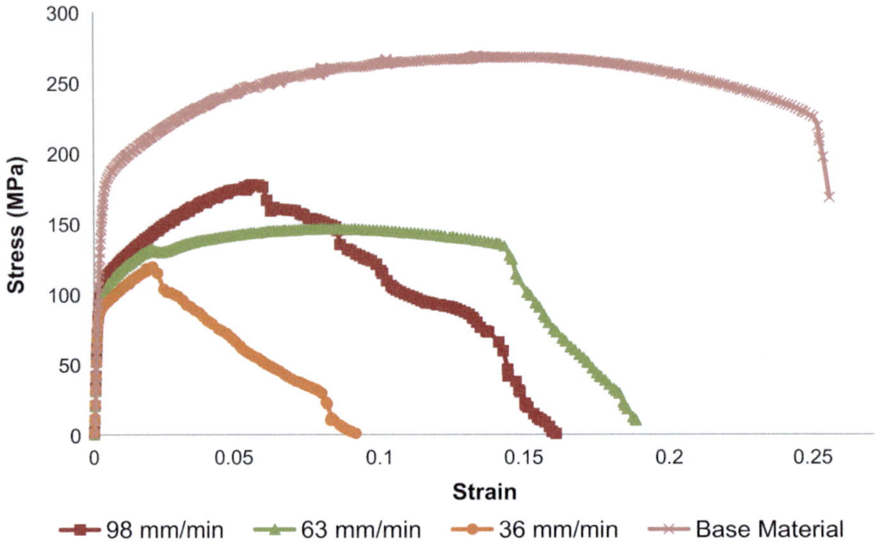

Fig. 12 Stress–strain curves for the base material and joints produced at different traverse speeds

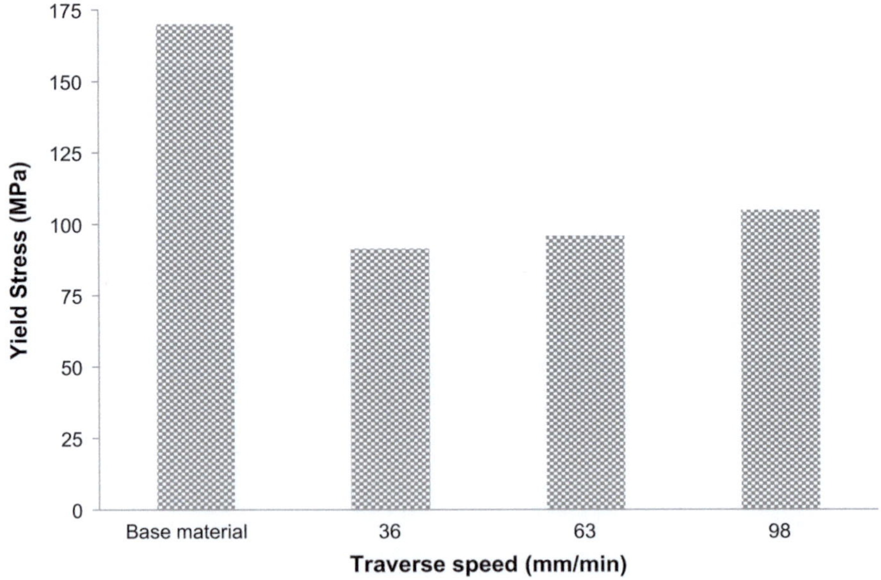

Fig. 13 Variation of yield stress for joints produced at various traverse speeds

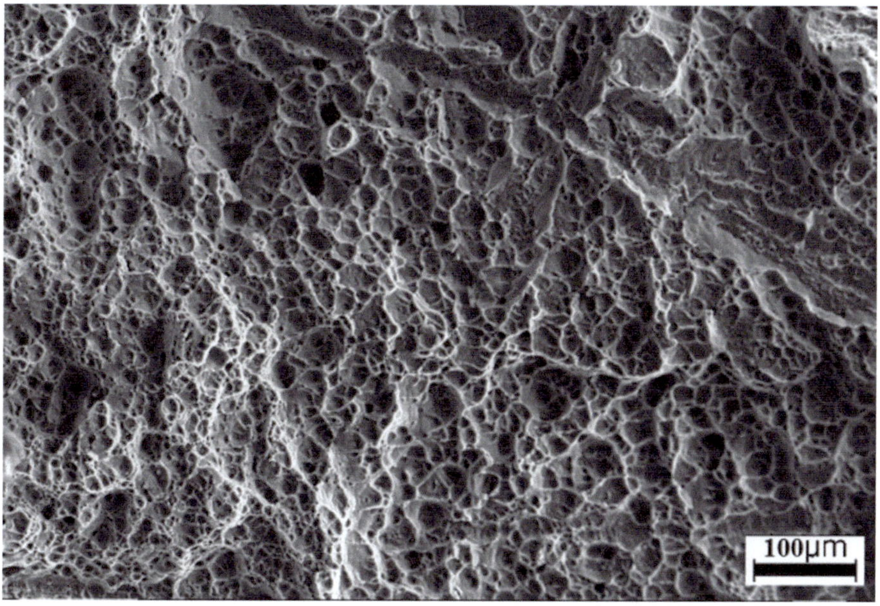

Fig. 14 Fracture surface of a joint prepared at 98 mm/min (magnification: 350×)

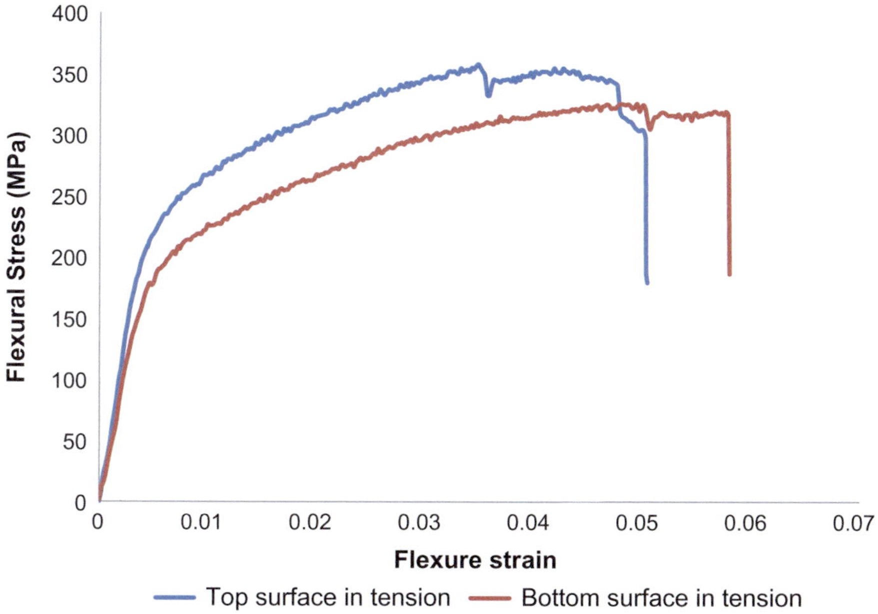

Fig. 15 Flexural stress–strain curve for joint produced at traverse speed of 63 mm/min

Fig. 16 Samples after three-point flexural test: **a** upper surface in tension, **b** lower surface in tension

Different zones were observed in the polished cross sections of welds. These zones were dependent on the thermo-mechanical conditions applied on the material. Traverse speed was found to have an influence on the formation of these zones. Figure 17 shows the characteristic saddle-shaped profile of BTFSW at 36 mm/min. Figures 18 and 19 show a magnified view of the saddle-shaped profile on the advancing side of the joints where it was more prominent at 36 and 98 mm/min respectively. A closer examination of the saddle-shaped profile revealed elongated grains formed due to the shearing of the material by the rotating tool.

In addition to the saddle-shaped profile, elliptical swirl zones were observed from the advancing side (AS) to the retreating side (RS) of the joints in polished samples with or without etching as shown in Figs. 20 and 21.

The temperature measurement using thermocouples at three locations at 600 rpm and 63 mm/min on the retreating side (15, 20 and 30 mm) were found in good agreement with the FEM simulation with an average error of 2.82%. The divergence between the experimental and FEM results increased as the distance from joint line increased. This may be due to the reason that the effect of the fixtures has been neglected in the FEM model. Figure 22a shows the variation of temperature

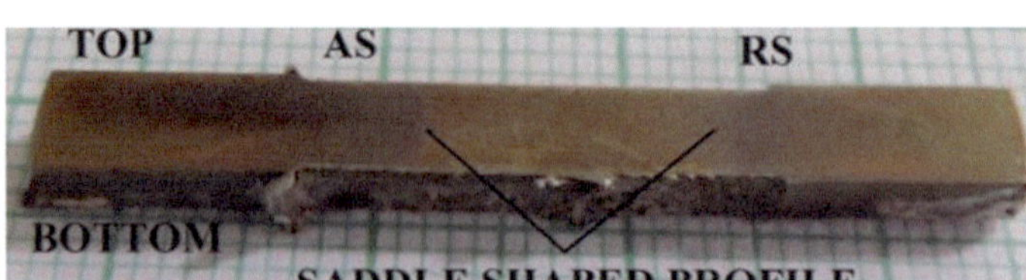

Fig. 17 Saddle-shaped profile in a welded joint after polishing and etching

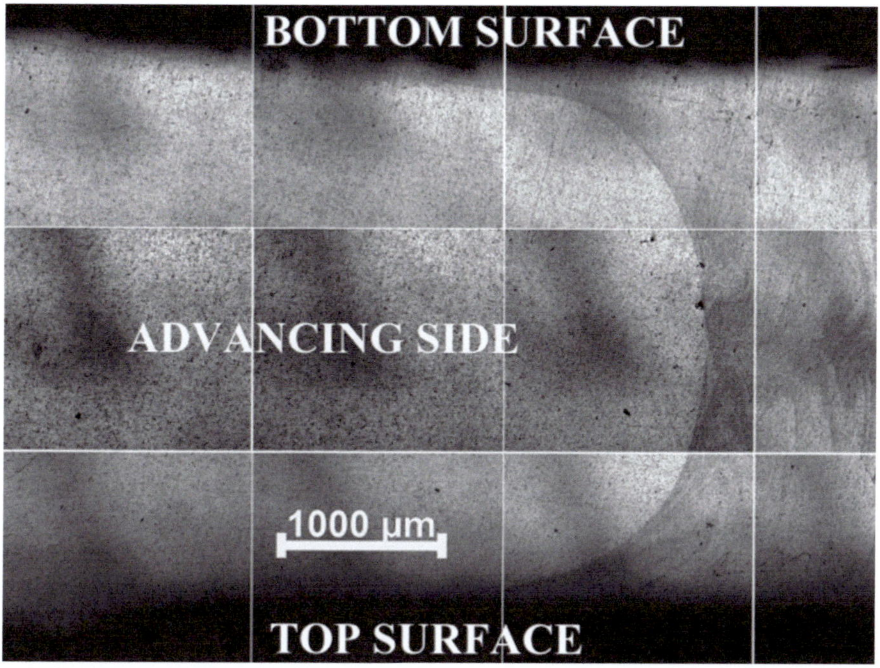

Fig. 18 Advancing side of a joint produced at traverse speed 36 mm/min showing the saddle shape and part of the swirl zone

Fig. 19 Magnified view of the saddle-shaped profile of a joint at traverse speed 98 mm/min (magnification: 50×)

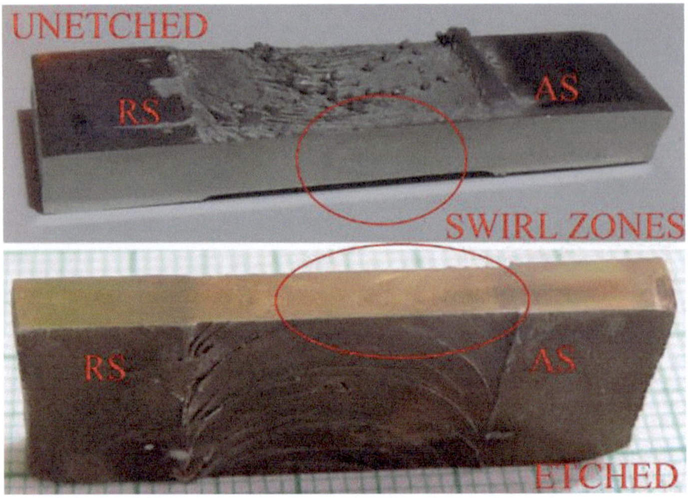

Fig. 20 Swirl zone observed in joints at 63 mm/min

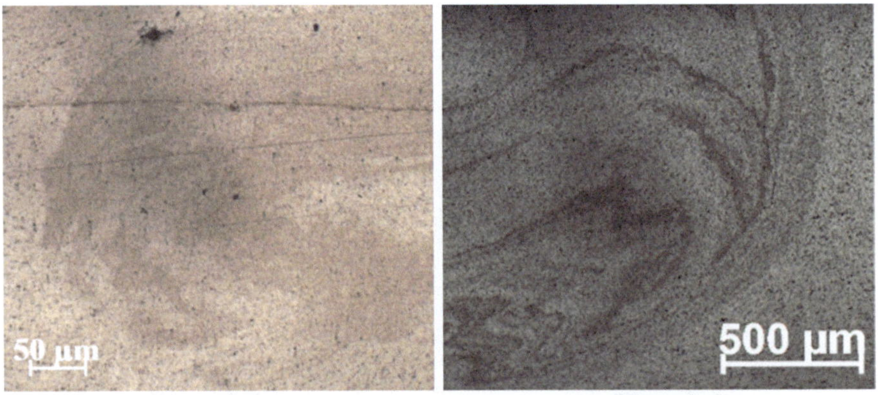

Fig. 21 Swirl zones observed at different traverse speeds (magnification: 10× and 20×)

with distance at the location shown in Fig. 22b. Figure 23 shows the velocity distribution as the material crosses the tool. Figure 23 shows temperature history at 600 rpm and 63 mm/min of the location 15 mm from the weld centreline on the retreating side and 2 mm from the top surface as it travels in the fluid region from the inlet to the outlet in the FEM model and transient temperature history provided by the thermocouple at the same location. The slope of the curve for the experimental results is steeper for the rising phase. Both the curves reach the peak at almost the same time. However, the cooling phase is very similar. The difference in the rising phase may be due to the quasi-steady state assumed in the model and also

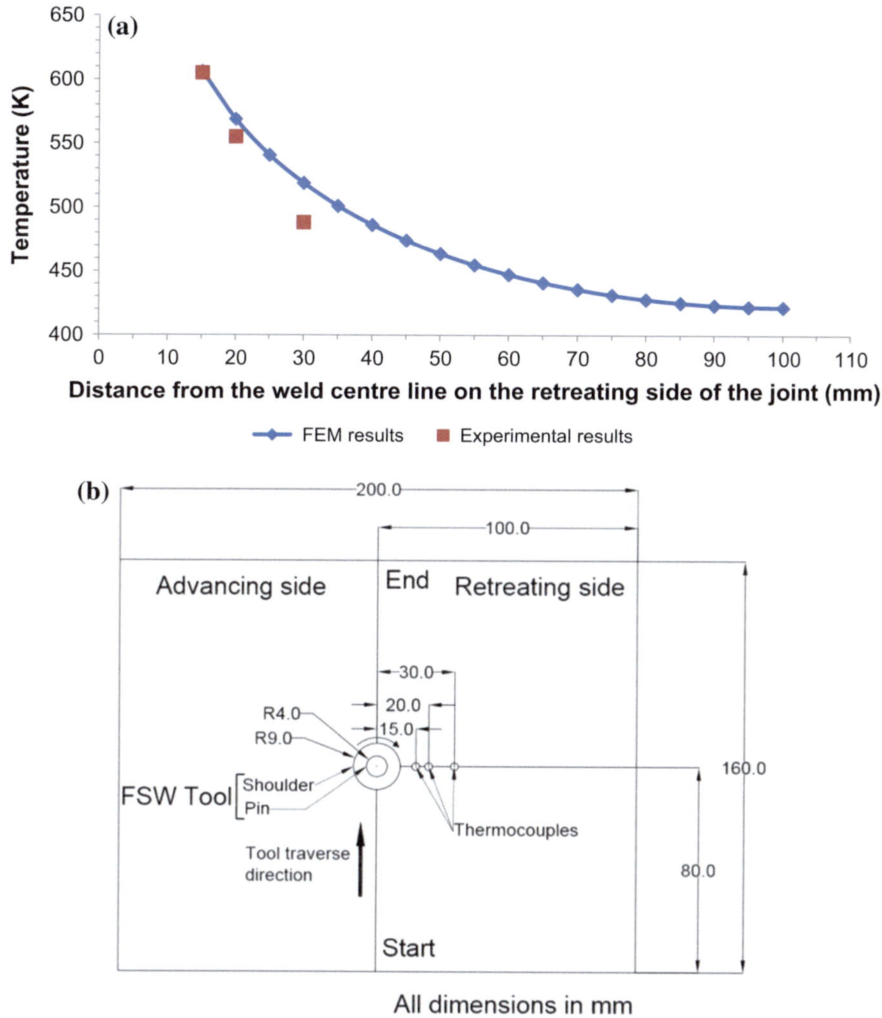

Fig. 22 **a** Temperature measurements from experiment and FEM simulation (600 rpm, 63 mm/min and 2 mm below top surface). **b** The locations at which temperature measurement was done

due to assumptions like not considering the fixtures. The slower cooling in case of BTFSW can be helpful in achieving better mechanical properties in the weld.

To understand the formation of these features, the results from the FEM model were analysed. The tool movement creates a void behind the pin which the rotating tool tries to fill by moving material from the retreating side to the advancing side. The cross section of this material which moves to fill the void appears as the swirl zone. The material flow is pin dominated and as the tool rotates the material is moved from the advancing side towards the retreating side. This shearing of the material leads to creation of a region which is visible as the saddle shape. The

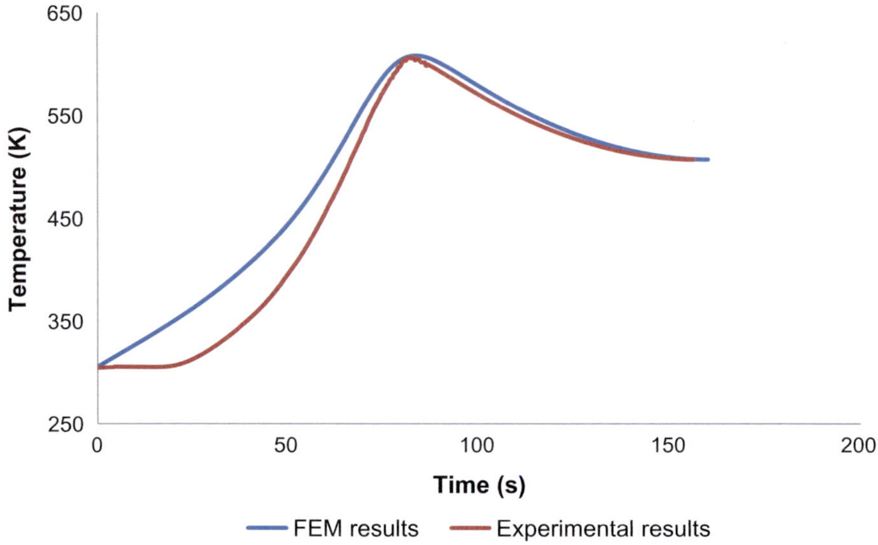

Fig. 23 Temperature history from the thermocouple and FEM model for the location 15 mm from the weld centreline on the retreating side and 2 mm below the top surface at 600 rpm and 63 mm/min

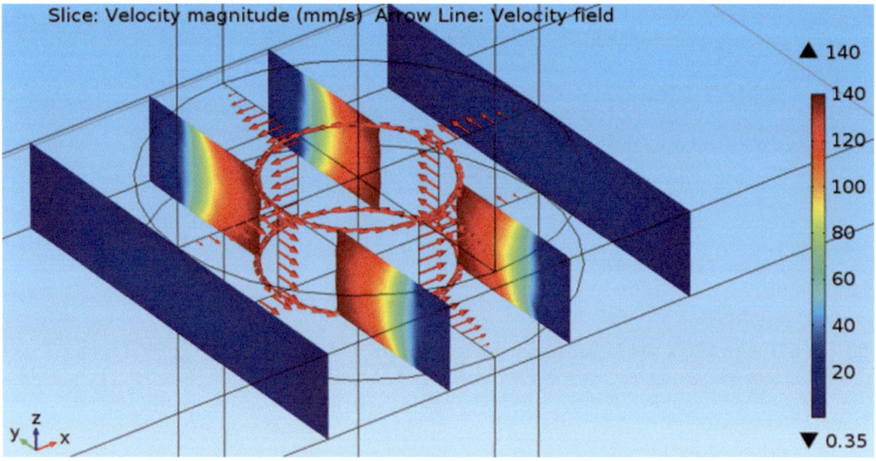

Fig. 24 Velocity distribution as the material moves across the tool at traverse speed 36 mm/min

velocity distribution and temperature history vary with the traverse speed which results in varying shape and size of the swirl zone. Figure 24 shows the velocity distribution as the material leaves the tool and the slice in Fig. 25 represents the swirl zone formation observed in the experimental results. Figure 26 shows the temperature distribution at the section at the centre of the tool. A similar distribution

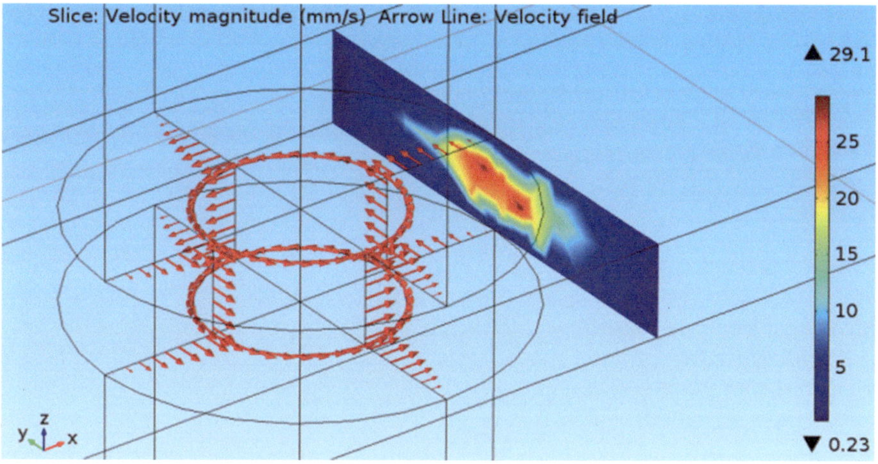

Fig. 25 Velocity distribution as the material leaves the tool at traverse speed 36 mm/min

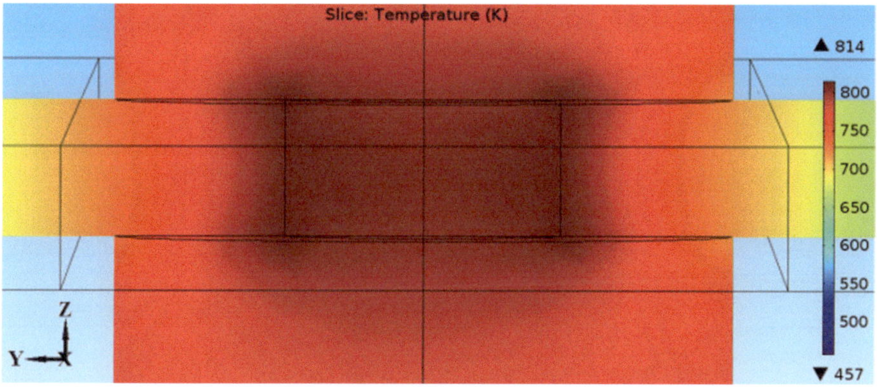

Fig. 26 Temperature distribution when the material passes the centre of the tool at traverse speed 36 mm/min

was observed near the shoulders, where the temperatures are higher and a slight drop towards the mid-thickness of the plate.

The temperature and velocity distribution (plastic deformation) together determine the resulting macrostructure and microstructural distribution. The nugget zone was found to have fine-grained equiaxed microstructure due to full dynamic recrystallization as shown in Fig. 27. In comparison, the base material has elongated large grain size due to rolling as shown in Fig. 28. Figure 29 shows the transition from the Heat Affected Zone (HAZ) to the Thermo-Mechanically Affected Zone (TMAZ). The HAZ has a microstructure similar to that of the base material.

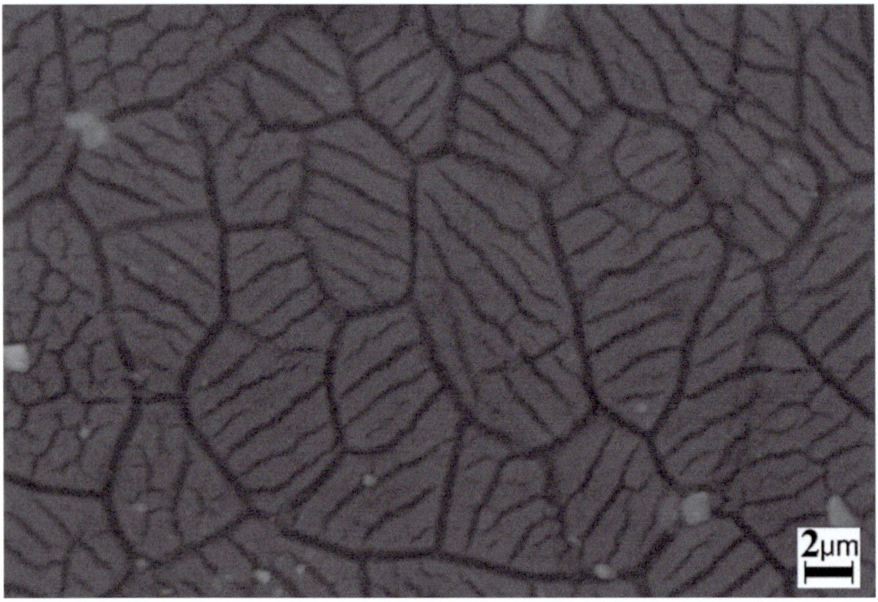

Fig. 27 Fine-grained equiaxed microstructure in the stir zone at traverse speed 36 mm/min (magnification: 8.22k×)

Fig. 28 Microstructure of the base material AA6061-T6 (magnification: 50×)

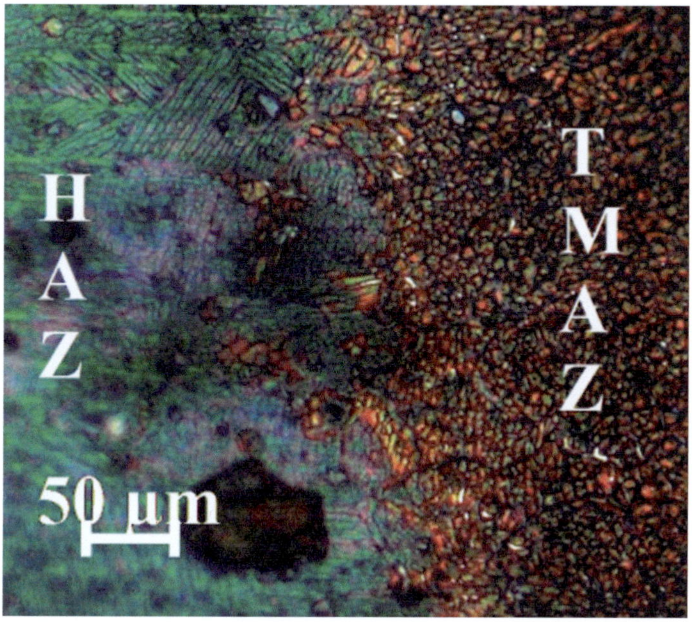

Fig. 29 Transition from the HAZ to the TMAZ on the advancing side of a joint produced at traverse speed 98 mm/min (magnification: 20×)

The resulting microstructure in the nugget zone due to dynamic recrystallization is independent of the initial grain structure and depends entirely on the temperature and plastic deformation the material is subjected to. The temperature distribution and material flow which are interrelated together determine the microstructure. The thermo-mechanical conditions can be represented by the Zener–Holloman parameter (refer Eq. 5). The microstructures in the different zones indicate that with increase in the Zener–Holloman Parameter the grain size reduces. The Zener–Holloman Parameter is dependent on the process parameters at which the welding has been carried out. Thus, the resulting microstructure can be related to the process parameters. Based on the experimental and FEM results, a regression model was developed

$$\log_{10}(\mathrm{GS}) = 2.28 - 0.08625\log_{10}(Z), \tag{9}$$

where GS is the Average Intercept length (μm) and Z is the Zener–Holloman Parameter.

Since the volume of data used for the model was small, the accuracy of the model was not very high. An average error of 9.85% with standard deviation of 1.27 was observed. This can be very useful for Friction Stir Processing of materials. Figure 30 shows the grain size distribution predicted in the nugget zone by the FEM model using the developed relationship. Similar to the measurements made on

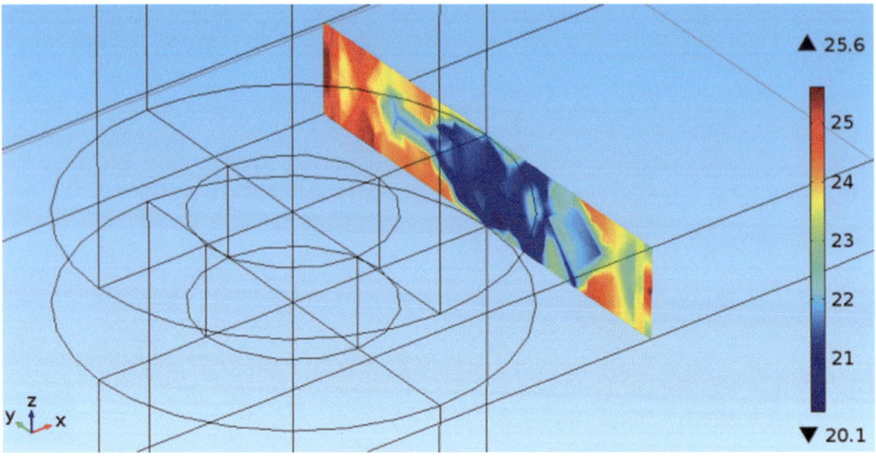

Fig. 30 Predicted average grain size (μm) at 600 rpm and 98 mm/min

the welded samples the average grain size predicted was found to be smaller near the centre where the pin had a greater influence. Similarly, Chang et al. [23] used Zener–Holloman parameter to develop a relationship between the resulting grain size and the applied working strain rate and temperature for the friction stir processing in AZ31 Mg.

5 Conclusions

From the results obtained in the present study the following conclusions can be drawn:

- The UTS increases significantly with the traverse speed. The highest UTS of the welded joint at traverse speed of 98 mm/min was 66% of that of the base material. The yield stress also increases with the increase in traverse speed but the increase was gradual. Ductile fracture was observed in the welded joints.
- The three-point flexural test points towards uniform mechanical properties across the thickness of the welded joint.
- The macrostructure of the joints had a saddle-shaped profile which was more prominent on the advancing side. In addition to this, elliptical swirl zones were observed.
- The formation of different features of the joints has been related to the results from the FEM model and the causes have been identified.

- The temperature distribution and plastic deformation result in full dynamic recrystallization. The base material which had an elongated coarse microstructure changed to fine equiaxed microstructure due to recrystallization. HAZ had a microstructure similar to the base material.
- A relationship between resulting grain size and process parameters was developed using Zener–Holloman parameters.

References

1. Thomas, W. M., Wiesner, C. S., Marks, D. J., & Staines, D. G. (2009). Conventional and bobbin friction stir welding of 12% chromium alloy steel using composite refractory tool materials. *Science and Technology of Welding and Joining, 14,* 247–253.
2. Lafly, A. L., Allehaux, D., Marie, F., Donne, C. Dalle, & Biallas, G. (2006). Microstructure and mechanical properties of the aluminium alloy 6056 welded by friction stir welding techniques. *Welding in the World, 50,* 98–106.
3. Threadgill, P. L., Ahmed, M. M. Z., Martin, J. P., Perrett, J. G., & Wynne, B. P. (2010). The use of bobbin tools for friction stir welding of aluminium alloys. *Materials Science Forum, 638–642,* 1179–1184.
4. Sued, M. K., Pons, D., Lavroff, J., & Wong, E. H. (2014). Design features for bobbin friction stir welding tools: Development of a conceptual model linking the underlying physics to the production process. *Materials and Design, 54,* 632–643.
5. Liu, H. J., Hou, J. C., & Guo, H. (2013). Effect of welding speed on microstructure and mechanical properties of self-reacting friction stir welded 6061-T6 aluminum alloy. *Materials and Design, 50,* 872–878.
6. Li, W. Y., Fu, T., Hütsch, L., Hilgert, J., Wang, F. F., dos Santos, J. F., et al. (2014). Effects of tool rotational and welding speed on microstructure and mechanical properties of bobbin-tool friction-stir welded Mg AZ31. *Materials and Design, 64,* 714–720.
7. Zhang, Huijie, Wang, Min, Zhang, Xiao, & Yang, Guangxin. (2015). Microstructural characteristics and mechanical properties of bobbin tool friction stir welded 2A14-T6 aluminum alloy. *Materials and Design, 65,* 559–566.
8. Hou, J. C., Liu, H. J., & Zhao, Y. Q. (2014). Influences of rotation speed on microstructures and mechanical properties of 6061-T6 aluminum alloy joints fabricated by self-reacting friction stir welding tool. *The International Journal of Advanced Manufacturing Technology, 73,* 1073–1079.
9. Wang, F. F., Li, W. Y., Shen, J. J., Hu, S. Y., Li, J. L., dos Santos, J. F., et al. (2015). Effect of tool rotational speed on the microstructure and mechanical properties of bobbin tool friction stir welding of Al-Li alloy. *Materials and Design*. https://doi.org/10.1016/j.matdes.2015.07.096.
10. Colegrove, Paul A., & Shercliff, Hugh R. (2005). 3-Dimensional CFD modelling of flow round a threaded friction stir welding tool profile. *Journal of Materials Processing Technology, 169,* 320–327.
11. Kang, S. W., & Jang, B. S. (2013). A study on computational fluid dynamics simulation of friction stir welding. In G. Soares & Romanoff (Eds.), *Analysis and design of marine structures* (pp. 433–439). ISBN 978-1-138-00045-2.
12. Hilgert, J., Huetsch, L. L., dos Santos, J. F., & Huber, N. (2010). Material flow around a bobbin tool for friction stir welding. In *Excerpt from the Proceedings of the COMSOL Conference*. Paris.
13. Hilgert, J. dos Santos, J. F., Huber, N. (2011). Investigation of the material shear layer in bobbin tool friction stir welding. In R. Mishra, M. W. Mahoney, Y. Sato, Y. Hovanski, R.

Verma (Eds.), *Friction Stir Welding and Processing VI.* TMS (The Minerals, Metals & Materials Society).

14. Hilgert, J., dos Santos, J. F., & Huber, N. (2012). Shear layer modelling for bobbin tool friction stir welding. *Science and Technology of Welding and Joining, 17,* 454–459.
15. ASM Material Data Sheet Al6061-T6. (2015). Available: http://asm.matweb.com/search/SpecificMaterial.asp?bassnum=MA6061t6.
16. ASM Material Data Sheet AISI Type H13 Hot Work Tool Steel. (2015). Available: http://www.matweb.com/search/datasheet_print.aspx?matguid=e30d1d1038164808a85cf7ba6aa87ef7.
17. Schlunder, E. U. (Ed.). (1983). *The international centre for heat and mass transfer, heat exchanger design handbook* (Vol. 1). Hemisphere Publishing Corporation.
18. Buyco', Edgar H., & Davis, Fred E. (1970). Specific heat of aluminum from zero to its melting temperature and beyond. *Journal of Chemical and Engineering Data, 15*(4), 518–523.
19. Nandan, R., Roy, G. G., Lienert, T. J., & Debroy, T. (2007). Three-dimensional heat and material flow during friction stir welding of mild steel. *Acta Materialia, 55,* 883–895.
20. Nandan, R., Roy, G. G., & Debroy, T. (2006). Numerical simulation of three-dimensional heat transfer and plastic flow during friction stir welding. *Metallurgical and Materials Transactions A, 37A,* 1247–1259.
21. Hamilton, C., Kopyscianski, M., Senkov, O., & Dymek, S. (2013). A coupled thermal/material flow model of friction stir welding applied to sc-modified aluminum alloys. *Metallurgical and Materials Transactions A, 44A,* 1730–1740.
22. Colligan, K. J., O'Donnell, A. K., Shevock, J. W., & Smitherman, M. T. (2012). *Friction Stir Welding of Thin Aluminum Using Fixed Gap Bobbin Tools, 9th International FSW Symposium*, Huntsville, AL, May 15–17, 2012.
23. Chang, C. I., Lee, C. J., & Huang, J. C. (2004). Relationship between grain size and Zener-Holloman parameter during friction stir processing in AZ31 Mg alloys. *Scripta Materialia, 51,* 509–514.

Thermomechanical Simulation of Friction Stir Welding Process Using Lagrangian Method

Rahul Jain, Surjya K. Pal and Shiv B. Singh

List of Symbols

A_{iN}	Area contribution of *i*th element to the node *N*
B_s	Strain rate matrix
C	Heat capacity matrix
D	Effective strain rate coefficient matrix
$\mathrm{E}(\dot{\varepsilon}_{ij})$	Work function
F_i	Traction
J_b	Jacobian matrix
K	Thermal conductivity
K_s	Global stiffness matrix
K_{wt}	Thermal conductance between the workpiece and tool
K_c	Heat conduction matrix
K_{wb}	Thermal conductance between the workpiece and backing plate
N^T	Transpose of shape function matrix
Q	Heat flux vector
Q_{wa}	Heat transfer between the workpiece and ambient
Q_{wt}	Heat transfer between the workpiece and tool
Q_{wb}	Heat transfer between the workpiece and backing plate

R. Jain · S. K. Pal (✉)
Friction Stir Welding Laboratory, Department of Mechanical Engineering, Indian Institute of Technology Kharagpur, Kharagpur 721302, West Bengal, India
e-mail: skpal@mech.iitkgp.ernet.in

R. Jain
e-mail: rahuljain214@gmail.com

S. B. Singh
Department of Metallurgical and Materials Engineering, Indian Institute of Technology Kharagpur, Kharagpur 721302, India
e-mail: sbs22@metal.iitkgp.ernet.in

U. S. Dixit and R. Kant (eds.), *Simulations for Design and Manufacturing*,
Lecture Notes on Multidisciplinary Industrial Engineering,
https://doi.org/10.1007/978-981-10-8518-5_4

S_F	Traction surface
T_t	Tool temperature
T_b	Backing plate temperature
T_w	Workpiece temperature
T_a	Ambient temperature
c	Specific heat capacity
f_R	Residual nodal point force
f	Yield function
g	Plastic potential
h_a	Convective heat transfer between the workpiece/tool and ambient
h_{wt}	Convective heat transfer between the workpiece and tool
h_{wb}	Convective heat transfer between the workpiece and backing plate
J_2, J_3	Second and third variant of stress tensor
k	Material constant
k_s	Static yield stress in shear
m	Shear factor
q_N	Shape function of element i at node N
q_α	Element shape function
$\dot{q}$	Heat generation rate
q_f	Surface heat flux due to friction
u_i^α	Velocity component at the αth node
u_i	Velocity component
v	Nodal velocity
v_s	Sliding velocity at the interface of tool workpiece
x, y, z	Cartesian coordinate system
V''	Viscosity constant

Greek Symbols

σ_b	Stefan–Boltzmann constant
ε_b	Emissivity
σ'_{ij}	Deviatoric stress
σ_{ij}	Cauchy stress
δ_{ij}	Kronecker delta
σ_m	Hydrostatic stress
σ_Y	Yield stress
ε_{ij}^p	Plastic strain
λ	Positive proportionality constant
$\dot{\gamma}_{xy}$	Shear strain rate
τ_{xy}	Shear stress
$\dot{\varepsilon}_x$	Strain rate in x-direction
$\dot{\bar{\varepsilon}}$	Effective strain rate
$\bar{\sigma}$	Effective stress or flow stress
$\dot{\varepsilon}_{ij}$	Strain rate component

$\bar{\tau}$	Contact stress
τ_{max}	Shear yield strength
λ_P	Penalty constant
$\dot{\varepsilon}_v$	Volumetric strain rate
$\dot{\bar{\varepsilon}}_0$	Limiting strain rate
$\bar{\varepsilon}_N$	Effective strain rate at node N
ξ, η, ζ	Natural coordinate system
φ	Inelastic heat fraction
ψ	Constant
ρ	Density

1 Introduction

Understanding the mechanics of the heat generation and material flow during friction stir welding (FSW) is of physical significance because it can minimize the number of experiments. Finite element method (FEM) is one of the efficient tools to do so. Availability of the commercial software and high computing facility has put this method ahead of the analytical analysis. Most of the finite element software is like a black box and the accuracy of the solution can be questioned until it is validated. At times, validation of the results takes huge time as the user is not familiar with various governing equations and solvers used in the software. In this chapter, FSW is modeled based on the Lagrangian method in DEFORM-3D. The chapter focuses on various steps of the model and different governing equations that are solved to achieve the final solution. Three different solvers, viz., conjugate gradient (CG), sparse, and combination of both (CG + sparse), are compared in terms of the computation time and output responses like temperature, forces, etc. Developed model shows a good correlation with the experimental data. Material flow path is investigated for different regions of the workpiece. Strain rate and material velocity are also studied in those zones.

Lightweight materials, especially aluminum alloys, are in high demand in industries for possible weight reduction and lower fuel consumption, because of their excellent strength-to-weight ratio. Aluminum alloys are difficult to weld, due to their high affinity for the oxygen that leads to the formation of aluminum oxide, high thermal conductivity, high coefficient of thermal expansion, etc. Though tungsten inert gas welding, commonly known as TIG, can weld the aluminum alloys, the weld strength is not up to the desired level [1]. Most of the researchers have reported that the weld strength with fusion welding ranges between 40 and 60% of the base material [1, 2]. FSW, developed in 1991, by The Welding Institute UK [3], is an efficient process to weld various aluminum alloys for high strength applications such as automobile, aerospace, etc. 2XXX series of aluminum alloys have the highest strength and it is preferred in the aerospace industry [2].

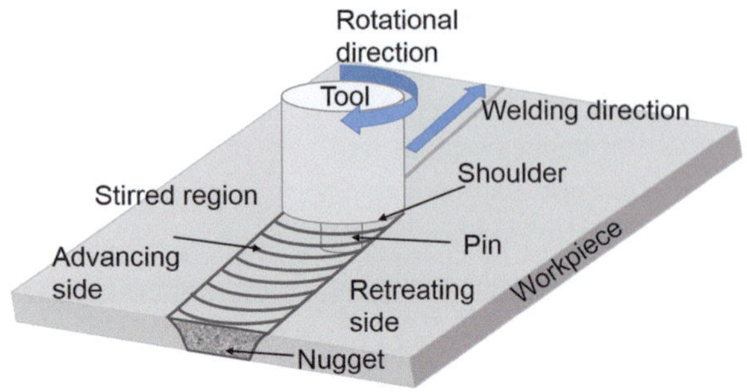

Fig. 1 Schematic of FSW with its nomenclature

Being a solid-state process, it is easier to weld aluminum alloys with FSW as compared to conventional fusion welding techniques. FSW possesses following advantages over conventional fusion welding process: no defects related to solidification of the material, low distortion, high weld strength, finer grain, etc. [4].

Friction stir welding is performed in three different stages, viz., plunging, dwelling, and welding. Each stage has its own significance. A non-consumable tool that is attached to the externally powered spindle rotates at a predefined speed and eventually impinges on the abutting edges of the clamped workpiece. Interaction of the two materials generates heat due to friction and plastic deformation. Once a contact between the tool and the workpiece is established, the tool is allowed to rotate in its position to further increase the temperature of the workpiece. Finally, the rotating tool is allowed to travel along the faying surface of the workpiece to stir the material and to weld the two material pieces. Schematic illustration of FSW during welding phase along with its nomenclature is shown in Fig. 1.

The tool has two special features, viz., shoulder and the pin. The ratio of shoulder to pin diameter ranges from 3 to 5 [5]. Efficient contact among the workpiece and tool is desired and that is achieved by defining a plunge depth that ranges from 0.1 to 0.4 mm. Plunge depth is the distance by which the shoulder of tool is impinged in the workpiece. Advancing side (AS) is the side of the workpiece where the direction of rotation of tool and the welding direction are same and the other side is known as the retreating side (RS). The difference in relative velocity on either side of the workpiece makes FSW asymmetric.

2 Literature Review

Analytical approach and numerical methods are the two well-known techniques to model any manufacturing process. Analytical method uses the appropriate process to break down the problems into the simpler ones and then solve them based on the set of fundamental equations. Analytical methods are accurate and provide an exact solution. The only drawback with this method is the difficulty in incorporating complex and nonlinear behaviors. Also, it is difficult to solve the equations analytically, when the number of unknown parameters is more than the number of equations. In such cases, numerical methods involving iterations are used to solve the mathematical relations in the form of differential equations. Various numerical methods like FEM, finite volume method (FVM), etc. can be used to solve the problem. With the increase in the availability of high-performance computer and commercial software, numerical methods have gained popularity in solving various manufacturing processes. Literature related to modeling are categorized as analytical, Lagrangian, Eulerian, Arbitrary Lagrangian and Eulerian.

2.1 Analytical Method

In the initial years, for the development of analytical models, researchers have used a moving heat source model to study the temperature distribution in FSW. Heat source equations were developed based on the different proposed possible mechanisms such as correlation with input machine power [6], frictional heating [7], the Rosenthal equation [8], etc. The gradual increment in the research leads to the coupling of the tool and workpiece to calculate the temperature distribution on the former [9]. The frictional contact condition at the workpiece and tool is different; therefore, researchers [10, 11] developed a heat generation equation based on sliding, sticking, and partial sliding/sticking contact conditions. They also reported that 86% of the heat is generated by the shoulder.

Analytical equations were used to solve the material flow and thermal analysis separately. The velocity field was modeled by considering three separate fields, viz., circumventing, vortex, and torsion velocity field based on the classical fluid mechanics concept. Thermal modeling was carried out based on the heat generation by Coulomb's law of friction and plastic deformation [12].

Solving analytical equations is a tedious task. Therefore, researchers have defined the moving heat source in FEM and solved them iteratively to study the thermal cycle [7, 13, 14]. Based on this method, the effect of tool geometries on temperature history [15] and thermal contact conductance between the workpiece and backing plate were studied. They found that clamping condition does not influence the temperature distribution [16].

Most of the researchers have explored the symmetric condition for the workpiece to study the thermal cycle. Also, heat generation through plastic deformation was not considered.

2.2 Lagrangian Method

In this method, nodes and their associated material points get displaced during the deformation of the domain. Throughout the problem, nodes on the boundary surfaces remain along the element edge; this eventually makes the definition of boundary and contact conditions easier [17]. Quadrature points (points at which results are calculated) coincide with the material points. Therefore, each time constitutive equations are evaluated at the same material points leading to the straightforward calculation of the constitutive equation. This makes the Lagrangian analysis best suited for solid mechanics or processes with lower deformation [18]. The Lagrangian method can be categorized into two types, viz., total Lagrangian and updated Lagrangian. In case of the former method, integral of weak form is performed over the initial configuration and derivatives are calculated with respect to the material coordinates. In case of the latter, integral of the weak form is carried out over the current or deformed configuration and the derivatives are calculated with respect to the spatial coordinates [19].

A two-dimensional model was proposed by Xu et al. to study the material flow during FSW [20]. They considered the thickness of the workpiece to be sufficiently high such that plane strain assumption was valid. Coulomb's law of friction was defined at the contact. The first three-dimensional model was reported by Ulysse to study the influence of welding speed on thermal history and force during FSW. Instead of defining the backing plate, a convective heat transfer coefficient was defined from the bottom face of the workpiece [21]. Residual stress was evaluated based on the three-dimensional thermomechanical model [22].

The workpiece symmetry was explored by the researchers, and a moving heat source was defined. Analysis was carried out in ANSYS software based on the Lagrangian analysis. The developed model was extended to study the influence of rotational and welding speed on the developed forces and thermal history [23, 24]. A comparative study between the uniform and nonuniform contact conductance was carried out by Soundarajan et al. [25]. They defined nonuniform contact conductance based on the stress distribution. Nonuniform contact conductance predicted better results during the plunging phase, while the peak temperature and cooling curve were the same for both the cases.

A group of researchers developed a three-dimensional, thermomechanical model based on the Lagrangian method using DEFORM 3D. They assumed the workpiece as a single continuum body and deformation as the rigid viscoplastic. Sticking frictional contact condition was defined at the interface [26]. The developed model was used to study the influence of included angle for the conical pin. Included angle between 20° and 30° was found to be the optimum value [27]. The model was further extended to predict the grain size based on dynamic recrystallization and compared with the experimental results [28, 29], modeling of the workpiece with different thickness [30, 31], and study of the material flow for cylindrical and conical pin [32]. The researchers did not consider the backing plate and restricted the maximum material velocity of the nodes to reduce the computation time.

A pseudo-steady-state FE model was developed in ABAQUS to study the various contact conditions at the interfaces. They found that surface heat flux predicts better result as the contribution of shoulder is higher than the pin [33]. Further, they also reported that a nonuniform contact conductance at the interface, i.e., higher conductance beneath the region near to the tool and zero elsewhere, provides better prediction of cooling rates [34].

A three-dimensional, thermomechanically coupled model based on Lagrangian implicit in DEFORM 3D was developed to simulate FSP of magnesium alloy. Sticking contact condition was defined. Flow stress was defined based on the Zener–Hollomon equation. Square pin shape was considered for the analysis. Pin height of 2.5 mm and the workpiece thickness of 5 mm were considered for the simulation. The predicted peak temperature was 470 °C. Particle tracking was used to study the material flow [35]. They have extended the model to study the material flow for the cylindrical pin. Al–Si as-cast aluminum alloy was used for the analysis. Experimentally measured temperature and axial force for the single parameter were used for the validation of the model. The material flow from the top region stretches toward the AS, and the same was found in experiment [36]. Similar analysis for the cylindrical pin was carried out by Pashazadeh et al. [37, 38] for copper alloys. They validated the model with the temperature obtained from the literature. Another researcher [39] used the Lagrangian model developed in DEFORM 3D to study the temperature and strain for threaded cylindrical pin. They considered Hill's anisotropy for the workpiece. Frictional contact was defined based on combined sliding/sticking criterion.

A Lagrangian model in the DEFORM 3D was developed [40] to compare the smooth and threaded pin. Instead of constraining the workpiece, a meshed die box was defined to arrest the rigid motion of the workpiece. The model was validated with the experimental force, torque, and peak temperature for a single parameter. Lower force and higher deformation were reported for the threaded pin. Also, the joint efficiency of the threaded pin was higher than the smooth cylindrical pin. The square and cylindrical pins were compared based on the model developed in DEFORM 3D [41]. They found that square pin produces higher strain, and hence finer grain size. In case of the square pin, material from AS and RS entered the stirred zone while for the cylindrical pin, only the material from AS entered the stirred zone. Zhang et al. [42] used the Lagrangian method to simulate the FSW in DEFORM 3D for the threaded cylindrical pin. They concluded that plastic flow is the driving force for the formation of banded texture and flash formation. They also studied the plastic flow of the material within the keyhole for one revolution of the tool. Shojaeefard et al. [43] studied the influence of the shoulder diameter on temperature and heat-affected zone. They also developed a neural network model and compared with the FEM results. The Lagrangian model has also been used to study the modeling of friction stir spot welding [44].

The developed model based on the Lagrangian method was further extended to predict the microstructure. Cellular automata were used to predict the developed microstructure during FSW of magnesium alloy for different pin shapes, viz., cylindrical [45] and square pin [46]. A simpler method based on Zener–Hollomon parameter was used to calculate the development of grain size [47].

2.3 Eulerian Method

Contrary to the Lagrangian analysis, the mesh nodes are fixed in space, and material points are allowed to flow through it in Eulerian method. Since the mesh is fixed in space, the problem associated with the mesh distortion is eliminated. Major drawback of this method is the difficulty to model free boundary surface because boundary nodes may not be coinciding with the element nodes. Therefore, Eulerian method can only be used when the boundaries of the deformed surface are known a priori [19]. Hence, plunging stage of the FSW cannot be modeled with the Eulerian analysis. Thus, the impression of the tool shape has to be created before initiating the simulation.

In case of the Eulerian analysis, all the researchers have considered the workpiece as a non-Newtonian, incompressible fluid [48]. The welding speed was defined as an inflow and outflow of the material. A three-dimensional model in FLUENT was developed to compare two different pin shapes, viz., triflute and Trivex. Trivex pin predicted lower forces as compared with the triflute [49, 50]. In order to study the influence of threaded cylindrical pin, an Eulerian model was developed in FLUENT. Steady-state condition was assumed for the modeling. The model predicted higher peak temperature and size of the deformation zone. Also, welding force was under-predicted [51].

A group of researchers developed a model to study the heat generation rate and temperature distribution during FSW [52]. They considered the material constant of Zener–Hollomon equation as a function of temperature. The heat flux was defined as a function of slip rate and coefficient of friction [53]. The developed model was further extended to predict the forces and torque requirement during FSW [54], strain and strain rate [55], and optimizing the shoulder diameter. The shoulder diameter of 18 mm was found as an optimum value [56]. Load bearing capacity of the pin was evaluated based on the shear stress experienced by the pin [57].

Two different contact conditions, viz., constant velocity and variable shear, were studied. They found that the latter provides better accuracy as compared to the former [58]. Researchers have used different frictional contact conditions for the analysis; a few have defined full sticking condition, i.e., material velocity was same as the tool velocity [59–61]; some defined the material velocity as a certain percentage of the tool velocity [62, 63] and partial sticking/sliding [56, 64–67]

Researchers have studied the influence of different types of pin and shoulder profiles along with the Eulerian analysis. Ji et al. [68] studied different shoulder geometries, viz., scroll and concave. The former shoulder provided higher material velocity. Further, they found that lower included angle of the pin and thread pitch, produced higher velocity. Su et al. [69] compared the conical and triflat pin. No significant difference in heat generation density was found for both the pins. The stirring action produced by the triflat pin was more dynamic than the conical pin because of the flat surfaces on the former. Kadian and Biswas [70] numerically compared seven different profiles having smooth, threaded, and flat surface. Peak temperatures for all the profiles were close and not much variation was reported. The flat surface pin produced swirl vortex flow.

Influence of the material models has been studied based on the two-dimensional Eulerian model. They reported that Zener–Hollomon equation predicted the saturation temperature at higher strain. Temperature variation of 21% was reported while performing the simulation with different material models [71]. Wormhole defect has been predicted by tracking the particle flow path and compared with the experimental defect [72]. A transient model to simulate all phases of FSW was reported. It has been shown that the temperature increases monotonically during plunging and sharply at the end of the plunging [73].

Simulations based on the Eulerian analysis were mostly carried out in FLUENT. Plunging has been neglected, and dwelling and welding phases have been modeled. Barring a few, most of the researchers have assumed steady-state analysis.

2.4 *Arbitrary Lagrangian and Eulerian Method*

Eulerian and Lagrangian methods have their own pros and cons. In the ALE method, the nodes can move arbitrarily, such that advantages of the Lagrangian and Eulerian methods are exploited to minimize their drawbacks. Freedom of moving mesh allows larger distortions of continuum that can be handled as compared with the Lagrangian method. However, mesh movement has its limitation and very large deformation is difficult to handle with this method.

A two-dimensional model was developed based on the ALE method in ABAQUS. They qualitatively compared the predicted and experimental material flows during FSW. Rotational and welding speeds were increased by 1000 times to reduce the computation time [74]. A three-dimensional thermomechanical model based on the ALE technique was developed. The workpiece was modeled as an Eulerian body with a circular shape. Pin height was larger than the workpiece thickness to achieve the convergence [75]. A good number of researchers have used ALE method to model the FSW process. All the models developed were in ABAQUS/Explicit. A few researchers have studied the effect of various phases of FSW on temperature, strain [76], effect of process parameters on material flow, velocity and temperature [77–79], grain size [80, 81], etc. Modified Johnson–Cook model has also been used to simulate FSW [82].

Plunging phase has been modeled based on ALE technique. Authors have compared the simulation predicted temperature and axial force with the experiments [83]. Two different frictional models, viz., Norton Hoff and Coulomb's law of friction, have been compared. It was found that the latter produced more realistic results as compared with the former [84]. The Johnson–Cook material and failure models have been used to model the different phases of FSW. The mesh on the stirred zone got deleted because of the failure model [85].

Re-meshing supported by the ALE method is not adequate to simulate the FSW process, since ALE does not support the generation of new mesh during the re-meshing procedure. Also, certain assumptions such as circular workpiece, pin height larger than the workpiece thickness, and pilot hole on the workpiece surface have been considered by most of the researchers.

3 Simulation Details

In an FSW process, the temperature reaches 70–90% of the melting point of the material. High temperature reduces the yield strength of the material. In the FSW process, temperature increases predominantly because of the friction between the tool shoulder and the work surface, along with the large-scale plastic deformation of the material due to the rotation of the pin. Therefore, deformation and temperature are interdependent and the same should be captured within the model for better accuracy of the results.

A three-dimensional coupled thermomechanical, nonlinear transient model based on the Lagrangian method is proposed to simulate the FSW. Displacement and temperature are coupled to capture their inter-dependency. Lagrangian method is generally associated with the low deformation problems, since mesh nodes are attached with the material points [19]. Hence, for the large deformation problems like FSW, mesh distortion is inevitable. In the proposed model, an adaptive re-meshing technique is defined to maintain the shape and integrity of the mesh. FSW is modeled in three stages, viz., plunging, dwelling, and welding with each having duration of 42, 10, and 25 s, respectively.

3.1 Geometric Modeling

The workpiece is modeled as a cuboid with a rectangular cross section within the simulation environment. It is modeled as a single continuum body, i.e., abutting edges are neglected. This assumption reduces the computation time by eliminating the contact instabilities [26]. AA2024 with a thickness of 5.9 mm is defined as the workpiece material. Length and width of each sheet are 120 and 60 mm, respectively. The tool is slightly complicated and DEFORM 3D does not have any inbuilt option to model it. Therefore, the tool with shoulder diameter of 15 mm and conical pin with top and bottom diameters of 6 and 2.5 mm, respectively, are modeled in CATIA V5R14 and imported in DEFORM 3D as a .STL file. Pin height used in the simulation is 5.5 mm. After modeling, both are assembled to replicate the initial plunging condition. Bottom of the pin must be in contact with the workpiece top surface prior to start of the simulation to avoid non-convergence as shown in Fig. 2. The tool is tilted by 2.5° toward its trailing edge during the assembly, and the same is also shown in Fig. 3.

Fig. 2 Geometric model of the tool and workpiece after assembly

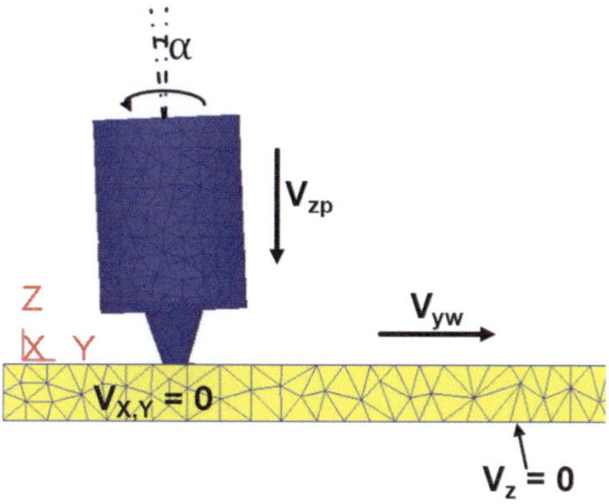

Fig. 3 Mechanical boundary conditions for different phases of FSW

3.2 *Boundary Conditions*

Accuracy of the results is greatly influenced by the boundary conditions. Boundary condition in FSW can be classified into two types, viz., mechanical and thermal. Mechanical boundary condition for the workpiece is required to constraint its rigid motion, so as to replicate the experimental clamping condition in the simulation. Tool rotation and translation are the two boundary conditions required to simulate the various phases of FSW. Thermal boundary conditions are required to define the thermal interactions of the workpiece and tool with the environment.

3.2.1 Mechanical Boundary Condition

Various mechanical boundary conditions defined on the workpiece tool are shown in Fig. 3.

(a) All degrees of freedom of the workpiece should be arrested to avoid any rigid rotation. Therefore, velocity of the workpiece bottom face was arrested in vertical (along *Z*-axis) direction. Since at the bottom surface of the workpiece material flow is in *XY* plane, therefore, velocity of the workpiece on side faces is constrained in *X*-and *Y*-directions.

(b) Rotational speed is defined about *Z*-axis. Velocity of the tool in plunging and welding phase was defined along −ve *Z*- and +ve *Y*-axes, respectively.

3.2.2 Thermal Boundary Condition

Initial temperature of the workpiece, tool, and environment is set at 20 °C.

(a) Both the workpiece and tool interact with the environment by two modes of heat transfer, i.e., convective heat transfer and radiation, as shown in Eq. 1. This is valid for all surfaces of the tool and workpiece, except the bottom face.
(b) Heat transfer between the workpiece and tool is due to the conduction but it is difficult to incorporate in simulation. Therefore, a convective heat transfer between them is defined such that the total heat transfer remains the same, as described in Eq. 2.
(c) Backing plate was neglected to reduce the computational time. Therefore, instead of conductive heat transfer, convective heat transfer was defined between the bottom face of the workpiece and environment, as described in Eq. 3.

$$Q_{\mathrm{wa}} = \sigma_{\mathrm{b}}\varepsilon_{\mathrm{b}}\left(T_{\mathrm{w}}^{4} - T_{\mathrm{a}}^{4}\right) + h_{\mathrm{a}}(T_{\mathrm{w}} - T_{\mathrm{a}}) \tag{1}$$

$$Q_{\mathrm{wt}} = K_{\mathrm{wt}}\frac{\partial T}{\partial z} = h_{\mathrm{wt}}(T_{\mathrm{w}} - T_{\mathrm{t}}) \tag{2}$$

$$Q_{\mathrm{wb}} = K_{\mathrm{wb}}\frac{\partial T}{\partial z} = h_{\mathrm{wb}}(T_{\mathrm{w}} - T_{\mathrm{b}}) \tag{3}$$

Values related to thermal boundary condition are mentioned in Table 1. Heat generation in FSW is also a thermal phenomenon, and it is discussed in Sect. 3.5.3.

3.3 *Material Model*

Deformation behavior of the material can be either defined as elastic–plastic or plastic. In case of the former, elastic deformation of the material is incorporated, while it is neglected for the latter. In case of plastic deformation, it is assumed that inception of deformation is from the yield point. This assumption allows better convergence and larger time increment. Plastic deformation can be classified into

Table 1 Constants related to thermal boundary condition

Variables	Values
Heat transfer from the workpiece/tool to environment (h_a) (N/mm s °C)	0.02
Heat transfer from the workpiece to tool (h_t) (N/mm s °C)	11
Heat transfer from the workpiece bottom surface to backing plate (h_b) (N/mm s °C)	11
Stefan Boltzmann constant (N/mm °C)	5.669×10^{-11}

two categories, viz., plastic and viscoplastic. In case of the former, strain hardening is more prominent, while strain rate is for the latter. FSW is a hot working process, and strain rate plays a vital role; hence, the material was defined as viscoplastic material. For viscoplastic material, the mechanical response can be defined with three relations, viz., constitutive law, yield criterion with hardening rule, and flow rule. These relations are discussed in the following subsections.

3.3.1 Yield Criterion

It is a hypothesis in the form of a mathematical equation to define the initiation (and continuation) of plastic deformation, as expressed in Eq. 4. Von Mises yield criterion including isotropic hardening law is expressed as [86]

$$f\left(\sigma_{ij}\right) = \text{Constant} \tag{4}$$

The material is defined as homogenous and isotropic, indicating that yielding is dependent on the magnitude of three principal stresses and independent of their directions. Also, yielding only depends on the principal component $(\sigma'_1, \sigma'_2, \sigma'_3)$ of the deviatoric stress tensor

$$\sigma'_{ij} = \sigma_{ij} - \delta_{ij}\sigma_m \tag{5}$$

where $\sigma_m = \frac{1}{3}\left(\sigma_x + \sigma_y + \sigma_z\right)$ is the hydrostatic component of the stress, and δ_{ij} is the Kronecker delta which results in value 1 for $i = j$ and 0 for $i \neq j$. Since the trace of deviatoric stress tensor is zero, the yield criterion reduces to

$$f(J_2, J_3) = \text{Constant} \tag{6}$$

where J_2 and J_3 are the second and third invariants of stress tensor

$$J_2 = -\left(\sigma_x\sigma_y + \sigma_y\sigma_z + \sigma_z\sigma_x\right) \tag{6a}$$

$$J_3 = \sigma_x\sigma_y\sigma_z \tag{6b}$$

In the current analysis, von Mises yield criterion is defined which states that yielding occurs when J_2 reaches a critical value or yielding is independent of J_3 and only depends on J_2. Therefore, Eq. 6 can be rewritten as

$$J_2 = \frac{1}{2}\left(\sigma_x'^2 + \sigma_y'^2 + \sigma_z'^2\right) = \frac{1}{2}\left(\sigma'_{ij}\sigma'_{ij}\right) = k^2 \tag{7a}$$

$$\left(\sigma_x - \sigma_y\right)^2 + \left(\sigma_y - \sigma_z\right)^2 + (\sigma_z - \sigma_x)^2 = 6k^2 \tag{7b}$$

where k is a material constant and depends on the material property. Equation 7 is written in suffix notation form, and Eq. 7b is by assigning the suffix value as x, y, and z. Material constants in Eq. 7 can be found through a simple tensile test as at yield point, $\sigma_x = \sigma_Y$ and $\sigma_y = \sigma_z = 0$ and we get $k = \frac{\sigma_Y}{\sqrt{3}}$ and therefore, Eq. 7 can be rewritten as [87]

$$\sqrt{\frac{3}{2}\left(\sigma'_{ij}\sigma'_{ij}\right)} = \sigma_Y = f(\sigma_{ij}) = \text{yield function} \tag{8}$$

Equation 8 is the yield criterion based on the von Mises principle. Also, yielding must incorporate the influence of prior plastic deformation (strain hardening of the material), and to incorporate this isotropic hardening rule with associated flow rule was defined. In case of isotropic hardening rule, the shape of the yield surface remains unchanged and only its size increases with the increase in stress, as shown in Fig. 4.

After incorporating hardening rule, Eq. 8 can be rewritten as

$$\sqrt{\frac{3}{2}\left(\sigma'_{ij}\sigma'_{ij}\right)} - \left(k\left(\varepsilon^p_{ij}\right)\right) = 0 \tag{9}$$

where $k\left(\varepsilon^p_{ij}\right)$ is a scalar which increases monotonically with the plastic deformation. Increase in deformation increases $k\left(\varepsilon^p_{ij}\right)$, and hence, the shape of yield surface.

3.3.2 Flow Rule

Hooke's law defines the relationship between the stress and strain within the elastic range but is invalid for the plastic region. In the plastic region, the stress–strain relation can be defined based on the concept of plastic potential as expressed in Eq. 10. Flow rule defines the rate of change of various plastic strain rate components during plastic deformation of the material. The associative flow rule is defined for this analysis in which plastic strain increment vector is normal to the yield surface [87].

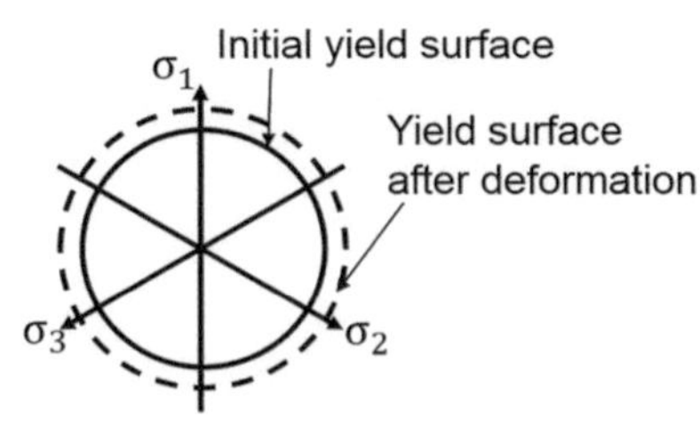

Fig. 4 von Mises isotropic hardening

$$\varepsilon_{ij}^{p} = \lambda \frac{\partial g}{\partial \sigma_{ij}} \tag{10}$$

where g is a scalar function of the invariants of the deviatoric stress and is called plastic potential and λ is a positive proportionality constant. When $g = f$(yield function), Eq. 10 is known as associated flow rule that is used to calculate the plastic strain and is expressed in Eq. 11.

$$\varepsilon_{ij}^{p} = \lambda \frac{\partial f}{\partial \sigma_{ij}} \tag{11}$$

Based on Eq. 7, above Eq. 11 can be rewritten in rate form, as shown in Eq. 12,

$$\varepsilon_{ij}^{p} = \sigma'_{ij} \dot{\lambda} \tag{12}$$

since

$$\frac{\partial f}{\partial \sigma_{ij}} = \frac{\partial f}{\partial \sigma'_{kl}} \frac{\partial \sigma'_{kl}}{\partial \sigma_{ij}} \tag{13}$$

In above equation, repeated subscripts k and l indicate summation with respect to these quantities. Now, Eq. 12 can be written as

$$\frac{\dot{\varepsilon}_{x}^{p}}{\sigma'_{x}} = \frac{\dot{\varepsilon}_{y}^{p}}{\sigma'_{y}} = \frac{\dot{\varepsilon}_{z}^{p}}{\sigma'_{z}} = \frac{\dot{\gamma}_{xy}^{p}}{2\tau_{xy}} = \frac{\dot{\gamma}_{yz}^{p}}{2\tau_{yz}} = \frac{\dot{\gamma}_{zx}^{p}}{2\tau_{zx}} \tag{14}$$

Here, the workpiece is defined as a viscoplastic body, i.e., elastic strain is neglected, and hence total strain rate will be equal to the plastic strain rate. Therefore, the Levy–Mises equation (Eq. 14) can be expressed in terms of stress components by three equations of the type

$$\dot{\varepsilon}_{x} = \left\{ \sigma_{x} - \frac{1}{3}\left(\sigma_{x} + \sigma_{y} + \sigma_{z}\right) \right\} \dot{\lambda} = \frac{2}{3}\dot{\lambda}\left\{ \sigma_{x} - \frac{1}{2}\left(\sigma_{y} + \sigma_{z}\right) \right\} \tag{15a}$$

Similar equation can be written for $\dot{\varepsilon}_{y}$ and $\dot{\varepsilon}_{z}$. Similarly, shear strain rate can be expressed as

$$\dot{\varepsilon}_{xy} = \frac{\dot{\gamma}_{xy}}{2} = \tau_{xy} \dot{\lambda} \tag{15b}$$

Similar equations can be written for $\dot{\varepsilon}_{yz}$ and $\dot{\varepsilon}_{zx}$. Equation 15 is known as Levy–Mises equation and the same is defined in the analysis. Proportionality factor $(\dot{\lambda})$ of Levy–Mises can be derived based on work hardening principle and is expressed as

$$\dot{\lambda} = \frac{3}{2}\frac{\dot{\bar{\varepsilon}}}{\bar{\sigma}} \tag{16}$$

with

$$\dot{\bar{\varepsilon}} = \sqrt{\frac{2}{3}\{\dot{\varepsilon}_{ij}\dot{\varepsilon}_{ij}\}} \tag{17}$$

where $\dot{\bar{\varepsilon}}$ is the effective strain rate.

3.3.3 Constitutive Law

This describes the change in the flow stress of the material as a function of plastic deformation, the rate of deformation, and temperature. In the current research, viscoplasticity principle based on the work of Perzyna is used. Workpiece material was defined as rigid viscoplastic with AA2024 material. Perzyna [88] introduced a function $F(\sigma_{ij})$ such that

$$F(\sigma_{ij}) = \frac{\sqrt{\frac{1}{2}\left(\sigma'_{ij}\sigma'_{ij}\right)}}{k_s} - 1 \tag{18}$$

where k_s is the static yield stress in shear. $F(\sigma_{ij})$ is a function similar to the plastic potential in Eq. 10, and therefore above can be expressed as

$$\dot{\varepsilon}_{ij} = V''\langle\phi(F)\rangle\frac{\partial F}{\partial\sigma_{ij}} \tag{19}$$

where V'' is a material viscosity constant and $\langle\phi(F)\rangle$ is a function of F such that

$$\langle\phi(F)\rangle = \begin{cases} 0, & \text{for } F \leq 0 \\ \phi(F), & \text{for } F > 0 \end{cases} \tag{20}$$

based on above equation, Eq. 19 can be rewritten as

$$\dot{\varepsilon}_{ij} = V''\left\langle\phi\left(\frac{\bar{\sigma}}{Y} - 1\right)\right\rangle\frac{\sigma'_{ij}}{\bar{\sigma}} \tag{21}$$

where $\bar{\sigma} = \sqrt{\frac{3}{2}\left(\sigma'_{ij}\sigma'_{ij}\right)}$ so that it is same as yield stress in tensile testing, and $\sigma_Y = \sqrt{3}k_s$ is the static yield stress in tension. By squaring above equation and using Eq. 17, we get

$$\dot{\bar{\varepsilon}}_{ij} = \frac{2}{3} V'' \left\langle \phi \left(\frac{\bar{\sigma}}{\sigma_Y} - 1 \right) \right\rangle \tag{22}$$

From Eqs. 21 and 22, the constitutive equation can be written as

$$\dot{\varepsilon}_{ij} = \frac{3}{2} \frac{\dot{\bar{\varepsilon}}}{\bar{\sigma}} \sigma'_{ij} \tag{23}$$

with $\bar{\sigma} = \sqrt{\frac{3}{2}\left(\sigma'_{ij}\sigma'_{ij}\right)}$ and $\dot{\bar{\varepsilon}} = \sqrt{\frac{2}{3}\left\{\dot{\varepsilon}_{ij}\dot{\varepsilon}_{ij}\right\}}$.

The effective stress in the above equation is dependent on strain rate dependent function ϕ and needs to be fed in the model in the form of the material model. In the current research work, it is defined as a function of temperature, strain, and strain rate as shown in Fig. 5 [89]. Flow stress is defined for a temperature range of 300 to 500 °C and strain rate of 0.3 and 10 s^{-1}. Log–log interpolation is defined to calculate the flow stress data within the defined range.

Yield strength of the tool material is much higher than the workpiece material and hence it undergoes negligible wear. Therefore, the tool is defined as a rigid body such that stress and strain are not calculated and computation time is reduced. Steel H13 is defined as the tool material and to calculate the temperature rise of the tool, thermal heat transfer is defined. Various physical properties of the workpiece and tool material are summarized in Table 2 [90].

3.4 Contact Condition

The contact interaction between the tool and workpiece has been defined through the surfaces. The mechanical interaction of this contact surface is described using the penalty contact algorithm [91]. In this algorithm, surface penetration is resisted

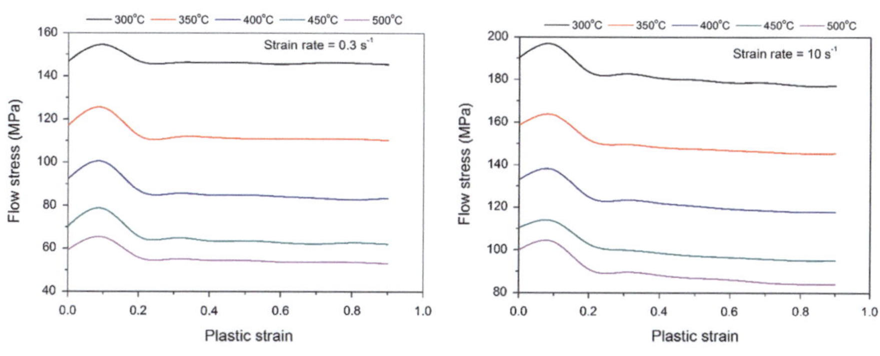

Fig. 5 Flow stress behavior of AA 2024 as a function of temperature and strain rate

Table 2 Physical properties of the tool and workpiece material

Properties	AA 2024	Tool steel H13
Young's modulus (N/mm^2)	68,900	210,000
Thermal conductivity (N/s °C)	181	24.5
Heat capacity (N/mm^2 °C)	2.43	2.78
Coefficient of thermal expansion (μmm/mm °C)	22	11.7
Poisson's ratio	0.33	0.3
Emissivity	0.05	0.7

by an assumption of a linear spring force whose value is proportional to the amount of penetration. In equilibrium condition, no penetration of the surface is assumed. Thus, in this condition, the contact pressure between the tool (master surface) and the workpiece (slave surface) does not get transmitted; however, during contact, there is no upper bound of the normal contact pressure. Sticking friction condition [92, 93] is assumed to define the shear stress between the said master and the slave contact surfaces, which is defined by Eq. 24. This condition is selected because it is best suited in most of the metal forming processes, and material deformation in the FSW operation can be assumed as one such operation [91].

$$\bar{\tau} = m\tau_{\max} \tag{24}$$

where $\bar{\tau}$ is the contact pressure, $\tau_{\max}$ is the shear yield strength of material (0.577 times of the yield strength as per von Mises yield criterion), and the shear factor, and m is defined as 0.4 [94].

3.5 Governing Equation and Finite Element Formulation for Lagrangian Method

The Lagrangian method is used for the current analysis. In this case, both material and nodal points deform along with the domain. Nodes along the boundary of the domain tend to remain along the element edges throughout the analysis, and hence simplifying the imposition of boundary condition and interaction. Lagrangian analysis is further classified into two categories, viz., total Lagrangian and updated Lagrangian. In case of the former, weak form involves integral over the initial configuration, and derivatives are taken with respect to the material coordinates. Dependent variable for the total Lagrangian method is displacement with respect to the reference configuration. In case of updated Lagrangian, the equations are formulated based on the current or deformed configuration and derivatives are with respect to spatial (Eulerian) coordinates. Dependent variables in this method are velocity and Cauchy stress as a function of time and current configuration. In the current research work, the updated Lagrangian method is used. In this section, a detailed discussion related to governing equation of the method along with the choice of the solver is discussed.

Governing equations are the mathematical relations which define the change in the dependent variables. FSW is a thermomechanical problem and the governing equations would be different for mechanical or plastic deformation and for the thermal analyses. In the subsequent section, governing equations used in the current work for mechanical and thermal analyses are laid down, along with the methodology used to solve them. Various methods are available to solve governing equations, viz., finite element method (FEM), finite difference method (FDM), finite volume method (FVM), etc. In the current work, FEM is used for the analysis.

3.5.1 Governing Equations for Plastic Deformation

Governing equations are the fundamental equations which determine the change in dependent variables. The governing equations for the plastic deformation of viscoplastic material are three equilibrium equations; the yield criterion and five strain rate ratios that are derived from the flow rule. Most of the relations are already discussed and are summarized below:

$$\text{Equilibrium condition} \quad \frac{\partial \sigma_{ij}}{\partial x_j} = 0 \quad (i,j = x,y,z) \tag{25}$$

$$\text{Yield condition} \quad f(\sigma_{ij}) = C, \quad \overline{\sigma} = \sqrt{\frac{3}{2}\left(\sigma'_{ij}\sigma'_{ij}\right)} \tag{26}$$

$$\text{Constitutive equation} \quad \dot{\varepsilon}_{ij} = \frac{3}{2}\frac{\dot{\bar{\varepsilon}}}{\bar{\sigma}}\sigma'_{ij} \quad \text{with } \dot{\bar{\varepsilon}} = \sqrt{\frac{2}{3}\left(\dot{\varepsilon}_{ij}\dot{\varepsilon}_{ij}\right)} \tag{27}$$

$$\text{Compatibility condition} \quad \dot{\varepsilon}_{ij} = \frac{1}{2}\left(\frac{\partial u_i}{\partial x_j} + \frac{\partial u_j}{\partial x_i}\right) \tag{28}$$

It is difficult to attain a complete solution that satisfies all governing equations in a single step. Therefore, FEM was used to approximate the results based on iteration techniques. The variational approach was used to derive the fundamental equation of FE formulation. It is based on one of the two variational principles. Velocity boundary conditions, incompressibility, and condition of compatibility must be satisfied by the admissible velocities u_i. The actual solution gives the following functional (function of functions) a stationary value [87]:

$$\pi = \int_{\mathrm{V}} E(\dot{\varepsilon}_{ij})\mathrm{d}V - \int_{S_F} F_i u_i \mathrm{d}S \tag{29}$$

where $F_i, V, S_F,$ and $E(\dot{\varepsilon}_{ij})$ are the surface traction, volume of the workpiece, traction surface, and work function, respectively. Above equation is based on the virtual work rate principle. Based on second extremum principle for the rigid viscoplastic material, work function is given by $\sigma'_{ij} = \frac{\partial E}{\partial \dot{\varepsilon}_{ij}}$. The solution to the original boundary value problem is obtained from the solution of the dual variational problem, where the first-order variation of the functional vanishes, as expressed in Eq. 30.

$$\delta\pi = \int_V \bar{\sigma}\delta\dot{\bar{\varepsilon}}\,\mathrm{d}V - \int_{S_F} F_i \delta u_i\,\mathrm{d}S = 0 \tag{30}$$

Incompressibility in penalized form is utilized to remove the incompressibility constraint on admissible velocity field. With this, Eq. 30 forms the fundamental equation for FE formulation.

$$\delta\pi = \int_V \bar{\sigma}\delta\dot{\bar{\varepsilon}}\,\mathrm{d}V + \lambda_{\mathrm{P}} \int_V \dot{\varepsilon}_{\mathrm{v}}\delta\dot{\varepsilon}_{\mathrm{v}}\,\mathrm{d}V - \int_{S_{\mathrm{F}}} F_i \delta u_i\,\mathrm{d}S = 0 \tag{31}$$

where λ_{P} and δu_i are the large penalty constant and arbitrary variation, respectively. $\delta\dot{\bar{\varepsilon}}$ and $\delta\dot{\varepsilon}_{\mathrm{v}}$ are the variations in strain rate, derived from δu_i. $\dot{\varepsilon}_{\mathrm{v}} = \dot{\varepsilon}_{ii}$ is the volumetric strain rate. Equation 31 is the fundamental equation for the FE formulation. The velocity field satisfying Eq. 31 is first obtained and the corresponding stresses can then be calculated based on flow rule.

In FSW process, deformation occurs only below the tool and rest of the domain remains almost undeformed. The undeformed region is termed as the rigid region in the simulation. Due to this, the value of the first term of Eq. 31 cannot be uniquely determined because the effective stress is not properly defined as effective strain approaches zero. This problem is eliminated by defining stress–strain relation expressed in Eq. 32 [87]

$$\dot{\varepsilon}_{ij} = \frac{3}{2}\frac{\dot{\bar{\varepsilon}}}{\bar{\sigma}}\sigma'_{ij} \quad \text{with } \bar{\sigma}_0 = \bar{\sigma}\left(\bar{\varepsilon}, \dot{\bar{\varepsilon}}_0\right) \text{ for } \dot{\bar{\varepsilon}} \leq \dot{\bar{\varepsilon}}_0 \tag{32}$$

where $\dot{\bar{\varepsilon}}_0$ is the limiting strain rate (0.001), i.e., region having strain rates lower than this will be treated as a rigid region. For this region, the first term of Eq. 31 is changed to

$$\int_V \frac{\bar{\sigma}_0}{\dot{\bar{\varepsilon}}_0}\delta\dot{\bar{\varepsilon}}\,\mathrm{d}V, \text{ for } \dot{\bar{\varepsilon}} \leq \dot{\bar{\varepsilon}}_0$$

Hence, Eq. 31 can be rewritten as

$$\delta\pi = \int_V \frac{\bar{\sigma}_0}{\dot{\bar{\varepsilon}}_0} \delta\dot{\bar{\varepsilon}} \mathrm{d}V + \lambda_\mathrm{P} \int_V \dot{\varepsilon}_\mathrm{v} \delta\dot{\varepsilon}_\mathrm{v} \mathrm{d}V - \int_{S_\mathrm{F}} F_i \delta u_i \mathrm{d}S = 0 \tag{33}$$

Thus, FE discretization process is based on Eq. 31 for regions having strain rate above the limiting value and Eq. 33 elsewhere.

3.5.2 Finite Element Formulation of Plastic Deformation

Domain in the FEM is the physical portion of the material, where loads and boundary conditions are applied and the solution is obtained based on the governing equations. There are several steps involved in FEM, viz., discretization of the domain, choice of interpolation function, evaluation of elemental stiffness matrix, assembly of elements, numerical integration, and solution. Sequence of the steps might change based on the implementation but all the steps need to be followed.

Discretization of the Domain or Mesh Generation

Selection of element type and size is a critical aspect of FEM analysis as accuracy of the results largely depends upon it. In the current analysis, the workpiece and tool need to be meshed. FSW is a coupled thermomechanical process; therefore, the workpiece was meshed with coupled tetrahedral elements, i.e., each node of the element calculates displacement and temperature simultaneously. The total number of elements influences the computation time. Therefore, biased meshing was defined on the workpiece with finer size near the interaction region and coarser elsewhere as shown in Fig. 6a, b. Mesh density window was used to define biased meshing. Interaction area between the workpiece and tool changes in different phases of FSW and size of mesh density window was selected accordingly. For plunging and dwelling, a square window of size 22 mm (larger than shoulder diameter) is defined, as shown in Fig. 6a, and in welding phase mesh density window of rectangular shape was defined, as shown in Fig. 6b.

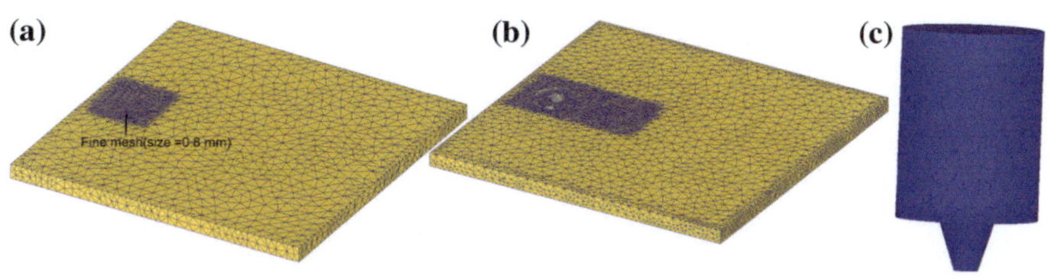

Fig. 6 Meshing of the domain during simulation. **a** Workpiece during plunging and dwelling, **b** welding phase, **c** tool

Though lower mesh size improves the accuracy of the result, at the same time it also increases the computation time. Mesh size of 0.8 mm is chosen for finer mesh (inside mesh density window), and for coarser mesh the size is 2 mm with a size ratio of 3. Total number of elements on the workpiece during plunging and welding phase is 33,886 and 67,098, respectively. Tool is also meshed with tetrahedral element with nodes having temperature degree of freedom only. On tool pin and shoulder, mesh size of 0.3 mm is defined. Tool shank is meshed with a mesh size of 1 mm with size ratio of 3. Total number of elements on the tool is 36,382.

Re-meshing Method

In the Lagrangian analysis, elemental nodes deform along with the material. Since FSW is a large deformation process with effective strain in the range of 80, mesh distortion is inevitable. Adaptive re-meshing is defined to tackle it. Re-meshing is triggered based on three criteria, and if any one of the criteria is achieved, the workpiece is then re-meshed.

(a) *Based on element penetration*: It is defined that re-meshing will initiate once the ratio of penetration distance from middle of the edge to the tool surface with original size of the element reaches 0.7 [89].
(b) *Shrinking of element*: In this criterion, re-meshing initiates when the size of the element shrinks by 40%. In this case, element size is 0.8; therefore, if element size reduces below 0.32 mm, re-meshing is initiated.
(c) *Stretching of element:* In this criterion, re-meshing will initiate once the element size stretches by 60%, i.e., if element size increases to 1.28 mm, re-meshing is then initiated.

The re-meshing is performed in two steps. First, generating a new mesh onto the workpiece and second transferring the results from old to new mesh through interpolation. New mesh is generated through the conventional mesh generator, which was used to mesh the workpiece initially. The output variables dependent on deformation history need to be interpolated on the new mesh, and for thermo-viscoplastic analysis those are temperature and effective strain. Temperature is calculated at the nodes and hence, its distribution is expressed using the shape function over the domain, i.e., workpiece. Hence, interpolation of temperature is straightforward, and it is simply evaluated at the locations of the new nodes.

Effective strain is calculated at the integration points and therefore, requires an additional step for the interpolation. Before interpolation, effective strain is calculated at nodes from the integration point of element and then interpolated on the new mesh system. Area-weighted averaging method is defined for the interpolation of data. In this method, the nodal value is obtained on the basis of the average of the adjacent element values weighted by the associated element size. For example, let node N is surrounded by the elements. The nodal value of the effective strain at node N is then calculated based on [87]

$$\bar{\varepsilon}_N = \frac{\sum_i \bar{\varepsilon}_i A_{iN}}{\sum_i A_{iN}} \quad (34)$$

where $\bar{\varepsilon}_N$ is the effective strain at node N, and $\bar{\varepsilon}_i$ is the effective strain value at the center of element i that surrounds the node N. A_{iN} is the area contribution of ith element to the node N and is defined by

$$A_{jN} = \int_{A_i} q_N(x, y)\mathrm{d}A \quad (35)$$

where q_N is the element shape function of element i at node N. The summation of Eq. 34 is performed over all the surrounding elements of node N. After calculating the effective strains at all the nodes, Eq. 36 is used to determine the strain distribution over each element

$$\bar{\varepsilon}(x, y) = \sum_e q_e \bar{\varepsilon}_e \quad (36)$$

where q_e is the element shape function. The interpolation from the old mesh to the new mesh is done by evaluating the effective strain at the location of the integration points of the new mesh. Comparative analysis of temperature and effective strain before and after re-meshing is mentioned in Table 3 at different time intervals during the welding phase. It shows that re-meshing has negligible effect on the output responses like temperature and effective strain.

Shape Function

Shape function is used to interpolate the results within the elements from discrete nodes. Therefore, appropriate shape functions are desired to be defined to achieve an approximate result that is closer to the exact solution. In the current analysis, linear shape function is defined for each tetrahedral element. Each node has three degrees of freedom which are nodal velocities. Node A_1 has coordinate and nodal velocities as (x_1, y_1, z_1) and u_{x1}, u_{y1}, u_{z1}, respectively. Coordinate and nodal velocities of point A_2 are (x_2, y_2, z_2) and u_{x2}, u_{y2}, u_{z2}, respectively, and so on for A_3 and A_4 (Fig. 7).

Table 3 Effect of re-meshing on output responses

	Temperature, °C			Effective strain		
S. No.	Before re-meshing	After re-meshing	% difference	Before re-meshing	After re-meshing	% difference
1	548.20	547.97	0.04	45.25	45.25	0
2	553.18	553.18	0	57.48	57.48	0
3	562.08	561.94	0.02	51.58	51.58	0

In matrix form, it can be written as

$$u = N^{\mathrm{T}} v \tag{37}$$

where v is the nodal velocity, N^{T} is the transpose of shape function matrix, and for tetrahedral element it is expressed as [95]

$$N^{\mathrm{T}} = \begin{bmatrix} q_1 & 0 & 0 & q_2 & 0 & 0 & q_3 & 0 & 0 & q_4 & 0 & 0 \\ 0 & q_1 & 0 & 0 & q_2 & 0 & 0 & q_3 & 0 & 0 & q_4 & 0 \\ 0 & 0 & q_1 & 0 & 0 & q_2 & 0 & 0 & q_3 & 0 & 0 & q_4 \end{bmatrix} \tag{38}$$

where q is the shape function and is given as

$$q_1 = 1 - \xi - \zeta - \eta, \quad q_2 = \xi, \quad q_3 = \eta, \quad q_4 = \zeta \tag{39}$$

It is important to develop a relation between the Cartesian coordinate system (x, y, z) and natural coordinate system (ξ, η, ζ). Based on the property of isoparametric elements, shape functions are correlated with Cartesian coordinate system, as expressed below [18]:

$$x = \sum_{i=1}^{4} q_i x_i; \quad y = \sum_{i=1}^{4} q_i y_i; \quad z = \sum_{i=1}^{4} q_i z_i \tag{40}$$

where (x, y, z) is the coordinate of point within the element, where solution needs to be interpolated. Using Eq. 39, above equation can be rewritten as a function of shape function as follows:

$$x = x_1 + \xi(x_2 - x_1) + \eta(x_3 - x_1) + \zeta(x_4 - x_1) \tag{41}$$

$$y = y_1 + \xi(y_2 - y_1) + \eta(y_3 - y_1) + \zeta(y_4 - y_1) \tag{42}$$

$$z = z_1 + \xi(z_2 - z_1) + \eta(z_3 - z_1) + \zeta(z_4 - z_1) \tag{43}$$

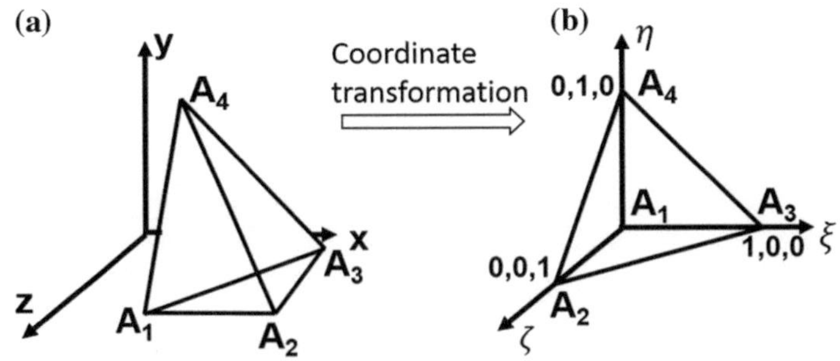

Fig. 7 Tetrahedral element in **a** original coordinate axis, **b** transformed coordinate axis

Calculation of dependent variables like strain rate will require differentiation of the shape function with respect to the Cartesian coordinate system. In Eq. 40, shape functions are defined in the natural coordinate system with chain rule. Differentiation of shape function with respect to natural coordinate can be expressed as

$$\begin{Bmatrix} \partial q/\partial \xi \\ \partial q/\partial \eta \\ \partial q/\partial \zeta \end{Bmatrix} = J_{\mathrm{b}} \begin{Bmatrix} \partial q/\partial x \\ \partial q/\partial y \\ \partial q/\partial z \end{Bmatrix} \tag{44}$$

where J is the Jacobian matrix and is given by

$$J_{\mathrm{b}} = \begin{bmatrix} \partial x/\partial \xi & \partial y/\partial \xi & \partial z/\partial \xi \\ \partial x/\partial \eta & \partial y/\partial \eta & \partial z/\partial \eta \\ \partial x/\partial \zeta & \partial y/\partial \zeta & \partial z/\partial \zeta \end{bmatrix} \tag{45}$$

Jacobian matrix for each element can be calculated from Eqs. 41–43. From Eq. 44, differentiation of shape function with respect to natural coordinate can be calculated as shown below:

$$\begin{Bmatrix} \partial q/\partial x \\ \partial q/\partial y \\ \partial q/\partial z \end{Bmatrix} = J_{\mathrm{b}}^{-1} \begin{Bmatrix} \partial q/\partial \xi \\ \partial q/\partial \eta \\ \partial q/\partial \zeta \end{Bmatrix} \tag{46}$$

Calculation of Strain Rate Matrix

Strain rate component is defined in Eq. 28. Also, admissible velocities within the element can be expressed as

$$u_i = \sum_e q_e u_i^e \tag{47}$$

where u_i^e is the velocity component at the eth node and summation is taken over all four nodes and is calculated by summation of nodal velocities over the element, as expressed in Eq. 47. Based on above equation, Eq. 28 can be rewritten as [96]

$$\varepsilon_{ij} = \frac{1}{2} \sum_e \left(\frac{\partial q_e}{\partial x_j} u_i^e + \frac{\partial q_e}{\partial x_i} u_j^e \right) \tag{48}$$

Hence, strain rate component can be evaluated if derivative of shape function is known and that can be evaluated from Eq. 46. For simplification of equation, let us assume

$$X_e = \frac{\partial q_e}{\partial x}; \quad Y_e = \frac{\partial q_e}{\partial y}; \quad Z_e = \frac{\partial q_e}{\partial z} \tag{49}$$

Based on Eqs. 48 and 49, six components of strain rate matrix can be written as

$$\dot{\varepsilon} = \begin{Bmatrix} \dot{\varepsilon}_x \\ \dot{\varepsilon}_y \\ \dot{\varepsilon}_z \\ \dot{\varepsilon}_{xy} \\ \dot{\varepsilon}_{yz} \\ \dot{\varepsilon}_{zx} \end{Bmatrix} = \begin{Bmatrix} \sum X_e u_x^e \\ \sum Y_e u_y^e \\ \sum Z_e u_z^e \\ \frac{1}{2}\sum Y_e u_x^e + X_e u_y^e \\ \frac{1}{2}\sum Z_e u_y^e + Y_e u_z^e \\ \frac{1}{2}\sum X_e u_z^e + Z_e u_x^e \end{Bmatrix} \tag{50}$$

In the matrix form, above equation can be written as

$$\dot{\varepsilon} = B_s v \tag{51}$$

where B_s is known as strain rate matrix, and for three-dimensional tetrahedral element, it is given as

$$B_s = \begin{bmatrix} X_1 & 0 & 0 & X_2 & 0 & 0 & X_3 & 0 & 0 & X_4 & 0 & 0 \\ 0 & Y_1 & 0 & 0 & Y_2 & 0 & 0 & Y_3 & 0 & 0 & Y_4 & 0 \\ 0 & 0 & Z_1 & 0 & 0 & Z_2 & 0 & 0 & Z_3 & 0 & 0 & Z_4 \\ Y_1 & X_1 & 0 & Y_2 & X_2 & 0 & Y_3 & X_3 & 0 & Y_4 & X_4 & 0 \\ 0 & Z_1 & Y_1 & 0 & Z_2 & Y_2 & 0 & Z_3 & Y_3 & 0 & Z_4 & Y_4 \\ Z_1 & 0 & X_1 & Z_2 & 0 & X_2 & Z_3 & 0 & X_3 & Z_4 & 0 & X_4 \end{bmatrix} \tag{52}$$

Similar to strain rate, it is necessary to express effective and volumetric strain rates in a matrix form. Effective strain rate is expressed in Eq. 17, and in a matrix form it can be rewritten as

$$\left(\dot{\bar{\varepsilon}}\right)^2 = \dot{\varepsilon}^{\mathrm{T}} D \dot{\varepsilon} \tag{53}$$

where D is the diagonal matrix with value of 2/3 as its components. Substituting Eq. 51 in Eq. 53, we get effective strain rate as

$$\left(\dot{\bar{\varepsilon}}\right)^2 = v^{\mathrm{T}} B_s^{\mathrm{T}} D B_s v = v^{\mathrm{T}} P v \tag{54}$$

where $P = B_s^{\mathrm{T}} D B_s$.

The volumetric strain rate is given by

$$\dot{\varepsilon}_v = \dot{\varepsilon}_x + \dot{\varepsilon}_y + \dot{\varepsilon}_z \tag{55}$$

and can be rewritten as

$$\dot{\varepsilon}_{\mathrm{v}} = C^{\mathrm{T}} v = C_I v_I \quad \text{with} \quad C_I = B_{1I} + B_{2I} + B_{3I} \tag{56}$$

where B_{IJ} is an element of the strain rate matrix B_{s}.

Elemental Stiffness Matrix

In Sect. 3.5.1, the basic equation for FEM formulation was derived and expressed in Eq. 31. Now based on the effective and volumetric strain rates, Eq. 31 should be modified to develop the elemental stiffness matrix. Equation 31 is solved from defined admissible velocity field based on shape function such that overreach element, there exists a continuous velocity field that can be uniquely defined with velocities of associated nodal points. Equation 31 can be written for nodal velocities v and their variation δv. From arbitrariness of δv_I, a set of algebraic equations or stiffness equation can be obtained as [87]

$$\frac{\partial \pi}{\partial v_I} = \sum_i \left(\frac{\partial \pi}{\partial v_I} \right)_i = 0 \tag{57}$$

where i indicates the quantity at the ith element. I is the nodal point number. Using the defined shape function and strain rate matrix, above equation can be rewritten as

$$\frac{\partial \pi}{\partial v_I} = \int\limits_V \frac{\bar{\sigma}}{\dot{\bar{\varepsilon}}} P_{IJ} v_J \mathrm{d}V + \int\limits_V \lambda_{\mathrm{P}} C_J v_J C_I \mathrm{d}V - \int\limits_{S_{\mathrm{F}}} F_j N_{jI} \mathrm{d}S \tag{58}$$

The third term on the right-hand side of the above equation is the applied nodal force and the other two are the reaction nodal point forces. The second derivative of above equation can be expressed as

$$\frac{\partial^2 \pi}{\partial v_I v_J} = \int\limits_V \frac{\bar{\sigma}}{\dot{\bar{\varepsilon}}} P_{IJ} \mathrm{d}V + \int\limits_V \left(\frac{1}{\dot{\bar{\varepsilon}}} \frac{\partial \bar{\sigma}}{\partial \dot{\bar{\varepsilon}}} - \frac{\bar{\sigma}}{\dot{\bar{\varepsilon}}^2} \right) \frac{1}{\dot{\bar{\varepsilon}}} P_{IK} v_K v_M P_{MJ} \mathrm{d}V + \int\limits_V \lambda_P C_J C_I \mathrm{d}V \tag{59}$$

Elemental level stiffness matrix can be formed using Eqs. 58 and 59. Equation 57 is nonlinear and the solution is obtained based on iterative method (viz., Newton–Raphson or direct integration). This method consists of linearization of the equation and achieving final solution based on convergence criterion. Taylor expansion near an assumed solution point $v = v_0$ (initial guess) is used for linearization, as expressed below:

$$\left[\frac{\partial \pi}{\partial v_I}\right]_{v=v_0} + \left[\frac{\partial^2 \pi}{\partial v_I v_J}\right]_{v=v_0} \Delta v_J = 0 \tag{60}$$

where Δv_J is the first-order correction of the velocity v_0. Once the elemental level matrix is calculated for each element, global stiffness matrix is then formed, as shown in Eq. 61 by assembling each elemental matrix in proper sequence

$$K_s \Delta v = f_R \tag{61}$$

where K_s is known as global stiffness matrix and f_R is the residual of the nodal point force vector. Choice of penalty constant (λ_P) is critical for the proper convergence of the solution. A large value of penalty constant is chosen such that volumetric strain rate remains close to zero. In the current work, penalty constant is 10^9.

Volume and surface integrals are required to solve Eqs. 58 and 59 and to represent them in stiffness matrix form. In implicit analysis, it is assumed that these integrations are exact but that is not possible every time. Therefore, approximations are made using numerical integration techniques. Numerical integration is performed over finite points, also known as integration points. Lower integration points help in faster solution and also reduce the possibility of volume locking. In the current analysis reduce integration method is chosen, i.e., one integration point is chosen at the centroid of the element to perform the integrals. Gauss quadrature formula is used to perform the integrals. Detailed study of numerical integration technique can be found in any FEM book [18, 95].

3.5.3 Governing Equation and Finite Element for Thermal Analysis

This section deals with the fundamental equations for the calculation of heat generation defined in the simulation. The energy balance equation for the thermal analysis is expressed below:

$$KT_{,ii} + \dot{q} - \rho c \dot{T} = 0 \tag{62}$$

$\dot{q}$ is the heat generation rate and $T_{,ii}$ is the heat transfer rate. $T_{,ii}$ is the Laplace operator (∇^2) applied on temperature (T). Volumetric heat generation rate due to plastic deformation is given by

$$\dot{q}_p = \varphi \sigma_{ij} \dot{\varepsilon}_{ij} \tag{63}$$

where φ is the inelastic heat fraction and is taken as 0.9. The remaining 0.1 causes change in dislocation density, grain boundaries, etc. [36]. Heat flux in terms of frictional heat is defined along the boundary of the workpiece and tool. Therefore, Eq. 62 can be rewritten for arbitrary variation in temperature δT [97]

$$\int_V KT_{,ii}\delta T \mathrm{d}V - \int_V \rho c\, \dot{T}\delta T \mathrm{d}V + \int_V \varphi \sigma_{ij}\dot{\varepsilon}_{ij}\delta T \mathrm{d}V = 0 \tag{64}$$

Applying divergence theorem in above equation, we get

$$\int_V KT_{,i}\delta T_{,i}\mathrm{d}V + \int_V \rho c\, \dot{T}\delta T \mathrm{d}V - \int_V \varphi \sigma_{ij}\dot{\varepsilon}_{ij}\delta T \mathrm{d}V - \int_{S_q} q_n \delta T \mathrm{d}S = 0 \tag{65}$$

where q_n is the heat flux across the boundary surface S_q, n indicates the normal to the boundary surface and

$$q_n = KT_{,n} \tag{66}$$

Temperature field needs to satisfy the boundary temperatures, and Eq. 65 for arbitrary perturbation δT.

3.5.4 Finite Element Formulation of Thermal Analysis

Approximation of temperature field is expressed in Eq. 65 [87]

$$T = \sum_e q_e T_e = N^{\mathrm{T}} T \tag{67}$$

where T_e is the temperature at eth node and q_e is the shape function. For the tetrahedral element, shown in Fig. 6, shape function matrix and temperature can be written in matrix form as

$$N^{\mathrm{T}} = \{q_1, q_2, q_3, q_4\} \quad \text{and} \quad T^{\mathrm{T}} = \{T_1, T_2, T_3, T_4\} \tag{68}$$

where $q_1 \ldots q_4$ are shape function, and are defined in Eq. 39. Now, putting

$$M_{ij} = N_{i,j} \quad \text{with} \quad M = \begin{bmatrix} X_1 & Y_1 & Z_1 \\ X_2 & Y_2 & Z_2 \\ X_3 & Y_3 & Z_3 \\ X_4 & Y_4 & Z_4 \end{bmatrix} \tag{69}$$

Elements of matrix M are already defined in Eq. 46. By putting Eqs. 68 and 69 in Eq. 65 and due to arbitrariness of δT, the following system of equations is obtained [87]

$$\sum_{j=1}^{m}\left[\int_{V_j} KMM^{\mathrm{T}}\mathrm{d}VT + \int_{V_j} \rho cNN^{\mathrm{T}}\mathrm{d}V\dot{T} - \int_{V_j} \varphi(\bar{\sigma}\dot{\bar{\varepsilon}})N\mathrm{d}V - \int_{S_{qj}} q_n N\mathrm{d}S\right] \tag{70}$$

Above equation can be expressed in the form of

$$C_{\mathrm{p}}\dot{T} + K_{\mathrm{c}}T = Q \tag{71}$$

where $C_{\mathrm{p}}, Q, K_{\mathrm{c}}, T, \dot{T}$ are the heat capacity matrix, heat flux vector, heat conduction matrix, vector of nodal point temperature and vector of nodal point temperature rates, respectively. Heat flux vector in Eq. 71 comprises various terms and is shown with interpolation function N by

$$\begin{aligned} Q = &\int_{V} \varphi(\bar{\sigma}\dot{\bar{\varepsilon}})N\mathrm{d}V + \int_{S_{\mathrm{r}}} \sigma_{\mathrm{b}}\varepsilon_{\mathrm{b}}\left(T_{\mathrm{w}}^4 - T_{\mathrm{a}}^4\right)N\mathrm{d}S + \int_{S_{\mathrm{a}}} h_{\mathrm{wa}}(T_{\mathrm{w}} - T_{\mathrm{a}})N\mathrm{d}S \\ &+ \int_{S_{\mathrm{c}}} h_{\mathrm{wt}}(T_{\mathrm{w}} - T_{\mathrm{t}})N\mathrm{d}S + \int_{S_{\mathrm{c}}} q_{\mathrm{f}}N\mathrm{d}S \end{aligned} \tag{72}$$

First term on the right-hand side of above equation is the heat generation due to plastic deformation within the workpiece. The second, third, and fourth terms are for defining the radiation from the workpiece to the environment, convection between the workpiece and environment, and convection between the workpiece and the tool, respectively. The last term defines the contribution of heat generation due to friction, where q_{f} is the surface heat generation rate and expressed as

$$q_{\mathrm{f}} = \bar{\tau}|v_{\mathrm{s}}| \tag{73}$$

where v_{s} is the velocity at the interface of the workpiece and tool. The convergence of the solution needs consistency and stability. Approximation based on Eq. 74 is required to achieve the consistency of the solution [97]

$$T_{t+\Delta t} = T_t + \Delta t\left[(1-\psi)\dot{T}_t + \psi\dot{T}_{t+\Delta t}\right] \tag{74}$$

where ψ is a constant and is taken as 0.75 for unconditional stability. t is the time.

3.5.5 Coupling the Plastic Deformation and Thermal Analysis

Friction stir welding is a coupled process, i.e., deformation and temperature are interdependent. Thus, simultaneous solution of finite element formulation is required. Let us take $T_{t+\Delta t}$ as a primary dependent variable. Therefore, at $t = 0$ from Eq. 74, we get [87]

$$\dot{T}_{\Delta t} = \frac{T_{\Delta t}}{\psi \Delta t} - \frac{T_0}{\psi \Delta t} - \left(\frac{1-\psi}{\psi}\right)\dot{T}_0 = \frac{T_{\Delta}}{\psi \Delta t} + \hat{\dot{T}} \tag{75}$$

where $\hat{\dot{T}} = \frac{T_0}{\psi \Delta t} - \left(\frac{1-\psi}{\psi}\right)\dot{T}_0$.

By putting Eqs. 75 in 71, we get

$$\left(K_c + \frac{C_p}{\beta \Delta t}\right) T_{\Delta t} = Q_{\Delta t} - C_p \hat{\dot{T}} \tag{76}$$

Equation 76 is used for coupling procedure and is shown in Fig. 8.

3.5.6 Solvers and Iterative Method

Solvers are the backbone of FEM technique. They are used to solve the set of equations and to obtain the final solution. FSW is a nonlinear problem, and a strong solver is required to achieve accurate solution with lower computation time. In this section, comparison between two solvers, viz., conjugate gradient (CG) and sparse

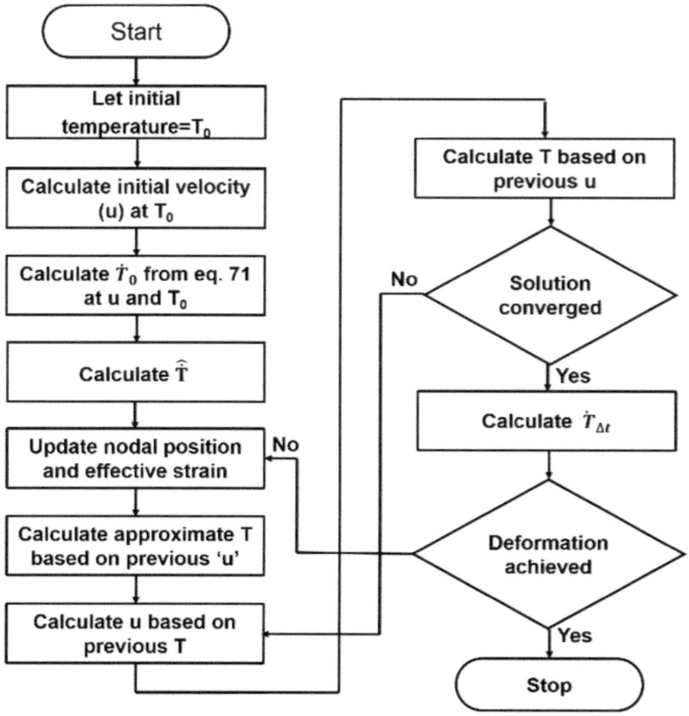

Fig. 8 Flowchart to depict coupling of deformation and temperature

method, is carried out. Simulation was carried out for three different cases; for the first and second cases, CG and sparse solvers are used, respectively, and for the third case both the solvers (CG + sparse) are used, such that if the CG solver is unable to achieve the convergence, simulation automatically switches to the sparse solver. The third case is the default option in the DEFORM 3D. To select only CG solver, user has to disable the sparse solver from "solver switch control" option in the simulation control module. Here, user needs to define the maximum number of the element above which solver will not switch from CG to sparse solver. Sparse solver is a direct solution method and makes use of sparseness of the stiffness matrix, i.e., instead of storing all the terms of stiffness matrix, only nonzero terms are stored. CG method obtains solution by iterative method. Total number of elements for plunging, dwelling phase is 33,886, and that for welding is 67,098. Due to dynamic re-meshing, the number of elements changes after every re-meshing step based on the extent of mesh distortion.

For plunging and dwelling phases, CG solver takes much higher time as compared with the other two, as shown in Table 4, indicating that higher time is required for convergence. This may be due to lower contact area during plunging between the tool and workpiece. While for sparse and combined solvers, computation time is almost similar. In welding phase, computation time is almost same for all the three cases with combined solver taking slightly lower time. In terms of total time, CG method takes the maximum computational time of 26 h 48 min and is followed by Sparse and combined (CG + sparse) solvers with 22 h 59 min and 21 h 41 min, respectively. All the simulations are carried out in 2.9 GHz, Intel i5 processor with 8 GB RAM. Apart from the highest computational time, CG solver also consumes the maximum memory (size of the file) of 33.09 GB. The other two methods take almost same memory size with a marginal difference of 700 MB. The results discussed above are contrary to the literature [98] which suggests that CG solver takes lower computation time and memory as compared to the sparse solver. But CG solver is prone to non-convergence if contact area is small as happened in the plunging phase.

Influences of solvers on output responses like temperature, axial force, and torque are studied and shown in Fig. 9. Figure 9a shows the temperature evolution using three different solvers. To study the temperature evolution, a point is defined

Table 4 Comparison of three different solvers in terms of computation time and memory

Solver	Computation time (hour:minute)				File size (GB)
	Plunging	Dwelling	Welding	Total time	
Conjugate gradient	9:52	4:50	12:06	26:48	33.09
Sparse	6:30	3:53	12:36	22:59	26.0
Conjugate gradient + sparse	7:08	3:13	11:20	21:41	26.7

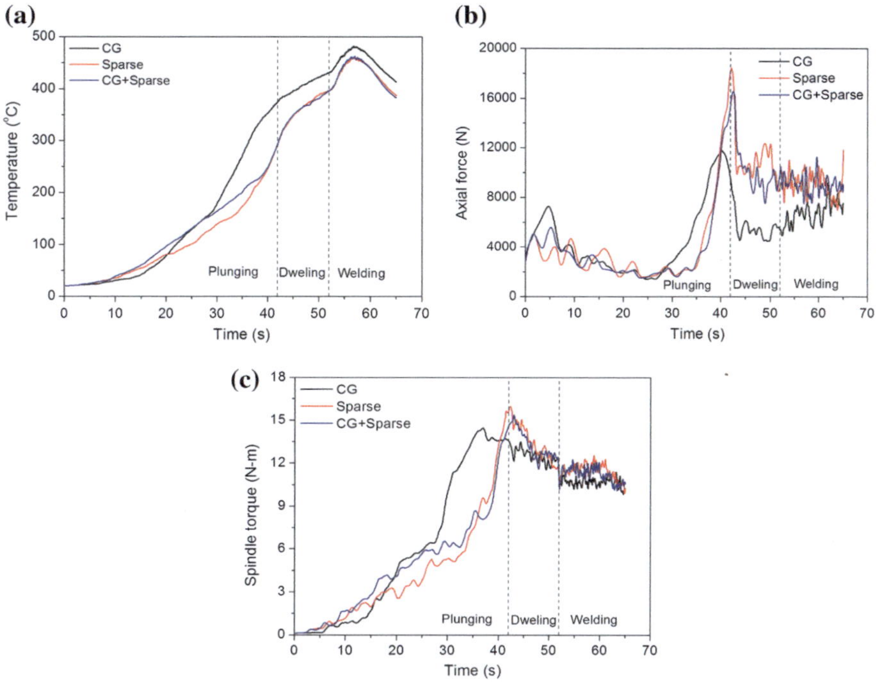

Fig. 9 Comparison of solvers with evolution of **a** temperature, **b** axial force, **c** spindle torque

on the workpiece at a distance of 1 mm from the periphery of the tool in both, welding and transverse directions. In FSW process, the temperature gradually rises during plunging phase till the shoulder comes in contact with the workpiece. At the start of the contact, temperature increases quickly. Afterward, during dwelling slope of the rise in temperature remains nearly constant. Lastly, temperature increases as the tool approaches the observation point and attains a peak value, followed by the cooling period as the tool passes. Sparse and combined solvers both predicted similar trend, and the temperature evolution for both is exactly the same in dwelling and welding phases. Peak temperatures predicted by sparse and combined solvers are 460.2 and 462.8 °C, respectively, which can be considered as same for all practical discussions. During the initial plunging phase, sparse solver has predicted lower temperature as compared with the combined solver and the difference between them is approximately 20 °C. CG solver is not able to predict the temperature evolution correctly. Initially, during the plunging phase, CG solver has under-predicted the temperature as compared with the other solver but after 28 s the gradient of temperature increases drastically and CG solver overpredicts the temperature for all phases of FSW. Peak temperature predicted by CG solver is 482 °C, i.e., 22 °C higher than that the other two solvers.

Comparison of axial force for three cases and evolution with time is shown in Fig. 9b. Initial peak force is observed during the initial contact of the tool with the workpiece. As the tool impinges, the temperature of the workpiece increases and leads to a drop in the axial force. Afterward, force gradually drops because of the conical shape of the pin; heat generation continuously increases leading to softening of the material till the shoulder comes in contact with the workpiece. A sudden spike in force is observed due to increase in the contact area with the tool shoulder but at the same time heat generation also increases; therefore, force suddenly drops and attains a steady value during the welding phase. All the three solvers show the similar trend, but CG solver predicts lower peak axial force of 12,533 N, while sparse and combined solvers predict 19,684 and 18,235 N, respectively. Similarly, average welding force for CG is much lower than that of the other two solvers. Also, axial force during welding continuously increases rather than attaining a steady value, but a different trend has been observed in case of the other two solvers. All three solvers show similar trend for spindle torque, as shown in Fig. 9c. The difference in the result predicted by the three solvers is not much except that CG solver predicted slightly lower peak in spindle torque. Average spindle torque for CG, sparse, and combine solver are 10.64, 11.44, and 11.19 N m, respectively.

Comparison of the solvers for three different cases suggests that for FSW process, sparse solver should be preferred or researcher can also choose CG + sparse. CG solver should not be preferred for friction stir welding process as it takes higher computation time, memory and also, predictions of temperature and axial force are inaccurate. All the results discussed below are with CG + sparse solver.

4 Results and Discussion

4.1 Forces and Torque

The model predictions of axial force (F_z) and welding force (F_y) match closely with the corresponding experimental values [99], as shown in Fig. 10a. Percentage difference between the experiment and the simulation results is close to 10%, except for the axial force at 40 mm/min welding speed. Forces increase with the increase in the welding velocity, because at higher velocity heat generation is reduced (as shown in Fig. 11). Lower heat generation leads to the reduced softening of the material and hence, higher forces. Figure 9b shows the comparison of the model predicted torque with the experiment values [99]. A good correlation is observed between experimental and the simulation results. Maximum percentage of error observed is 11%. Similar to forces, torque increases with the increase in the welding velocity. Torque is a direct indicator of power consumption during the process. Therefore, a higher capacity machine will be required to perform FSW at a higher velocity.

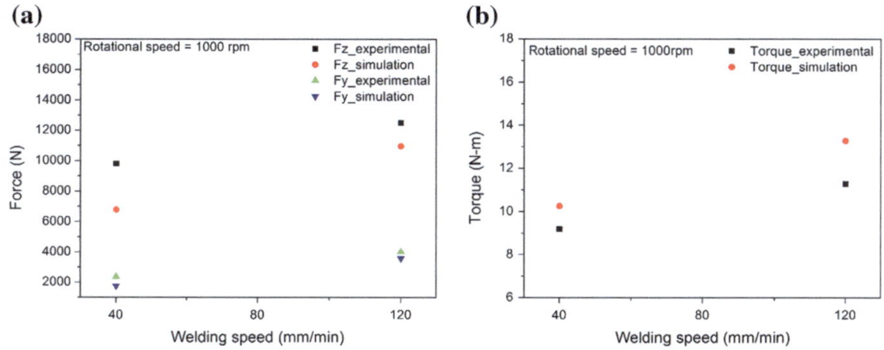

Fig. 10 Validation of simulation results with experiment. **a** Force, **b** spindle torque

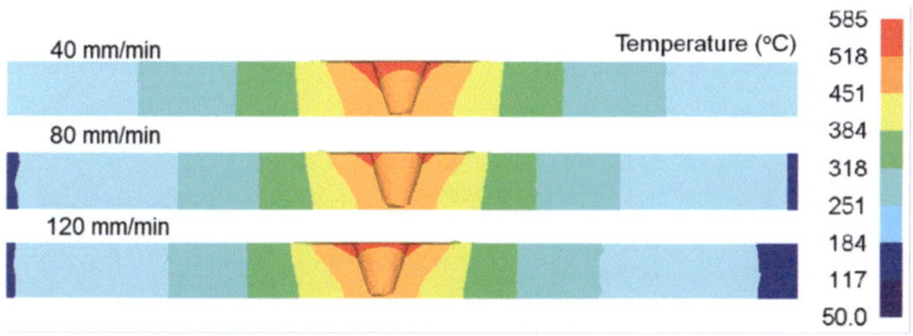

Fig. 11 Effect of welding speed on temperature distribution

4.2 Temperature Distribution

Temperature distribution in the workpiece is compared for three different welding speeds, i.e., 40, 80, and 120 mm/min and for a rotational speed of 1000 rpm, as shown in Fig. 11.

Maximum temperature is observed on the top surface of the workpiece which is in contact with the tool shoulder. This is because of the two factors: first, shoulder diameter is 2.5 times that of the pin diameter; and therefore, it is the primary source of heat generation in the process. Second, heat transfer from the bottom face of workpiece to the backing plate is higher (heat transfer increases with pressure) as compared with the top surface. Hence, temperature distribution forms an inverted V shape. Increase in welding velocity reduces the temperature. Figure 10 shows that the temperature range of 518–585 °C has occupied a larger area for 40 mm/min as compared with 120 mm/min. This is because numbers of tool rotation per mm movement of the tool at 40 and 120 mm/min are 30 and 10, respectively. Higher rotation/mm of tool indicates larger deformation of the material and hence higher temperature.

4.3 Material Flow

Material flow influences the defect formation in the weld. Efficient material flow leads to the defect-free weld, and vice versa. It is affected by a number of parameters like rotational speed, tilt angle, pin shape, etc. [94]. Material flow is investigated for conical pin based on the point tracking method. Figure 12a shows the initial position of the points from top view (*XY* plane) and back view (*XZ* plane). Distance between two consecutive points in *X*- and *Z*-directions is 1 mm. Points are selected to cover the pin area and to study the material flow in the stir zone. Initially, points are 6 mm ahead of the tool pin. Study has been performed at a rotational speed of 1000 rpm and a welding velocity of 120 mm/min.

Flow of the material in the stirred zone at intermittent locations is shown in Fig. 12b, and final location is shown in Fig. 12c. Material stirs through the RS and gets deposited on the trailing side of the tool behind its original location. Material points from centreline and RS are deposited as a hive (shown in black color

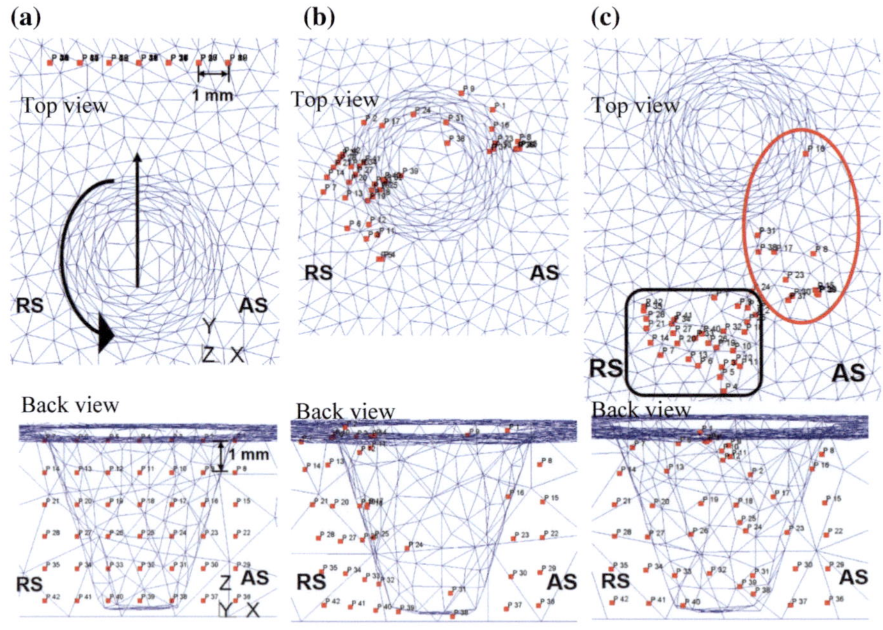

Fig. 12 Material flow from stirred zone. **a** Initial location, **b** intermittent location, **c** final deposition for a rotational speed of 1000 rpm and 120 mm/min welding speed

rectangle). It confirms that rather than stirring from the pin, the material from this region is mostly displaced and gets deposited on the trailing edge of the tool. In comparison to RS, material from AS is deposited in a scattered way (shown in the red-color ellipse) indicating material is stirred repeatedly by the pin before getting deposited in the AS. Scattered depositions of the material on AS could be one of the reasons for defect generation only in AS. Further investigation in AS reveals that material points closer to the tool shoulder (P1-3) stir once, as they are influenced more by the shoulder as compared with the pin, and are deposited either on centreline or on RS. It leads to surface defects like line defects. Material points away from the shoulder are only influenced by the pin. Hence, material flow study with one or two-marker material will not give thorough information.

Figure 13 shows the three-dimensional as well as two-dimensional graphs obtained by plotting the coordinates of the material points, P17-19. All three points are located 2 mm below the top surface of the workpiece, as shown in Fig. 12a. Points P17 and P19 are 1 mm toward AS and RS, respectively, and P18 is on the CL. WD, TD, and ND are welding direction, transverse direction (along the width of the workpiece), and normal direction (along the thickness), respectively. Three-dimensional material flow path of P17 is shown in Fig. 13a. Vertical movement of material is observed for all the three cases. This could be due to the tapered pin shape since the material mostly travels in the horizontal plane for cylindrical pin and no vertical movement is observed. Figure 13d shows the comparison of material path for AS, CL, and RS on TD-WD plane. Material path for CL and RS is circular, and it confirms that material is displaced by the tool rather than participating in stirring action. Flow path of AS is elliptical because material travels larger distance and at the same time tool is traveling along welding direction. This generates elliptical movement of the material. Evolution of strain rate is plotted for the three points (P17–19) in Fig. 14a. Strain rate gradually evolves as the tool pin approaches the material and attains a peak value during stirring of the material and finally reduces to zero. Maximum strain rates for AS, CL, and RS are 18.6, 15, and 11 s^{-1}, respectively. AS has higher strain rate due to higher deformation of the material on this side as compared with the other. Material velocity shows similar evolution as strain rate, as shown in Fig. 14b. Maximum velocities for AS, CL, and RS are 14.6, 11.56, and 9.14 mm s^{-1}, respectively. There are two peaks of material velocity for the material on AS, while single peak is observed for the RS and CL. This is because the material on AS enters the stirred region and some regions stick with the tool pin and stir for a longer time. Hence, it gets deposited in a scattered fashion (points marked with red ellipse in Fig. 12c). On the other hand, the material on the RS and CL side gets just displaced by the pin. This behavior also leads to multiple peaks for the strain rate.

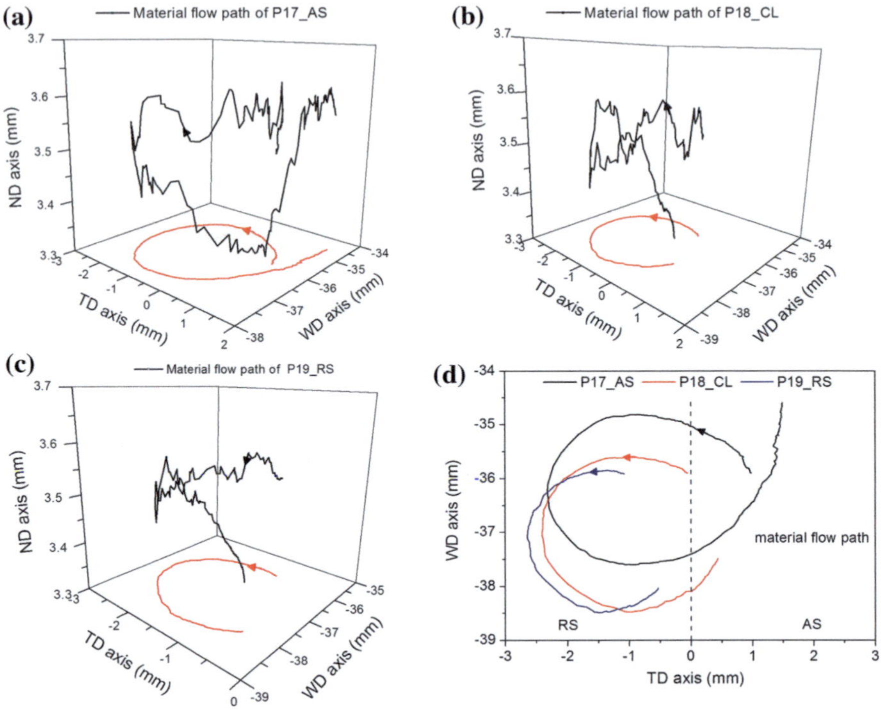

Fig. 13 Three-dimensional material flow path. **a** P17 on AS, **b** P18 on CL, **c** P19 on RS, **d** two-dimensional material path for AS, CL, and RS

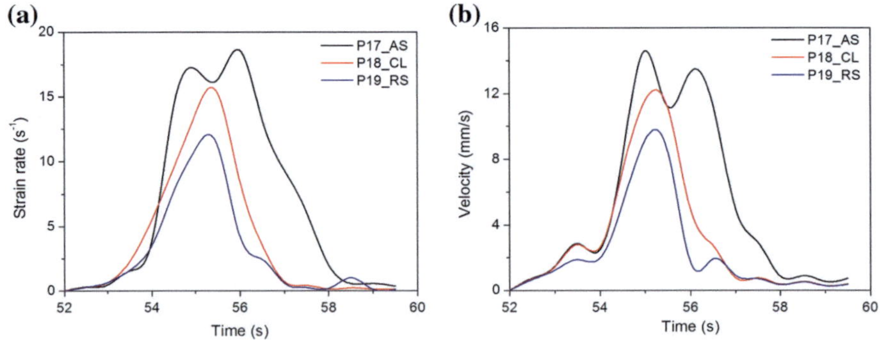

Fig. 14 Comparison of output responses for AS, CL, and RS. **a** Strain rate, **b** material velocity

5 Conclusion

A methodology to develop a three-dimensional coupled thermomechanical, non-linear transient model based on Lagrangian formulation is discussed. Formulation of governing equations and their derivation are thoroughly discussed. Shear friction model is defined to formulate the contact between the tool and the workpiece. Three different solvers (conjugate gradient, sparse, and conjugate gradient + sparse) are compared based on computation time, memory used, and output responses. Adaptive re-meshing is used to tackle mesh distortion. Following are the conclusions of the present study:

(1) Sparse solver and combined sparse and conjugate solver are suited to simulate friction stir welding.
(2) Model predictions show a good correlation with experimental results and have successfully predicted axial force and torque. Model predictions are consistent with the overall mechanism of FSW.
(3) The shoulder is the primary source of heat generation in FSW. Increase in welding speed decreases the temperature of the workpiece.
(4) Material flows in a circular path around the pin and is displaced maximum by one pin diameter. Material from centreline and RS is deposited in a smaller region as compared with AS.

References

1. Munoz, A. C., Ruckert, G., Huneau, B., Sauvage, X., & Marya, S. (2008). Comparison of TIG welded and friction stir welded Al-4.5Mg-0.26Sc alloy. *Journal of Materials Processing Technology, 197*(1–3), 337–343.
2. Jain, R., Kumari, K., Kesharwani, R. K., Kumar, S., Pal, S. K., Singh, S. B., et al. (2015). Friction stir welding: Scope and recent developement. In P. J. Davim (Ed.), *Mordern manufacturing engineering* (pp. 179–228). Swizerland: Springer International Publishing.
3. Thomas, W. M., Nicholas, E. D., Needham, J. C., Murch, M. G., Temple Smith, P., & Dawes, C. J. (1991). International patent number PCT/GB92/02203 and GB patent application number 9125978.9.
4. Mishra, R. S., & Ma, Z. Y. (2005). Friction stir welding and processing. *Materials Science and Engineering: R: Reports, 50*(1–2), 1–78.
5. Nandan, R., Debroy, T., & Bhadeshia, H. (2008). Recent advances in friction-stir welding—Process, weldment structure and properties. *Progress in Materials Science, 53*(6), 980–1023.
6. Khandkar, M. Z. H., Khan, J. A., & Reynolds, A. P. (2003). Prediction of temperature distribution and thermal history during friction stir welding: input torque based model. *Science and Technology of Welding and Joining, 8*(3), 165–174.
7. Song, M., & Kovacevic, R. (2003). Thermal modeling of friction stir welding in a moving coordinate system and its validation. *International Journal of Machine Tools and Manufacture, 43*(6), 605–615.
8. Mandal, S., & Williamson, K. (2006). A thermomechanical hot channel approach for friction stir welding. *Journal of Materials Processing Technology, 174,* 190–194.

9. Song, M., & Kovacevic, R. (2004). Heat transfer modelling for both workpiece and tool in the friction stir welding process: A coupled model. *Proceedings of the Institution of Mechanical Engineers, Part B: Journal of Engineering Manufacture, 218*(1), 17–33.
10. Schmidt, H., Hattel, J., & Wert, J. (2004). An analytical model for the heat generation in friction stir welding. *Modelling and Simulation in Materials Science and Engineering, 12,* 143–157.
11. Schmidt, H. B., & Hattel, J. H. (2008). Thermal modelling of friction stir welding. *Scripta Materialia, 58*(5), 332–337.
12. Heurtier, P., Jones, M. J., Desrayaud, C., Driver, J. H., Montheillet, F., & Allehaux, D. (2006). Mechanical and thermal modelling of Friction stir welding. *Journal of Materials Processing Technology, 171,* 348–357.
13. Chao, Y. J., Liu, S., & Chien, C. (2008). Friction stir welding of al 6061–T6 thick plates: Part II—numerical modeling of the thermal and heat transfer phenomena. *Journal of the Chinese Institute of Engineers, 31*(5), 769–779.
14. Tufaro, L. N., Manzoni, I., & Svoboda, H. G. (2015). Effect of heat input on AA5052 friction stir welds characteristics. *Procedia Materials Science, 8,* 914–923.
15. Biswas, P., & Mandal, N. R. (2011). Effect of tool geometries on thermal history of FSW of AA1100. *Welding Journal, 90*(July), 129–135.
16. Contuzzi, N., Campanelli, S. L., Casalino, G., & Ludovico, A. D. (2016). On the role of the thermal contact conductance during the friction stir welding of an AA5754-H111 butt joint. *Applied Thermal Engineering, 104,* 263–273.
17. Donea, J., Huerta, A., Ponthot, J., & Rodr, A. (1999). *Arbitrary Lagrangian–Eulerian methods*. Wiley, New York.
18. Bathe, K.-J. (2006). *Finite element procedures* (2nd ed.). United States of America: Prentice Hall, Pearson Education.
19. Priyadarshini, A., Pal, S. K., & Samantaray, A. K. (2012). Finite element modeling of chip formation in orthogonal machining. In P. J. Davim (Ed.), *Statistical and computational techniques in manufacturing* (pp. 101–144). Berlin: Springer.
20. Xu, S., Deng, X., Reynolds, A. P., & Seidel, T. U. (2001). Finite element simulation of material flow in friction stir welding. *Science and Technology of Welding and Joining, 6*(3), 191–193.
21. Ulysse, P. (2002). Three-dimensional modeling of the friction stir-welding process. *International Journal of Machine Tools and Manufacture, 42,* 1549–1557.
22. Chen, C. M., & Kovacevic, R. (2003). Finite element modeling of friction stir welding—thermal and thermomechanical analysis. *International Journal of Machine Tools and Manufacture, 43*(13), 1319–1326.
23. Chen, C., & Kovacevic, R. (2004). Thermomechanical modelling and force analysis of friction stir welding by the finite element method. *Proceedings of the Institution of Mechanical Engineers, Part C: Journal of Mechanical Engineering, 218*(5), 509–519.
24. Chen, C. M., & Kovacevic, R. (2006). Parametric finite element analysis of stress evolution during friction stir welding. *Proceedings of the Institution of Mechanical Engineers, Part B: Journal of Engineering Manufacture, 220*(8), 1359–1371.
25. Soundararajan, V., Zekovic, S., & Kovacevic, R. (2005). Thermo-mechanical model with adaptive boundary conditions for friction stir welding of Al 6061. *International Journal of Machine Tools and Manufacture, 45*(14), 1577–1587.
26. Buffa, G., Hua, J., Shivpuri, R., & Fratini, L. (2006). A continuum based fem model for friction stir welding—model development. *Materials Science and Engineering A, 419*(1–2), 389–396.
27. Buffa, G., Hua, J., Shivpuri, R., & Fratini, L. (2006). Design of the friction stir welding tool using the continuum based FEM model. *Materials Science and Engineering A, 419*(1–2), 381–388.
28. Buffa, G., Fratini, L., & Shivpuri, R. (2007). CDRX modelling in friction stir welding of AA7075-T6 aluminum alloy: Analytical approaches. *Journal of Materials Processing Technology, 191*(1–3), 356–359.

29. Fratini, L., & Buffa, G. (2005). CDRX modelling in friction stir welding of aluminium alloys. *International Journal of Machine Tools and Manufacture, 45*(10), 1188–1194.
30. Fratini, L., Buffa, G., & Shivpuri, R. (2007). Improving friction stir welding of blanks of different thicknesses. *Materials Science and Engineering A, 459*(1–2), 209–215.
31. Buffa, G., Fratini, L., & Shivpuri, R. (2008). Finite element studies on friction stir welding processes of tailored blanks. *Computers & Structures, 86*(1–2), 181–189.
32. Fratini, L., Buffa, G., Palmeri, D., Hua, J., & Shivpuri, R. (2006). Material flow in FSW of AA7075–T6 butt joints: Numerical simulations and experimental verifications. *Science and Technology of Welding and Joining, 11*(4), 412–421.
33. Simar, A., Pardoen, T., & de Meester, B. (2007). Effect of rotational material flow on temperature distribution in friction stir welds. *Science and Technology of Welding and Joining, 12*(4), 324–333.
34. Simar, A., Lecomte-Beckers, J., Pardoen, T., & de Meester, B. (2006). Effect of boundary conditions and heat source distribution on temperature distribution in friction stir welding. *Science and technology of welding and joining, 11*(2), 170–177.
35. Asadi, P., Mahdavinejad, R. A., & Tutunchilar, S. (2011). Simulation and experimental investigation of FSP of AZ91 magnesium alloy. *Materials Science and Engineering A, 528*(21), 6469–6477.
36. Tutunchilar, S., Haghpanahi, M., Besharati Givi, M. K., Asadi, P., & Bahemmat, P. (2012). Simulation of material flow in friction stir processing of a cast Al-Si alloy. *Materials and Design, 40,* 415–426.
37. Pashazadeh, H., Masoumi, A., & Teimournezhad, J. (2013). A study on material flow pattern in friction stir welding using finite element method. *Proceedings of the Institution of Mechanical Engineers, Part B: Journal of Engineering Manufacture, 227*(10), 1453–1466.
38. Pashazadeh, H., Teimournezhad, J., & Masoumi, A. (2014). Numerical investigation on the mechanical, thermal, metallurgical and material flow characteristics in friction stir welding of copper sheets with experimental verification. *Materials and Design, 55,* 619–632.
39. Gok, K., & Aydin, M. (2013). Investigations of friction stir welding process using finite element method. *International Journal of Advanced Manufacturing Technology, 68*(1–4), 775–780.
40. Trimble, D., Monaghan, J., & O'Donnell, G. E. (2012). Force generation during friction stir welding of AA2024-T3. *CIRP Annals-Manufacturing Technology, 61*(1), 9–12.
41. Marzbanrad, J., Akbari, M., Asadi, P., & Safaee, S. (2014). Characterization of the influence of tool pin profile on microstructural and mechanical properties of friction stir welding. *Metallurgical and Materials Transactions B Process Metallurgy Materials Processing Science, 45*(5), 1887–1894.
42. Zhang, P., Guo, N., Chen, G., Meng, Q., Dong, C., Zhou, L., et al. (2014). Plastic deformation behavior of the friction stir welded AA2024 aluminum alloy. *International Journal of Advanced Manufacturing Technology, 74*(5), 673–679.
43. Shojaeefard, M. H., Khalkhali, A., Akbari, M., & Asadi, P.: Investigation of friction stir welding tool parameters using FEM and neural network. *Proceedings of the Institution of Mechanical Engineers, Part L: Journal of Materials: Design and Applications, 229*(3), 209–217.
44. Zhao, Y., Liu, H., Yang, T., Lin, Z., & Hu, Y. (2015). Study of temperature and material flow during friction spot welding of 7B04-T74 aluminum alloy. *International Journal of Advanced Manufacturing Technology, 83*(9), 1467–1475.
45. Asadi, P., Besharati Givi, M. K., & Akbari, M. (2015). Microstructural simulation of friction stir welding using a cellular automaton method: a microstructure prediction of AZ91 magnesium alloy. *International Journal of Mechanical and Materials Engineering, 10*(1), 20.
46. Asadi, P., Besharati Givi, M. K., Akbari, M. (2015). Simulation of dynamic recrystallization process during friction stir welding of AZ91 magnesium alloy. *The International Journal of Advanced Manufacturing Technology, 83*(1), 301–311.
47. Wan, Z. Y., Zhang, Z., & Zhou, X. (2016). Finite element modeling of grain growth by point tracking method in friction stir welding of AA6082-T6. *International Journal of Advanced Manufacturing Technology, 90*(9), 3567–3574.

48. Seidel, T. U., & Reynolds, A. P. (2003). Two-dimensional friction stir welding process model based on fluid mechanics. *Science and Technology of Welding and Joining, 8*(3), 175–183.
49. Colegrove, P. A., & Shercliff, H. R. (2004). Development of Trivex friction stir welding tool Part 1—two-dimensional flow modelling and experimental validation. *Science and Technology of Welding and Joining, 9*(4), 352–361.
50. Colegrove, P. A., & Shercliff, H. R. (2004). Development of Trivex friction stir welding tool Part 2—three-dimensional flow modelling. *Science and Technology of Welding and Joining, 9*(4), 352–361.
51. Colegrove, P. A., & Shercliff, H. R. (2005). 3-Dimensional CFD modelling of flow round a threaded friction stir welding tool profile. *Journal of Materials Processing Technology, 169,* 320–327.
52. Nandan, R., Roy, G. G., Debroy, T., & Introduction, I. (2006). Numerical simulation of three-dimensional heat transfer and plastic flow during friction stir welding. *Metallurgical and Materials Transactions A: Physical Metallurgy and Materials Science, 37*(4), 1247–1259.
53. Nandan, R., Roy, G. G., Lienert, T. J., & Debroy, T. (2007). Three-dimensional heat and material flow during friction stir welding of mild steel. *Acta Materialia, 55*(3), 883–895.
54. Arora, A., Nandan, R., Reynolds, A. P., & DebRoy, T. (2009). Torque, power requirement and stir zone geometry in friction stir welding through modeling and experiments. *Scripta Materialia, 60*, 13–16.
55. Arora, A., Zhang, Z., De, A., & DebRoy, T. (2009). Strains and strain rates during friction stir welding. *Scripta Materialia, 61*(9), 863–866.
56. Arora, A., De, A., & Debroy, T. (2011). Toward optimum friction stir welding tool shoulder diameter. *Scripta Materialia, 64*(1), 9–12.
57. Arora, A., Mehta, M., De, A., & DebRoy, T. (2011). Load bearing capacity of tool pin during friction stir welding. *International Journal of Advanced Manufacturing Technology, 61*(9–12), 911–920.
58. Liechty, B. C., & Webb, B. W. (2008). Modeling the frictional boundary condition in friction stir welding. *International Journal of Machine Tools and Manufacture, 48*(12–13), 1474–1485.
59. Kim, D., Badarinarayan, H., Kim, J. H., Kim, C., Okamoto, K., Wagoner, R. H., et al. (2010). Numerical simulation of friction stir butt welding process for AA5083-H18 sheets. *European Journal of Mechanics-A/Solids, 29*(2), 204–215.
60. Wu, C., Zhang, W., Shi, L., & Chen, M. (2012). Visualization and simulation of plastic material flow in friction stir welding of 2024 aluminium alloy plates. *Transactions of the Nonferrous Metals Society of China, 22*(6), 1445–1451.
61. Chen, G., Shi, Q., Li, Y., Sun, Y., Dai, Q., Jia, J., et al. (2013). Computational fluid dynamics studies on heat generation during friction stir welding of aluminum alloy. *Computational Materials Science, 79,* 540–546.
62. Mohanty, H., Mahapatra, M. M., Kumar, P., Biswas, P., & Mandal, N. R. (2012). Study on the effect of tool profiles on temperature distribution and material flow characteristics in friction stir welding. *Proceedings of the Institution of Mechanical Engineers, Part B: Journal of Engineering Manufacture, 226*(9), 1527–1535.
63. Zhang, J., Shen, Y., Li, B., Xu, H., Yao, X., Kuang, B., et al. (2014). Numerical simulation and experimental investigation on friction stir welding of 6061-T6 aluminum alloy. *Materials and Design, 60,* 94–101.
64. Shi, L., Wu, C. S., & Liu, H. J. (2015). The effect of the welding parameters and tool size on the thermal process and tool torque in reverse dual-rotation friction stir welding. *International Journal of Machine Tools and Manufacture, 91,* 1–11.
65. Tang, J., & Shen, Y. (2016). Numerical simulation and experimental investigation of friction stir lap welding between aluminum alloys AA2024 and AA7075. *Journal of Alloys and Compounds, 666,* 493–500.
66. Al-moussawi, M., Smith, A. J., Young, A., Cater, S., & Faraji, M. (2017). Modelling of friction stir welding of DH36 steel. *The International Journal of Advanced Manufacturing Technology.*

67. Su, H., Wu, C. S., Pittner, A., & Rethmeier, M. (2014). Thermal energy generation and distribution in friction stir welding of aluminum alloys. *Energy, 77,* 720–731.
68. Ji, S. D., Shi, Q. Y., Zhang, L. G., Zou, A. L., Gao, S. S., & Zan, L. V. (2012). Numerical simulation of material flow behavior of friction stir welding influenced by rotational tool geometry. *Computational Materials Science, 63*, 218–226.
69. Su, H., Wu, C. S., Bachmann, M., & Rethmeier, M. (2015). Numerical modeling for the effect of pin profiles on thermal and material flow characteristics in friction stir welding. *Materials and Design, 77,* 114–125.
70. Kadian, A. K., & Biswas, P. (2017). Effect of tool pin profile on the material flow characteristics of AA6061. *Journal of Manufacturing Processes, 26,* 382–392.
71. Kuykendall, K., Nelson, T., & Sorensen, C. (2013). On the selection of constitutive laws used in modeling friction stir welding. *International Journal of Machine Tools and Manufacture, 74,* 74–85.
72. Zhu, Y., Chen, G., Chen, Q., Zhang, G., & Shi, Q. (2016). Simulation of material plastic flow driven by non-uniform friction force during friction stir welding and related defect. *Materials and Design, 108*(July), 400–410.
73. Shi, L., & Wu, C. S. (2017). Transient model of heat transfer and material flow at different stages of friction stir welding process. *Journal of Manufacturing Processes, 25,* 323–339.
74. Deng, X., & Xu, S. (2004). Two-dimensional finite element simulation of material flow in the friction stir welding process. *Journal of Manufacturing Processes, 6*(2), 125–133.
75. Schmidt, H., & Hattel, J. (2005). A local model for the thermomechanical conditions in friction stir welding. *Modelling and Simulation in Materials Science and Engineering, 13*(1), 77–93.
76. Guerdoux, S., & Fourment, L. (2009). A 3D numerical simulation of different phases of friction stir welding. *Modelling and Simulation in Materials, 17*(7), 75001.
77. Zhang, Z., & Zhang, H. W. (2008). A fully coupled thermo-mechanical model of friction stir welding. *International Journal of Advanced Manufacturing Technology, 37*(3–4), 279–293.
78. Zhang, H. W., Zhang, Z., & Chen, J. T. (2007). 3D modeling of material flow in friction stir welding under different process parameters. *Journal of Materials Processing Technology, 183,* 62–70.
79. Zhang, Z., & Zhang, H. W. (2009). Numerical studies on the effect of transverse speed in friction stir welding. *Materials and Design, 30*(3), 900–907.
80. Grujicic, M., He, T., Arakere, G., Yalavarthy, H. V., Yen, C.-F., & Cheeseman, B. A. (2010). Fully coupled thermomechanical finite element analysis of material evolution during friction-stir welding of AA5083. *Proceedings of the Institution of Mechanical Engineers, Part B: Journal of Engineering Manufacture, 224*(4), 609–625.
81. Grujicic, M., Ramaswami, S., Snipes, J. S., Avuthu, V., Galgalikar, R., & Zhang, Z. (2015). Prediction of the grain-microstructure evolution within a friction stir welding (FSW) joint via the use of the Monte Carlo simulation method. *Journal of Materials Engineering and Performance, 24*(9), 3471–3486.
82. Grujicic, M., Pandurangan, B., Yen, C. F., Cheeseman, B. A. (2012). Modifications in the AA5083 Johnson-Cook material model for use in friction stir welding computational analyses. *Journal of Materials Engineering and Performance, 21*(11), 2207–2217.
83. Mandal, S., Rice, J., & Elmustafa, A. A. (2008). Experimental and numerical investigation of the plunge stage in friction stir welding. *Journal of Materials Processing Technology, 203* (1–3), 411–419.
84. Assidi, M., Fourment, L., Guerdoux, S., & Nelson, T. (2010). Friction model for friction stir welding process simulation: Calibrations from welding experiments. *International Journal of Machine Tools and Manufacture, 50*(2), 143–155.
85. Yu, M., Li, W. Y., Li, J. L., & Chao, Y. J. (2012). Modelling of entire friction stir welding process by explicit finite element method. *Materials Science and Technology, 28*(7), 812–817.
86. Chakrabarty, J. (2006). *Theory of plasticity* (3rd ed.). Oxford: Butterworth-Heinemann.
87. Kobayashi, S., OH, S.-I., & Altan, T. (1989). *Metal forming and the finite element method, Oxford series.* Oxford: Oxford University Press.

88. Perzyna, P. (1966). Fundamental problems in viscoplasticity. *Advances in Applied Mechanics, 9,* 243–377.
89. DEFORM. (2015). v11 Documentation.
90. Jain, R., Pal, S. K., & Singh, S. B. (2017). Finite element simulation of pin shape influence on material flow, forces in friction stir welding. *The International Journal of Advanced Manufacturing Technology.*
91. Jain, R., Pal, S. K., & Singh, S. B. (2017). Numerical modeling methologies for friction stir welding process. In P. J. Davim (Ed.), *Computational methods and production engineering: Research and development* (1st ed., pp. 125–170). United Kingdom: Woodhead Publishing, Elsevier.
92. Jain, R., Pal, S. K., & Singh, S. B. (2012). Finite element simulation of temperature and strain distribution in Al2024 aluminum alloy by friction stir welding. In *5th International & 26th All India Manufacturing Technology, Design and Research Conference (AIMTDR 2014), 98,* 3–7.
93. Jain, R., Pal, S. K., & Singh, S. B. (2017). Finite element simulation of temperature and strain distribution during friction stir welding of AA2024 aluminum alloy. *Journal of The Institution of Engineers (India): Series C, 98*(1), 37–43.
94. Jain, R., Pal, S. K., & Singh, S. B. (2016). A study on the variation of forces and temperature in a friction stir welding process: A finite element approach. *Journal of Manufacturing Processes, 23,* 278–286.
95. Zienkiewicz, O. C., & Taylor, R. L. (2000). *The finite element method volume 1 : The basis*, 5th ed.
96. Zienkiewicz, O. C., Taylor, R. L. (2000). *The finite element method volume 2: Solid mechanics*, 5th ed.
97. Rebelo, N., & Kobayashi, S. (1980). A coupled analysis of viscoplastic deformation and heat transfer-I. *International Journal of Mechanical Sciences, 22,* 699–705.
98. Rout, M., Pal, S. K., & Singh, S. B. (2017). Finite element modeling of hot rolling. In P. J. Davim (Ed.), *Computational methods and production engineering: Research and development* (pp. 83–124). Woodhead Publishing, .
99. Su, H., Wu, C. S., Pittner, A., & Rethmeier, M. (2013). Simultaneous measurement of tool torque, traverse force and axial force in friction stir welding. *Journal of Manufacturing Processes, 15,* 495–500.

Studies on Spring Back Effect of TIG Welded Ti-6Al-4V Sheets

A. Karpagaraj, Abhilash, N. Siva Shanmugam, B. Suresha and S. Arungalai Vendan

1 Introduction

Sheet metal forming is primary and inevitable fabrication method employed for mass production in industries. Stamping processes in general involve bending operation. During stamping operation, sheet metal is bent beyond its yield strength to cause plastic deformation of the desired shape. Upon unloading, a portion of the energy is recovered elastically which results in the divergence from the target shape. This deviation in shape is termed the 'spring back' effect. The degree of deviation relies on the material property, geometry and tooling dimensions. Forecasting the extent of occurrence of spring back and its compensation scheme to solve the additional dimensional variation is crucial in the design of forming tools and dies. Figure 1 explains the overview of spring back phenomenon where the sheet is punched to achieve the target shape with target angle α_i. Once the applied force is removed, the stress redistribution allows the material to release elastic part

A. Karpagaraj · N. Siva Shanmugam (✉)
Department of Mechanical Engineering, National Institute of Technology Tiruchirapalli, Tiruchirapalli, Tamil Nadu, India
e-mail: nsiva@nitt.edu

A. Karpagaraj
e-mail: karpaganit@gmail.com

Abhilash · B. Suresha
Department of Mechanical Engineering, National Institute of Engineering, Mysuru, Karnataka, India
e-mail: abhilash.ashok.d@gmail.com

B. Suresha
e-mail: sureshab2004@yahoo.co.in

S. A. Vendan
School of Electrical Engineering, Vellore Institute of Technology, Vellore, Tamil Nadu, India
e-mail: arungalaisv@yahoo.co.in

U. S. Dixit and R. Kant (eds.), *Simulations for Design and Manufacturing*, Lecture Notes on Multidisciplinary Industrial Engineering,
https://doi.org/10.1007/978-981-10-8518-5_5

of the energy which results in the final shape denoted by α_f. Thus, there will be difference between the target angle and the final actual angle. This angular difference between target and actual final position is known as spring back denoted by α_s. The spring back angle is calculated as

$$\alpha_s = \alpha_f - \alpha_i \tag{1.1}$$

It is known that starting from small-scale industries to industries of high-end products like automobiles, aero-structures, aero-engines, etc., bending and stamping are the most common operations used in their manufacturing process chain and spring back undermines the benefits of these processes. Thus, it is essential to predict the spring back to devise compensation for accomplishing the desired final shape.

2 Spring Back Phenomenon

When sheet metal undergoes forming, the tool bends the sheet into required bend angle with given radius. Once the tool is removed, the metal will release elastic stress eventually widening the angle by increasing the radius. This ability or phenomenon of material to recover its elastic part of the state after removing the applied load is known as spring back. In order to understand the spring back phenomenon more clearly, review of material stress–strain curve is important. Figure 2a represents a simple bending process for which the typical stress–strain curve is shown in Fig. 2b. The stress–strain curve consists of both elastic stress and plastic stress region. During bending, the material is overstressed from point *A* to *B*, i.e. beyond its yield point in order to induce permanent deformation.

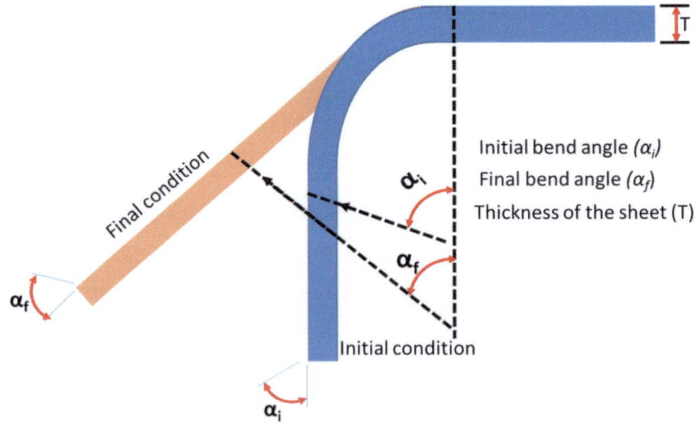

Fig. 1 Overview of spring back phenomenon in sheet metal [1]

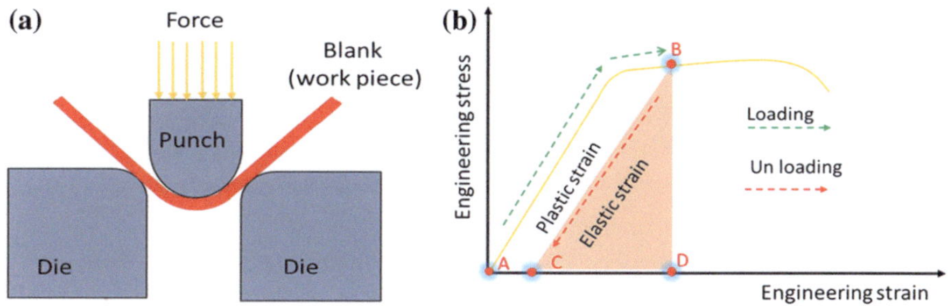

Fig. 2 **a** Typical sheet metal forming process [2] and **b** stress–strain curve representing elastic and plastic stress region [3]

After reaching point *B* when the load is removed, the stress value will return to zero along the path parallel to the elastic modulus which is represented as *BC*. Due to this, the material will release part of the energy stored which results in deviation of bending angle, and the permanent deformation will usually be lesser than the designer intended deformation of the strip. The spring back will be equal to the amount of elastic strain recovered when the die is removed and is represented as the area under the curve *C*–*B*–*D*. The plastic strain which is responsible for the deformed shape of the strip is denoted by the area under the curve *A*–*B*–*C*.

It is noted that during bending, the effect of bending moment is more at the top and bottom surfaces of the strip than at the neutral axis. Hence, the stress values at the top and bottom surfaces will be the highest and zero at the neutral axis. The outer surface of the strip experiences pure tensile stress and the inner surface of the strip experiences the compressive stresses. Hence, the interior of the strip would like to straighten out the bend when the load is removed, while the outer edge tends to resist the straightening. Due to permanent deformation, the strip cannot come back to its original shape. However, the spring back force tries to bring the workpiece to equilibrium position, where all internal stresses will be balanced. In order to obtain the desired part without deviation in angle, it is required to compensate for spring back of the material. Caution must be adopted for the design of die and tooling for forming processes. Prediction of the spring back in sheet metals becomes inevitable.

3 Literature Survey

Titanium alloys can be formed to tolerances similar to those obtained in forming of stainless steels. Joseph et al. have worked on forming of titanium and titanium alloys. It observed very high effect of spring back in cold forming cases as well as Bauschinger effect resulted in hysteresis stress–strain loops. To overcome the effect of spring back, they have recommended ageing or stress relief operations for cold-formed titanium alloy sheets [4]. Gupta et al. conducted a study on strain hardening in aerospace alloys. The strain hardening coefficient was found to be dependent on strain rate and the value was high for cold forming operations [5]. Kwei Leu has studied the effect of normal anisotropic value (*R*) and the strain hardening exponent (*n*) on the pure bending of sheet metal. A simple analytical model incorporating the normal anisotropic value and hardening exponent was developed to estimate spring back. It was concluded that spring back was almost proportional to *R*. It decreases sharply with respect to smaller *n* values [6]. Micari et al. have worked on effective spring back prediction in three-dimensional finite element model of stamping process. A prediction was based on combined approach in which explicit FE code was used to simulate the forming process, whereas the implicit code was used to analyse the spring back phase. An excellent agreement with the experimental tests has been found out [7]. Annika Nilsson et al. tested the ability of finite element method to predict the spring back in aluminium, stainless steel, low-carbon steel and hot-rolled steel. The material properties of both base and weld metals were taken from the corresponding true stress–strain curves. Spring back has been studied for eight different materials of variable thickness. The process has been simulated with Nike2D code, and good correlation was obtained between the simulations and the experimental results [8]. Zhang and Lin have developed an analytical solution for the spring back of sheet metals stamped by a rigid punch and an elastic die under plain stress deformation. The workpiece material was considered to be elastic/perfectly plastic. Based on the solutions obtained, the effects of the most important process factors, namely, the friction between the workpiece and die and the elastic properties of the die, were examined. The solution shows that spring back stamping process can be achieved by selecting optimal processing parameters [9]. Lee and Yang developed a numerical model to simulate the spring back effect on sheet metal during forming operation. They inferred that analytical techniques were highly impossible to study the spring back effect in a three-dimensional geometry. However, the results of spring back simulated by FEM may vary greatly according to numerical factors like element size, punch velocity, penalty contact, etc. In this work, such numerical factors influencing spring back have been evaluated quantitatively using the Taguchi method. U-draw bending process was chosen as an evaluation problem because of its large spring back. FE numerical simulations were carried out under various combinations of numerical factors. Optimal element sizes for the workpiece and dies corner radius have been found. In addition, it was found that kinematic hardening should be included for more realistic simulations [10].

Narasimhan and Michael Lovell outlined a coupled explicit to implicit finite element procedure for predicting spring back deformations in sheet metal forming processes [11]. An actual automotive component was selected for analysis and the results indicated that the coupled finite element procedure can be utilised to significantly reduce the number of die prototype designs. Ling et al. have conducted a parametric study in order to reduce time spent on corrections of the die, and various simulations were conducted to understand the effect of die parameters like die clearance, die radius, step height and step distance on spring back [12]. Tekaslan et al. further considered the subject of bending dies and spring back in the

V-bending process [13]. A 'V'-shaped die was designed to find out the energy absorbed by steel sheet metal material in various angles and also to study the spring back effect. The conclusions drawn from the study eliminate the effect of punch stroke, type of material and sheet thickness of spring back phenomenon.

Panthi et al. have employed finite element method software based on the total elastic incremental plastic strain that was capable of handling large rotation and elastic recovery or spring back [14]. In this investigation, the effect of load and thickness on spring back has been examined. Firat worked based on kinematic hardening plasticity model in order to improve the predicted sheet metal deformation [15]. The performance of the model in forming sheet metal is analysed, and FE analysis of U-channel was performed using an implicit method based on kinematic hardening plasticity model. Relative improvements were found in the predicted deformed channel profile due to the capability of proposed material model that considers Bauschinger effect. Further, Thipprakmas has selected finite element method to investigate the effects of punch stroke on spring back in V-bending [16]. The FEM simulation results revealed that the effect of punch stroke on the bending angle was clearly based on material and stress distribution analysis. The application of small punch stroke resulted in a large gap between the punch and die as well as small reversed bending zone. When too-large punch stroke was given more, bending angle was obtained than required.

Nanu and Brabie investigated an interaction between residual stress generated in the forming parts and spring back [17]. The obtained model of interaction was extended for the case out of plane forming of metal sheets. The study concludes the dependency of interaction model on the initial state of stress and remarks that the spring back depends on the level and distribution of residual stresses generated in the samples by in-plane forming before spring back. Thus, if the residual stresses had high values the spring back parameters also show high values. Schwarze et al. worked with the hexahedral solid-shell finite element [18]. The analysis covered the field starting from single integration point in case of shell element to an arbitrary number of integration points in 3D hexahedron element to incorporate highly stress–strain behaviour in the model. The key point of the formulation was the construction of hourglass stabilisation by means of different Taylor expansions. The hourglass stabilisation includes the assumed natural strain concept which eliminates transverse shear locking and volumetric locking in elements. Using ABAQUS/Standard implicit time integration scheme, the simulations were carried out for spring back predictions. For performing the forming simulations, reduced integration method was utilised due to its superior performance and stability.

Xu et al. developed a finite element solver based on solid-shell element dynamic explicit and implicit algorithms [19]. The solid-shell model considers bending effects along thickness direction multi-point integration approach. The comparison of results demonstrated that the solid-shell model is competitive enough to simulate medium thick plate metal forming and their spring back prediction. Maati et al. have analysed the influence of constitutive modelling on the prediction of the degree of spring back in stretch bending [20]. Tests have been carried out on titanium sheets which have a wide range of applications for high-tech industries. At the same time, the dependence of some process parameters such as clamping force is investigated. The study concludes the influence of parameters such as the radius of curvature of tool and velocity of punch on spring back.

Rojek et al. presented various methods which can be used to determine mechanical properties of weld zone in tailor welded blanks [21]. Methods based on experimental tests as well as those combining experimental procedures with numerical studies are described. The presented methods include uni-axial tension tests, micro-hardness tests and indentation tests combined with inverse numerical analysis. The results prove that the tensile test can be considered as the most trustworthy method to determine stress–strain curve as the curve is a direct result of the test. Comparison of results obtained using standard and substandard specimens indicates similar results as the yield stress and hardening of the material in a weld zone are concerned.

From available literature study, it is clear that many researchers have put effort to study spring back in sheet metals by analytical, experimental and numerical methods. However, the scope of previous works is limited to pure base metals, and inadequate number of literature are available on titanium alloys. The study on spring back effect of welded joints is a new area which covers majority applications in today's industries. Hence, the present study involves the prediction of spring back in titanium alloy and welded butt joints under three-point bending by experimental and numerical techniques. In numerical technique, inputs like reduced integration method and frictional interaction properties are taken from available literature.

4 Methodology

The methodology followed for this work is shown in Fig. 3. Initially, a detailed literature review is carried out to understand the problem statement. It was followed by two different procedures, experimental and finite element methods to predict spring back in both base metal and welded specimens. Tensile tests of base metal, as well as weld zone metal, are carried out to define material models as input to FE package. Three-point bend tests were performed on all specimens, and spring back angle is observed experimentally. The FE simulation results are then compared with experimental ones to validate the model. The validated model is then used further for parametric analysis.

5 Materials and Experimental Producers

5.1 Research Material

Titanium is the most desirable material for aerospace, chemical and medical applications and fabrication industries due to its high strength-to-weight ratio, good corrosion resistance and stability at higher temperatures. Welding has the greatest

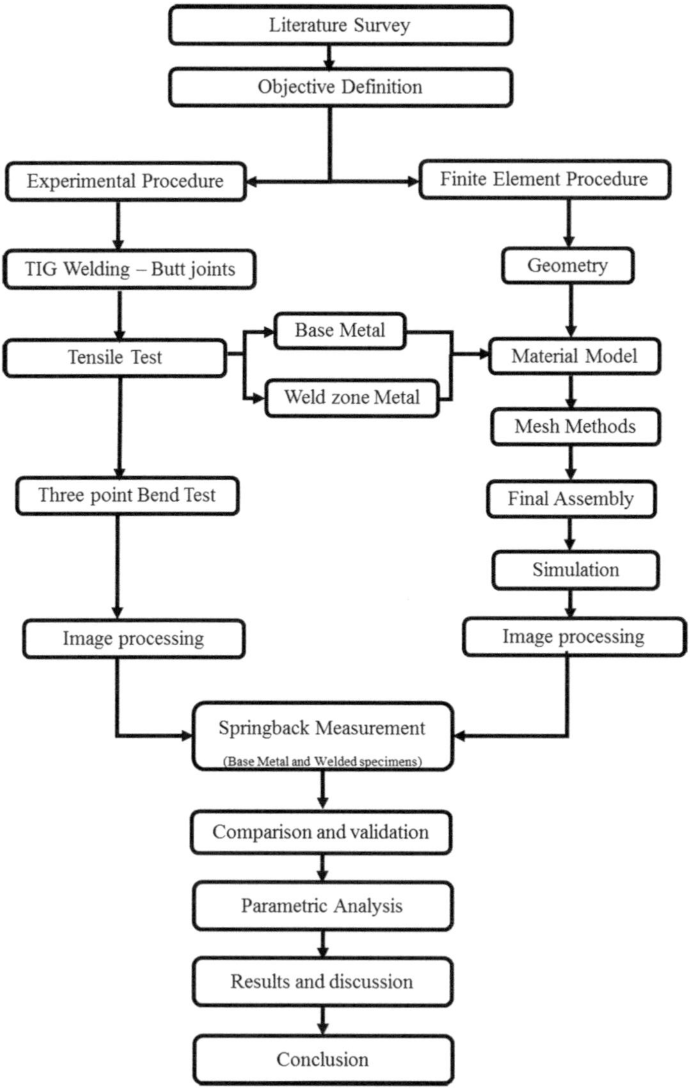

Fig. 3 Flow diagram of experiment and simulation analysis

potential for affecting material properties. For titanium in all types of welds contamination by impurities such as nitrogen, oxygen is common and must be minimised to maintain useful ductility in the weldment. Welding generally increases strength and hardness but decreases tensile and bend ductility. Thus, it is required to characterise the weld zone material. According to industry statistics, 90% of the titanium parts intended to be used in final assembly are rejected at various stages of the product development. This is due to either excess oxidation or distortion caused during welding process and shape imperfections produced during the forming operations [22]. Titanium exists in two crystallographic forms. At room temperature, unalloyed titanium has a hexagonal closely packed (HCP) crystal structure referred as alpha phase and body-centred cubic (BCC) structure known as the beta phase at transus temperature of 890 °C.

Based on the phases, present titanium alloys can be categorised into three categories as alpha alloys, beta alloys and alpha + beta alloys [23]. Ti-6Al-4V alloy is the most common alpha + beta alloy with aluminium as an alpha stabiliser and vanadium as a beta stabiliser. Alpha stabiliser defines the strength, whereas beta stabiliser improves ductility. Thus, Ti-6Al-4V alloy shows balanced advantage of both high strength and good ductility properties. Titanium alloys exhibit poor plastic deformation which leads to significant elastic recovery of the material after removal of the forming force. This has an important influence on the desired shape and accuracy of the final product. The quantity needs to be accounted and hence, it is selected for this study. Ti-6Al-4V is a two-phase alloy with aluminium as an alpha stabiliser and vanadium as a beta stabiliser. It is alloyed material with 6% of aluminium and 4% of vanadium and is commonly known as Ti-6Al-4V. It is used in the annealed condition. Table 1 shows the composition of Ti-6Al-4V.

The corrosion resistance of the Ti-6AL-4V is due to continuous and spontaneous formation of oxide layer upon exposure to oxygen. It is resistant to general corrosion but may be quickly attacked by environments that cause breakdown of the oxide protective layer including hydrofluoric, hydrochloric, sulphuric and phosphoric acids.

Ti-6Al-4V alloy is easily welded in the solution-treated or in the annealed condition and partially aged condition, with ageing being completed during the post-weld heat treatment. Further, titanium is highly reactive material and easily affected by the environment once the processing temperature crossed above 400 °C and care must be taken to prevent oxygen, nitrogen and hydrogen contamination. These contaminations could result in a deterioration in ductility which would adversely affect notch sensitivity and forming characteristics. Some of the applications of Ti-6Al-4V include compressor blades, discs and rings for jet engines, airframe and space capsule components, pressure vessels, rocket engine cases, helicopter rotor hubs, fasteners and critical forging applications [24].

5.2 *TIG Welding*

Tungsten inert gas (TIG) welding is a type of arc welding process that uses non-consumable tungsten electrode to produce the weld. The weld area is protected

Table 1 Composition of Ti-6Al-4V

Component	Al	Fe	O	V	Ti
Weight %	6	0.25	0.20	4	Balance

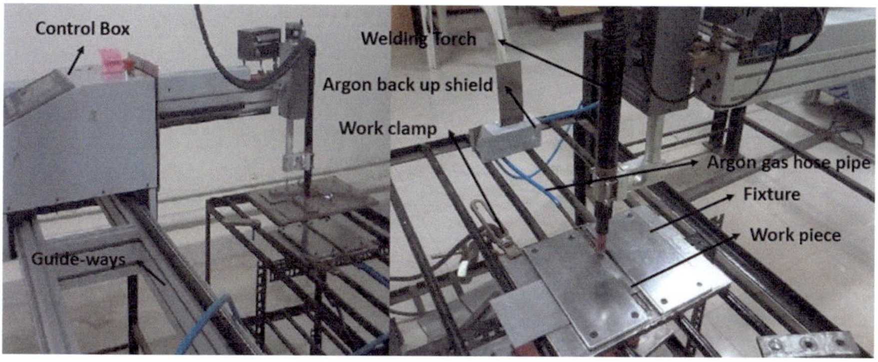

Fig. 4 Tungsten inert gas welding setup

from atmospheric contamination by a continuous supply of inert shielding gas such as argon. Normally, a filler material is used though some welds do not require it.

A constant current welding power supply produces electrical energy, which is conducted across the arc through a column of highly ionised gas and metal vapours known as plasma. TIG welding is most commonly used for joining titanium and titanium alloys. As discussed earlier, at high temperatures, titanium is highly reactive with oxygen and nitrogen available in the atmosphere. Due to this reason, welding of titanium is observed to be difficult and need more attention. Figure 4 shows the completely numerically controlled automatic TIG welding setup used in the work. As the thickness of the sheet is less than 2.4 mm, there is no need for filler material [25]. Argon is used as shielding gas to maintain an inert atmosphere in order to prevent oxygen and nitrogen contamination during welding. Initially, to optimise the welding parameters namely welding speed and welding current for quality weld, bead on plate trials is carried out on test samples [26].

A test sample of same material and thickness is taken and welded with parameters obtained from literature [27]. Once the quality weld with good depth of penetration is obtained, then the same parameters are recorded as optimum values and are used further. The optimum parameters are obtained out of weld trials. The welding is carried out with optimum parameters mentioned in Table 2. The final weld joints are shown in Fig. 5.

Table 2 Optimized welding parameters

Parameters	Arc length (mm)	Welding speed (mm/min)	Current (A)	Gas flow (l/mm) [25]	Electrode diameter (mm)	Sheet thickness (mm)
Values	3	250	125	15/25/10	2.4	2

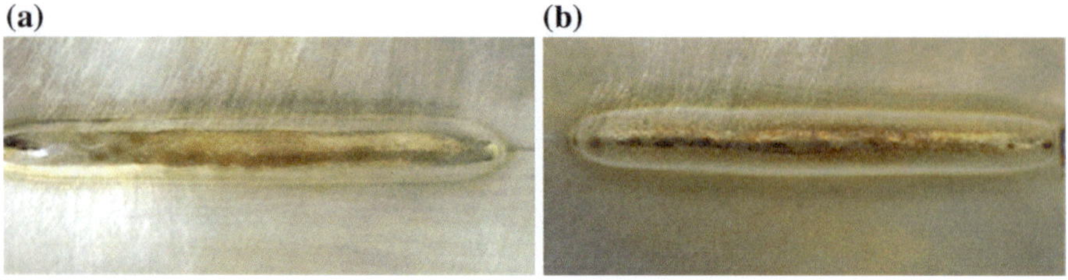

Fig. 5 The final weld joints are shown in **a** face side and **b** root side

5.3 Tensile Test

The tensile test is the most common test and it is an effective way to characterise material's response to loading. The test results help to calculate the mechanical properties such as strength, ductility, toughness and elastic modulus for the material. In this case, in order to prepare material models for finite element simulations, it is required to get the material elastic and plastic data. The test is conducted for both titanium base metal and weld zone metal. For the tensile test of zone metal, two plates of dimension 150 mm × 25 mm × 2 mm are taken. The specimens for base metal and weld zone metal are prepared according to ASTM E8 standard [28]. Both the specimens are cut to the standard dimensions by Wire Electric Discharge Machining (EDM). The stress–strain curve is recorded by conducting tensile test on both base metal and welded joints. To get an effective property of weld metal, longitudinal tensile test specimen is prepared as per ASTM standard and it is subjected to tensile. The fractured specimen after the tensile test and the stress–strain curve for base metal and longitudinal tensile test are shown in Fig. 6.

5.4 Three-Point Bend Test

In the present study, three-point bend test is carried out on both base metal and welded joints to study the spring back effect. The three-point bend test represents the simplest form of bending process. This test is selected due to the ease of setup. Bend test may be carried out on a Universal Testing Machine with the help of certain attachments. Any cracks at the weld indicate defect in the weld. The stretching of the metal indicates its ductility to some extent. Bend tests are used primarily to determine the integrity of weld. During the test, the mandrel or punch is depressed until the specimen becomes U-shaped as in the punch or mandrel. Further, the downward moving punch deforms the sheet into required bending angle. The bend test specimen prepared as per ASTM E190/192 standard is shown in Fig. 7a [29].

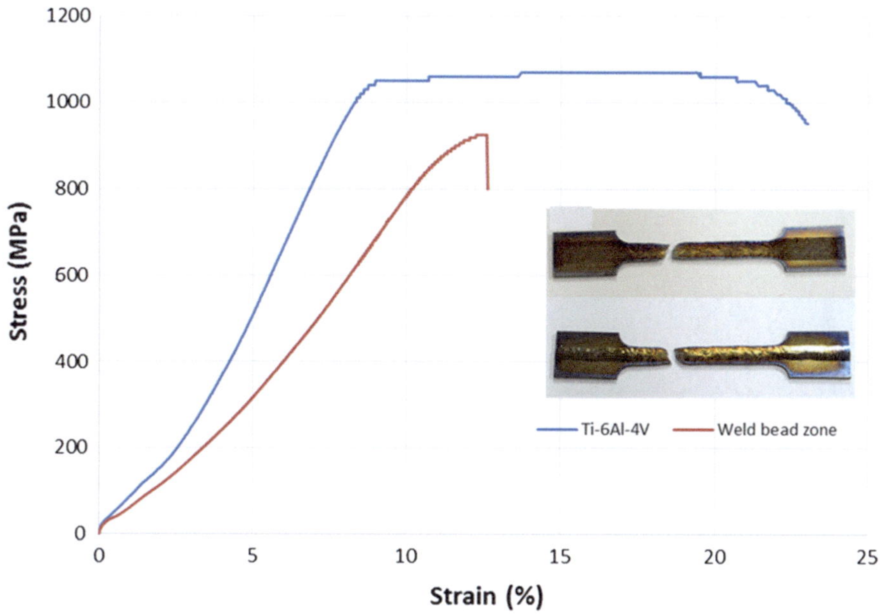

Fig. 6 Engineering stress–strain curve

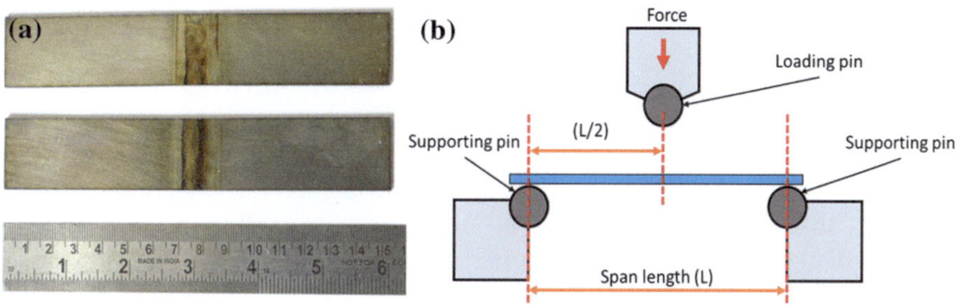

Fig. 7 **a** Three-point bending test specimen, **b** schematic representation of three-point bending test [30]

Figure 7b shows the schematic of general test setup used for testing of materials. The punch and supports are of roller shape with 10 mm diameter. The span length of 100 mm is given between the supports to ensure the punch movement to the required distance.

6 Finite Element Method

The chapter involves the finite element simulation of three-point bend test using ABAQUS/CAE FE package. It gives the various steps involved in the modelling starting from geometry, boundary conditions, material modelling to simulation procedure and validation. ABAQUS is a general-purpose simulation tool which works on finite element method. It can solve problems ranging from simple linear to complex nonlinear types. It contains extensive element library which enables us to model any type of geometry. It has an equally extensive list of material models namely metals, rubber, polymers, composites, reinforced concrete, foams, soils and rock to simulate problems in diverse areas including mechanical structural analysis, heat transfer problems, coupled thermal–mechanical problems, coupled thermal–electrical problems, acoustic problems, soil mechanics and fluid dynamics. ABAQUS includes three main analysis products—ABAQUS/Standard, ABAQUS/Explicit and ABAQUS/CFD. In this study, ABAQUS/ Standard is used to simulate three-point bend test. It is based on implicit time integration technique.

ABAQUS/CAE is an interactive graphical interface which allows us to create and discretise the geometry into a number of finite elements. It has very powerful options to preprocess the model. It includes nine main modules namely part, property, assembly, step, interaction, load, mesh, job and visualisation.

6.1 Material Model

The study involves spring back prediction in both base metal and welded specimens. Hence, there is a need to develop material models for both the cases. A nonlinear behaviour of material is taken into account; the nonlinear analysis step created in ABAQUS will be broken into number of time increments. In general, ABAQUS solves the nonlinear problems by incremental method. According to the geometrical changes or material property changes, it develops the stiffness matrix and then a new stiffness matrix is constructed and the next increment of load (or displacement) is applied to the system. At each incremental level, it solves for system of equilibrium equations to write output results. It follows Newton's numerical method and takes several iterations to determine solution to each time increment. Similar type of analysis and step methods is utilised from ABAQUS for solving nonlinear models [31, 32].

In order to solve the problem numerically, ABAQUS performs calculations by the incremental method. Thus, the material model developed should be good at predicting the variables like deformation at each incremental step. Thus, it is required to feed the material data in terms of true stress versus true strain [33]. The results recorded from the tensile test for both base metal and weld zone metal are processed in terms of true stress and true strain data and are given as input to develop the material models. The initial gauge length and initial area of the

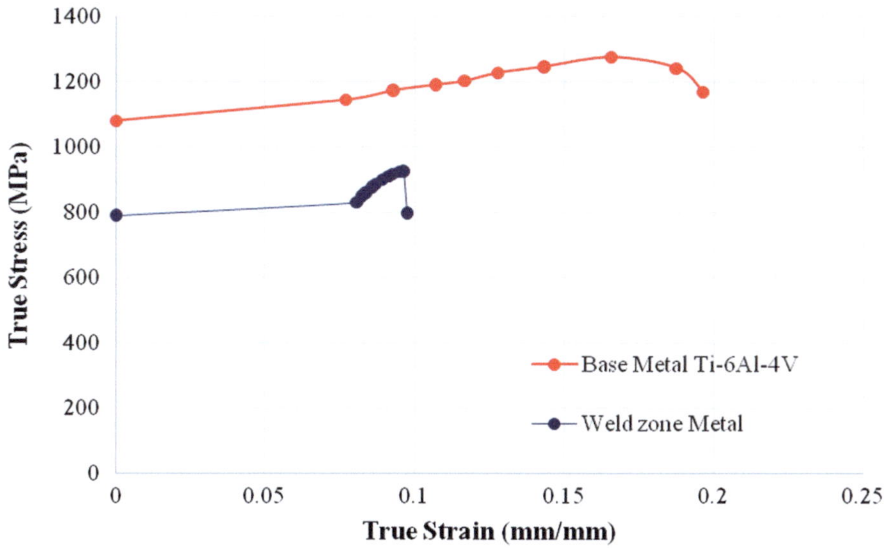

Fig. 8 Material input for base and weld zone metal

specimen used to measure the deflection are recorded before the test is begun. The specimen is then mounted in the universal test machine, and the upper end of the specimen is loaded under tension with the input strain rate of 1 mm/min. The test is conducted till the fracture point of the material. The test results in the form of load vs. displacement are acquired. The further load is converted to stress by the calculation [34]:

$$\text{True stress } (\sigma) = \frac{P}{A_o}, \quad \text{True strain } (\varepsilon) = \frac{G - G_0}{G_0}$$

where A_o is the original cross-sectional area. The deflection or extension of the gauge length is given by $G - G_0$, where G is the gauge length corresponding to the load P and G_0 is the original gauge length. The processes data is plotted as true stress–strain curve and is shown in Fig. 8.

6.2 *Geometry and Meshing*

The base metal geometry considered in the present study is uniform throughout. But welded specimen is characterised by two sections namely unaffected base metal and weld zone. The sketch of the butt joint is shown in Fig. 9a. For welded specimen, it is required to know the weld zone dimensions to precisely represent the original experimental conditions. Hence, the cross section of the welded sheets is cut and

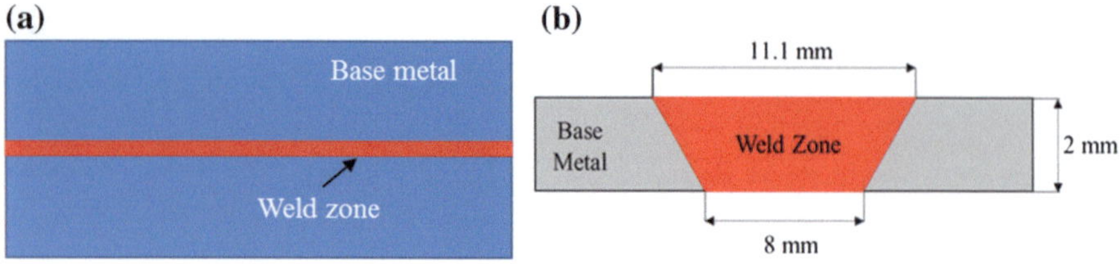

Fig. 9 Schematic diagram of **a** weld sample and **b** bead geometry with measurement

metallographic process is carried out. For measuring and identifying the weld zone, macroscopic examination is done by 20× magnification.

The magnified image of the weld zone is recorded and measured to know the exact dimensions. The sketch of the measured dimension of weld zone is shown in Fig. 9b. The 3D model representing the exact test setup consists of four parts. Two roller supports, a cylindrical punch of 10 mm diameter and specimen. All the models are prepared in ABAQUS/CAE using the part module. The discretization is done using continuum 8-noded 3D stress hexahedron elements with reduced integration (C3D8R) available in element library. The parts prepared along with the dimensions and mesh are shown in Figs. 10 and 11. The sample final assembly setup for three-point bending simulation is shown in Fig. 12.

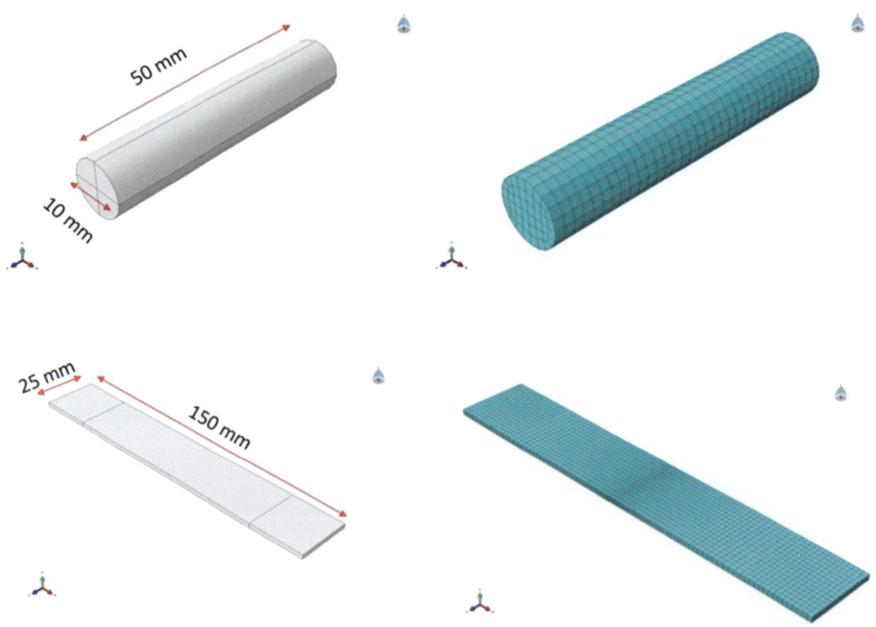

Fig. 10 3D model and mesh details of support/punch and base metal specimen with mesh

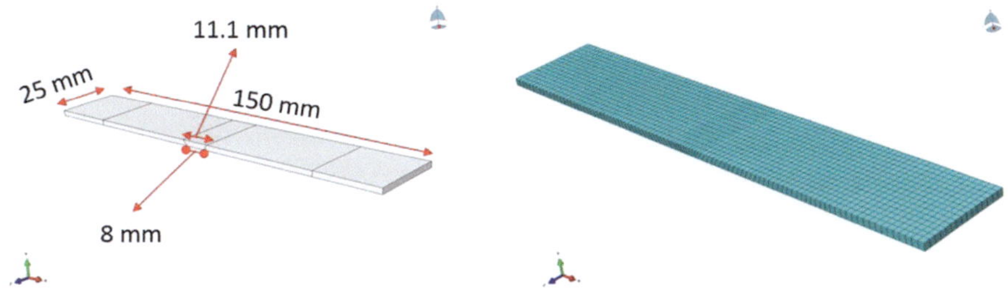

Fig. 11 3D model and mesh details of welded specimen

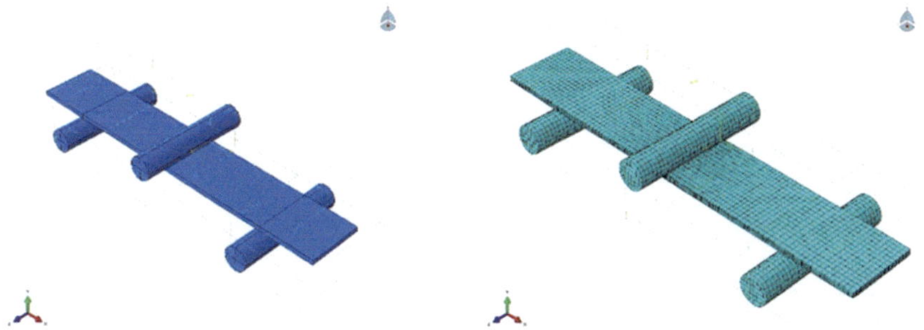

Fig. 12 3D Assembly setup for three-point bending simulation

6.3 *Boundary Conditions and Constraints*

In finite element simulations, boundary conditions and constraints are given for degrees of freedom of parts play a major role. In this case, for support all the six degrees of freedom are constrained, i.e. $Ux = Uy = Uz = URx = URy = URz = 0$. The punch is constrained in all directions except Uy direction. For specimen, only Uz direction is constrained. The interactions between specimen punch and specimen supports are defined as a surface-to-surface contact with friction coefficient of 0.15 [22]. In order to make post-processing easy, a reference point is created for punch with coupled interaction with all nodes. This will help us further to extract results from the simulation. For simulation, the input punch stroke is given in the form of vertical displacement, i.e. Uy. Figure 13 gives all the boundary conditions and constraints of the simulation model.

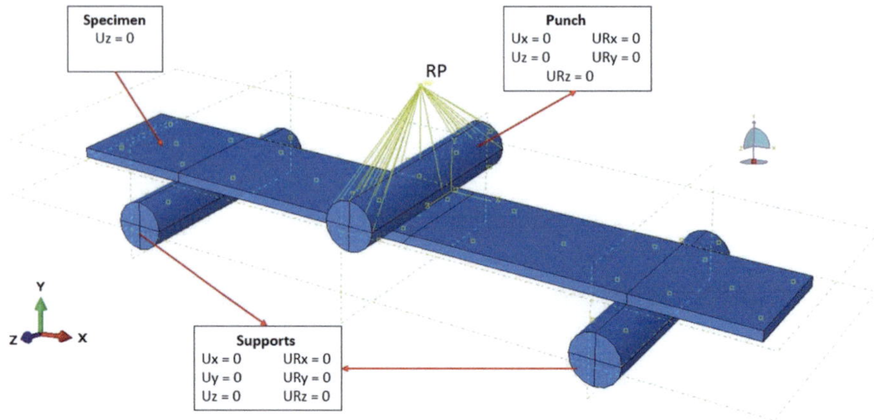

Fig. 13 Boundary conditions and constraints

7 Result and Discussion

7.1 General Procedure for Spring Back Measurement

As the main concern of the test is to study spring back effect, the sheet must be stressed beyond its yield stress to achieve the forming effect. The specimen is placed as shown in the test setup. The input is given in terms of punch stroke at a rate of 1 mm/min. The mid-portion of the sheet experiences maximum stress and stressed beyond its yield point which makes it deform plastically. Once considerable plastic deformation is achieved in the specimen, the punch is retracted to observe the spring back effect. To calculate the total spring back angle, the specimen images at maximum stroke position and final position are recorded. The bending angle is measured from specimen position at a maximum stroke and final bent angle from final position. Similar type of spring back measurements is carried out by Davoodi and Zareh-Desari [35]. The difference between bending angle and the bent angle is recorded as spring back angle for each case. For image processing, i.e. measuring, the angle AutoCAD software is used. Images are imported into the auto CAD software and measured without any errors.

7.2 Spring Back Measurement

In base metal, as the specimen material is unaffected and being ductile, the stroke is continued to the displacement of 28 mm which is the maximum limit for the test setup to induce plastic deformation in the specimen. In next step, the punch is taken back and the base metal experiences a noticeable amount of spring back. The measured angles for bend and final bent conditions are tabulated in Table 3.

Table 3 Experimental results for base metal bending, face bend specimen and root bend specimen

Parameter	Bent angle at maximum stroke (α_i)	Bent angle at final position (α_f)	Spring back angle (α_s)
Base bend specimen	99°	144°	45°
Face bend specimen	111°	153°	42°
Root bend specimen	102°	145°	43°

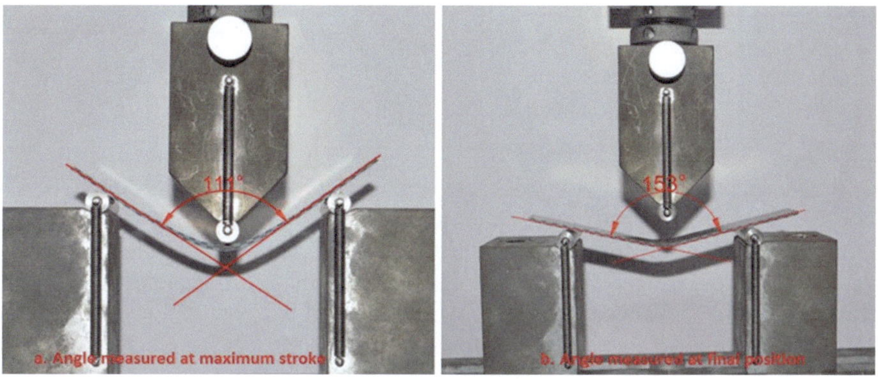

Fig. 14 Experimental spring back prediction for face bend

In the case of welded joints for face bend, the specimen is placed in such a way that the face side of the weld is kept inverted and bending stroke is applied on the root side of the weld. The specimen consists of both base material and weld zone material. Experimentally, it has been observed that the bending specimen fails at the punch stroke of 27.5 mm. Hence, only 25 mm stroke is used for the study. The specimen positions measured at the maximum punch and final cases are shown in Fig. 14. For root bend, bending stroke is applied on the face side of the weld.

7.3 Convergence and Validation

The results obtained are only approximate solutions of engineering problems. Many uncertain parameters and assumptions made during modelling will account for some differences between predicted results and actual ones. Even though FEM method eliminates the process tryouts and cycle time of product design, the final models must be validated in comparison with experimental or analytical procedures. The models thus validated can further be used for the parametric study. A sequence of convergence tests has been conducted to select suitable number of elements particularly at the area closer to the weld zone (where the load is going to

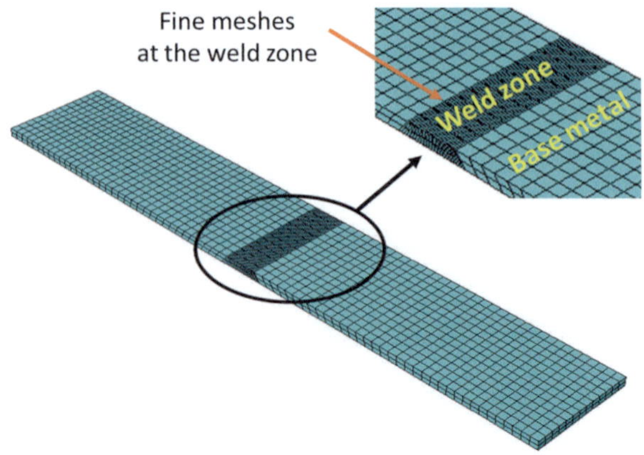

Fig. 15 Finite element model

act on the specimen) and in the thickness direction for three specimens/models, i.e. base metal, face bend and root bend with 1.6 mm thickness. Based on the convergence test, it is inferred that geometry with 13,356 numbers of elements gives reasonable convergence and less solution time with good accuracy. Figure 15 shows the finite element model of the weld zone and base metal region.

The final results of the simulation are drawn in the form of reaction force versus displacement. The results obtained in all the three cases are compared with that of experimental results to validate the effectiveness of the simulation model. The comparison made for all three cases is shown in Fig. 16.

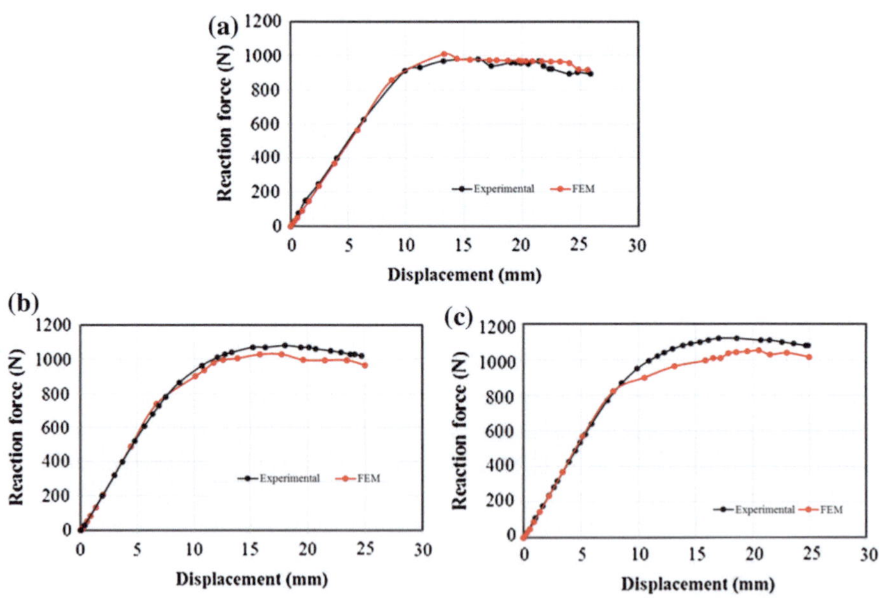

Fig. 16 Validation of FE model for **a** base metal, **b** face bend and **c** root bend simulation

7.4 Three-Point Bending Simulation

The simulation of three-point bending test is carried out in two major steps namely loading and unloading. The required input from literature and experimental data recorded is processed to develop the complete numerical model. The simulations are carried out for both base metal and welded specimens. The numerical modelling is done by the following step-by-step procedure as discussed below. The deform shape of the specimen at maximum displacement position and final position are measured in all the cases. The measured bend angle and spring back angle are recorded, and the comparison is made with that of experimental results as represented in Table 4. From Table 4, it is clear that the results obtained from numerical simulations are in good correlation with that of experimental results. However, the variation is observed only in the nonlinear behaviour of materials in all the three cases. The deviation observed is well within the acceptable range. The comparisons between experimental and numerical procedure in each case are presented below.

7.5 Parametric Analysis

The validated FE model is further considered to study the effect of various parameters on the spring back angle. FE simulations are run for base metal with three different thicknesses of 1.2, 1.6 and 2 mm for the stroke of 28 mm and for three different punch strokes 20, 25 and 30 mm with a constant base metal sheet thickness of 2 mm to study the effect of thickness and punch stroke in the spring back.

Table 4 Comparison of results for base metal

Method	Material condition	Bend angle (α_i)	Bent angle (α_f)	Spring back angle (α_s)
Experimental	Base metal bending	99°	144°	45°
FEM		103°	150°	47°
Error (%)		3.8	4.1	4.4
Experimental	Face bending	111°	153°	42°
FEM		112°	154°	42°
Error (%)		0.90	0.653	Nil
Experimental	Root bending	102°	145°	43°
FEM		106°	146°	40°
Error (%)		3.9	0.68	6.9

7.5.1 Effect of Stress Distribution

During three-point bending process, material near the neutral axis experiences less moment effect than that of the material away from the neutral axis and the mid-portion of beam undergoes a maximum amount of stress. The inner portion of the beam towards punch undergoes compression, whereas outer portion undergoes tensile stresses. When the applied bending moment is removed, the elastic part of the total deformation tends to return to its original shape. This can be observed from FE result plots for each case. Figure 17a, b shows the stress state plot for the case of base metal bending with the stroke of 25 mm. For base metal, at maximum punch stroke, the maximum stress value of 1089 MPa at the outer surfaces and zero stress value at the neutral axis are recorded. After removing punch, the maximum stress value induced reduces to 474.3 MPa. In the case of welded specimen, the weld zone material experiences majority part of bending moment as it is at the point of action. Figure 17c, d shows the von Mises stress plots for face bend operation. The maximum stress values of 997 and 377.6 MPa are recorded for maximum stroke and final cases.

Figure 17e, f shows the stress plots for root bend specimen at maximum stroke position and final position, respectively. At maximum stroke position, the maximum stress observed is about 995.6 MPa and at the final position, the maximum stress is 420.3 MPa.

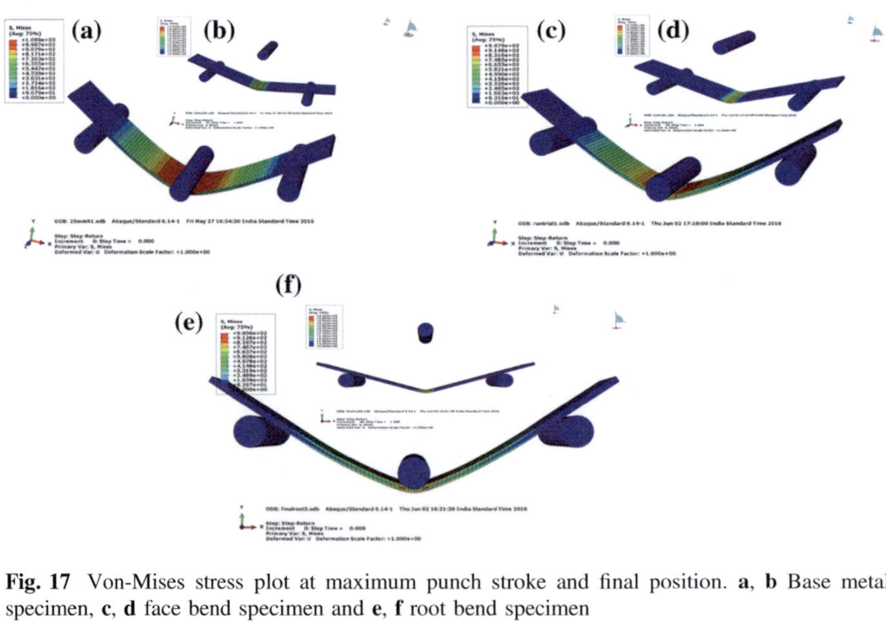

Fig. 17 Von-Mises stress plot at maximum punch stroke and final position. **a**, **b** Base metal specimen, **c**, **d** face bend specimen and **e**, **f** root bend specimen

7.5.2 Effect of Sheet Thickness

To understand the effect of sheet thickness on spring back angle, the simulations are run for three different thicknesses with constant punch stroke of 28 mm. The results are presented in Table 5. Figure 18 shows the reaction force versus displacement plot from variable thickness FE simulations. For the same value of punch stroke, the bend angle is less for least thickness and more for higher thickness. This is attributed to the fact that an increase in sheet thickness offers higher resistance. As the thickness increases, material availability away from the neutral axis increases and the amount of plastic deformation will increase. Hence, the spring back angle reduces.

Table 5 FE results for variable sheet thickness

Spring back measurement for variable thickness			
Thickness (mm)	Bend angle (α_i)	Bent Angle(α_f)	Spring back angle (α_s)
1.2	94°	170°	76°
1.6	101°	157°	56°
2.0	103°	150°	47°

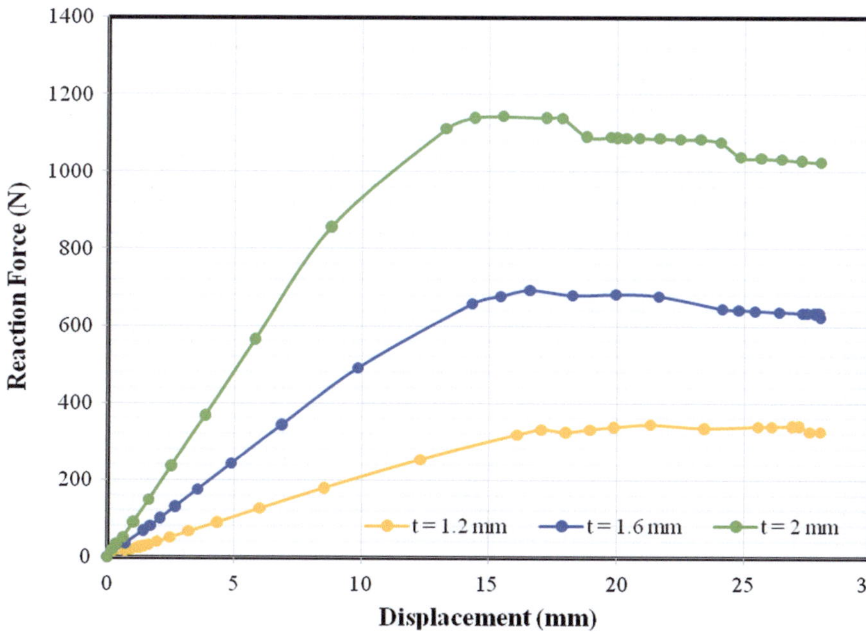

Fig. 18 Reaction force versus displacement curve for thickness variable

Table 6 FE results of spring back measurement for variable punch stroke

Punch stroke (mm)	Bend angle (α_i)	Bent angle (α_f)	Spring back angle (α_s)
20	121°	164°	43°
25	110°	155°	45°
30	98°	147°	49°

7.5.3 Effect of Punch Stroke

The effect of punch stroke on spring back is studied, and the results are shown in Table 6. From Table 6, it is clear that with an increase in punch stroke the spring back angle also increases. As the stroke value (length) increases, the bend angle at maximum stroke position goes on reducing which in turn increases the number of induced stresses. The compressive stresses on the inner side of the specimen, which induce elastic deformation, tend to release their energy. On the other hand, tensile stresses are induced in the outer part of the specimen, which deforms the material plastically. However, due to the domination of elastic deformation over plastic, the spring back angle increases.

8 Conclusions

In this study, the parameters required to analyse spring back effect in sheet metal forming are measured by both experimental and numerical methods. The optimum parameters for better quality of weld with good depth of penetration are identified by conducting multiple trials. The optimised welding parameters are tabulated. The experimental procedures for tensile and three-point bend tests are given. Three-point bending test is conducted for both base metal and TIG welded Ti-6Al-4V specimens. The phenomenon of spring back is quantified experimentally for base metal and welded samples. The stress condition before and after spring back for all the three cases including base metal and welded specimens are discussed based on the developed FE model.

Based on the investigations performed, the following main conclusions are drawn:

- In the case of base metal, spring back of 45° is recorded for punch stroke of 28 mm. The spring back angles of 42° and 43° are recorded for face bend and root bend cases, respectively, with 25 mm as punch stroke.
- The amount of spring back increases in proportion to an increase of the inside radius. If the bend radius is slender, the material is thin (1.6 and 2 mm) and the expected spring back is minor. Besides, it further depends on the yield strength, elastic modulus and strain hardening exponent. Ti-6Al-4V is having low yield strength than other materials (Al, SS304, SS316 and Commercially Pure Titanium) the spring back angle (α_s) is maximum.

- The maximum stress induced after forming is found to be more in the case of base metal specimens compared to welded ones for the same stroke.
- Spring back effect of welded sample is comparable with that of the parent metal which proves the strength of weld joint. Hence, welded sheets can be used in forming processes. From results of the parametric simulation, it is clear that spring back effect is inversely dependent on the thickness of sheet for same punch stroke and is directly proportional to punch stroke for the case of constant sheet thickness.
- The spring back angle measured for sheet metal in both FE model and experimental procedures is in good agreement with each other.
- Spring back angle in sheet metal bending depends on punch stroke as well as sheet thickness. Spring back angle reduces with the increase in sheet thickness. Spring back angle increases with the increase in punch stroke.

References

1. Nebosky, P. S., Schmid, S. R., & Sellés, M. A. (2011). The spring back characteristics of a porous tantalum sheet-metal. *Journal of Manufacturing Science and Engineering, 133*(6), 061022.
2. Kim, H., Nargundkar, N., & Altan, T. (2007). Prediction of bend allowance and spring back in air bending. *Journal of Manufacturing Science and Engineering, 129*(2), 342–351.
3. Andrade, E. C., Bernardi, H. H., & Otubo, J. (2014). Microstructural evaluation on shape recovery in stainless Fe-Mn-Si-Cr-Ni-Co SMA processed by wire drawing. *Materials Research, 17*(3), 583–587.
4. Beal, J. D., Boyer, R., & Sanders, D. (2006). Forming of titanium and titanium alloys. In *Metalworking: Sheet Forming (ASM Handbook)* (vol. 14, pp. 656–669).
5. Gupta, R. K., Mathew, C., & Ramkumar, P. (2015). Strain hardening in aerospace alloys. *Frontiers in Aerospace Engineering, 4*(1), 2325–6796.
6. Leu, D. K. (1997). A simplified approach for evaluating bendability and spring back in plastic bending of anisotropic sheet metals. *Journal of Materials Processing Technology, 66*(1–3), 9–17.
7. Micari, F., Forcellese, A., Fratini, L., Gabrielli, F., & Alberti, N. (1997). Spring back evaluation in fully 3-D sheet metal forming processes. *CIRP Annals-Manufacturing Technology, 46*(1), 167–170.
8. Nilsson, A., Melin, L., & Magnusson, C. (1997). Finite-element simulation of V-die bending: a comparison with experimental results. *Journal of Materials Processing Technology, 65* (1–3), 52–58.
9. Zhang, L. C., & Lin, Z. (1997). An analytical solution to spring back of sheet metals stamped by a rigid punch and an elastic die. *Journal of Materials Processing Technology, 63*(1–3), 49–54.
10. Lee, S. W., & Yang, D. Y. (1998). An assessment of numerical parameters influencing spring back in explicit finite element analysis of sheet metal forming process. *Journal of Materials Processing Technology, 80,* 60–67.
11. Narasimhan, N., & Lovell, M. (1999). Predicting spring back in sheet metal forming: an explicit to implicit sequential solution procedure. *Finite Elements in Analysis and Design, 33* (1), 29–42.

12. Ling, Y. E., Lee, H. P., & Cheok, B. T. (2005). Finite element analysis of spring back in L-bending of sheet metal. *Journal of Materials Processing Technology, 168*(2), 296–302.
13. Tekaslan, Ö., Şeker, U., & Özdemir, A. (2006). Determining spring back amount of steel sheet metal has 0.5 mm thickness in bending dies. *Materials and Design, 27*(3), 251–258.
14. Panthi, S. K., Ramakrishnan, N., Pathak, K. K., & Chouhan, J. S. (2007). An analysis of spring back in sheet metal bending using finite element method (FEM). *Journal of Materials Processing Technology, 186*(1), 120–124.
15. Firat, M. (2007). U-channel forming analysis with an emphasis on spring back deformation. *Materials and Design, 28*(1), 147–154.
16. Thipprakmas, S. (2010). Finite element analysis of punch height effect on V-bending angle. *Materials and Design, 31*(3), 1593–1598.
17. Nanu, N., & Brabie, G. (2011). Influence of material properties on the interaction between residual stress and spring back in the case of in plane sheets forming. *Archives of Civil & Mechanical Engineering, 11*(4), 979–991.
18. Schwarze, M., Vladimirov, I. N., & Reese, S. (2011). Sheet metal forming and spring back simulation by means of a new reduced integration solid-shell finite element technology. *Computer Methods in Applied Mechanics and Engineering, 200*(5), 454–476.
19. Xu, H. J., Liu, Y. Q., & Zhong, W. (2012). Three dimensional finite element simulation of medium thick plate metal forming and spring back. *Finite Elements in Analysis and Design, 51,* 49–58.
20. Maati, A., Tabourot, L., Balland, P., Ouakdi, E. H., Vautrot, M., & Ksiksi, N. (2015). Constitutive modelling effect on the numerical prediction of spring back due to a stretch-bending test applied on titanium T40 alloy. *Archives of Civil & Mechanical Engineering, 15*(4), 836–846.
21. Rojek, J., Hyrcza-Michalska, M., Bokota, A., & Piekarska, W. (2012). Determination of mechanical properties of the weld zone in tailor-welded blanks. *Archives of Civil & Mechanical Engineering, 12*(2), 156–162.
22. ASM International, ASM International. Handbook Committee, & ASM International. (1990). *Metals handbook: Properties and selection* (vol. 2). ASM International.
23. Welsch, G., Boyer, R., & Collings, E. W. (Eds.). (1993). *Materials properties handbook: Titanium alloys*. ASM International.
24. Peters, M., & Leyens, C. (2009). Aerospace and space materials. *Materials Science & Engineering, 3,* 1–11.
25. Karpagaraj, A., Shanmugam, N. S., & Sankaranarayanasamy, K. (2015). Some studies on mechanical properties and microstructural characterization of automated TIG welding of thin commercially pure titanium sheets. *Materials Science and Engineering A, 640,* 180–189.
26. Karpagaraj, A., Shanmugam, N. S., & Sankaranarayanasamy, K. (2016). Studies on mechanical behavior and microstructural analysis of tailor welded blanks of Ti–6Al–4V titanium alloy sheet. *Journal of Materials Research, 31*(14), 2186–2196.
27. Ravikumar, B. V. R., & Harshitha, V. (2015). Experimental investigation into the effect of gas tungsten arc welding on Ti-6al-4v. *International Journal of Advances in Production & Mechanical Engineering (IJAPME), 1,* 2394–6202.
28. Standard test methods for tension testing of metallic materials, ASTM E8/E8M-11.
29. Standard test method for guided bend test for ductility of welds, ASTM E190-92 (Reapproved 2008).
30. Mujika, F., Arrese, A., Adarraga, I., & Oses, U. (2016). New correction terms concerning three-point and four-point bending tests. *Polymer Testing, 55,* 25–37.
31. Baskar, K., Shobana, R., & Cyril Thomas, A. (2016). Non-linear analysis of castellated beams. In *Proceedings of 6th International Conference on Computational Mechanics* (pp. 894–897).
32. Sun, B., Liu, R., & Gu, B. (2012). Numerical simulation of three-point bending fatigue of four-step 3-D braided rectangular composite under different stress levels from unit-cell approach. *Computational Materials Science, 65,* 239–246.

33. Cyril Thomas, A., & Baskar, K. (2016). Assessment of load carrying capacity of thin-webbed castellated beams. In *Proceedings of Structural Engineering Convention* (pp. 322–327).
34. Faridmehr, I., Osman, M. H., Adnan, A. B., Nejad, A. F., Hodjati, R., & Azimi, M. (2014). Correlation between engineering stress-strain and true stress-strain curve. *American Journal of Civil Engineering & Architecture, 2*(1), 53–59.
35. Davoodi, B., & Zareh-Desari, B. (2014). Assessment of forming parameters influencing spring-back in multi-point forming process: a comprehensive experimental and numerical study. *Materials and Design, 59,* 103–114.

Numerical Prediction of Fracture in Parts Formed with Incremental Sheet Forming Process

Divyansh Dwivedi, Abhishek Kumar, Satwik Priyadarshi, Prashant K. Jain and Puneet Tandon

1 Introduction

For half a century, mass manufacturing played a significant role in developing the world. Mass manufactured products ruled market and economy. With the rise of globalization and capital market economies, products ill-suited for mass manufacturing were lost into oblivion. Nowadays, almost every product or artefact used in daily life is a mass manufactured product. Mass manufacturing also has its flaws and fallacies. Standard size products seldom have a perfect fit and a few special purpose products like prosthetics require heavy investment and customized tooling for every piece. In this dynamic market, where products are changing rapidly, it is becoming increasingly difficult to make huge investments on fabrication of specialized dies and tools. These tools serve no purpose once the product changes. The prominence of mass manufacturing also hinders the ability to experiment with innovative products because it makes specialized tools difficult to afford.

It is also obligatory for the sheet metal forming industry to adopt new and evolving techniques to manufacture diverse products with reduced time and cost.

D. Dwivedi · A. Kumar · S. Priyadarshi · P. K. Jain · P. Tandon (✉)
PDPM Indian Institute of Information Technology, Design and Manufacturing,
Jabalpur, India
e-mail: ptandon@iiitdmj.ac.in

D. Dwivedi
e-mail: d.divyansh@iiitdmj.ac.in

A. Kumar
e-mail: k.abhishek@iiitdmj.ac.in

S. Priyadarshi
e-mail: satwik.p@iiitdmj.ac.in

P. K. Jain
e-mail: pkjain@iiitdmj.ac.in

U. S. Dixit and R. Kant (eds.), *Simulations for Design and Manufacturing*,
Lecture Notes on Multidisciplinary Industrial Engineering,
https://doi.org/10.1007/978-981-10-8518-5_6

The conventional sheet metal forming process requires dedicated tooling and dies, which are designed for mass production. Thus, to reduce cost, lead time and difficulties in making dies, a new manufacturing technology called 'Incremental Sheet Forming' (ISF) is under the process of development. In ISF process, the sheet is locally deformed by a simple hemispherical tool (punch), which follows the tool path of finishing geometry. Moreover, the dexterity of the process is further augmented through rationale provided by pre-fabrication simulations. The Finite Element Analysis (FEA) is one of the most important Computer-Aided Engineering (CAE) tools that helps in analysing the forming process under any imaginable condition. In the process, it allows the forming process to be further refined, prior to any real-time experiment. This reduces the number of experiments to be performed and hence, the lead time to manufacture products with ISF process.

In its present form, ISF is appropriate for rapidly producing sheet metal parts in small batch sized productions. Reduction in lead time, forming forces and removal of dies make this process superior to any other conventional forming processes, viz. deep drawing, punching, stretching, pressing, etc. However, non-uniform thinning and unpredictable onset of fracture accompanied by undesired bending and spring back in the formed components limit the applicability of ISF process. To make this process more efficient and prolific, thorough investigation of material deformation mechanism of the ISF process is essentially required before communicating it to the shop floor. A comprehensive experimental study can be done for this purpose but it is time consuming as well as inadequate. Thus, CAE analysis is needed to eliminate the superfluous experimentation. CAE analysis also helps to examine the flexibility and adaptability of ISF for a particular application. This makes CAE analysis desirable before actual manufacturing. As per available literature, some work had been carried out in the domain of FE of ISF process; however, most of the work reports lack of accuracy. This, thus, motivates us to conduct some research to improve the accuracy of fracture prediction through finite element analysis due to the inherent applications and advantages of FEA of ISF.

2 Incremental Sheet Forming

The origin of ISF lies in the realm of conventional shear forming or spinning process. Mason proposed a variant of spinning process in 1978 at the University of Nottingham, which is often regarded as the origin of ISF [1]. However, the elementary concept of single-point ISF was initially epitomized through the US patent, US3342051, filed in 1964 by Leszak [2]. The resurgence of ISF as a technology of interest was due to work done by Iseki [3], particularly in developing a process of deforming steel and titanium sheets of 0.7 mm thickness through a robotic arm to form non-symmetric shapes. In the last two decades, ISF matured as a technology with capability to form the complex 3D geometry. The foremost attribute of ISF is that it does not require any dedicated tooling or dies to form components. The dieless nature of this process has helped to establish ISF as an efficient alternative

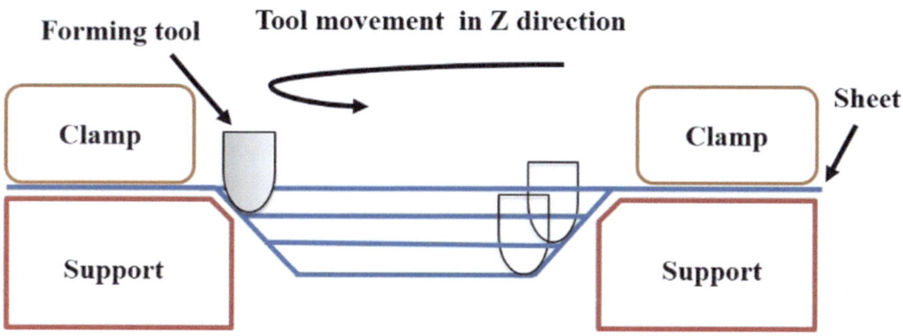

Fig. 1 Incremental sheet metal forming process

for low-volume production of the sheet metal components. The complete description of ISF process is shown in Fig. 1 in which the sheet is clamped on its outer periphery. The tool moves in *Z*-direction and makes a highly localized deformation which is dragged as described by the tool path.

Earlier, ISF was classified on the basis of the shape of geometry being formed, viz. convex and concave [4], but with new developments in ISF, it is now possible to form many complex combinations of concave and convex shapes. Nowadays, the classification of ISF is normally done on the basis of contact point between tool and sheet, i.e. Single-Point Incremental Forming (SPIF), Two-Point Incremental Forming (TPIF) and Double-sided Incremental Forming (DSIF) [5] (Fig. 2). In SPIF, tool deforms the sheet with only one point of contact as shown in Fig. 2a. On the other hand, in TPIF there are two contact points as shown in Fig. 2b. The second contact point in TPIF is made by the tool with the full die or partial die. Double-sided Incremental Forming (DSIF), also known as Duplex Incremental Sheet Forming, is characterized by use of a counter tool which moves simultaneously with the forming tool, as shown in Fig. 2c.

3 Deformation Mechanism and Material Modelling

The mechanism in ISF is a complex combination of three deformation mechanisms, namely bending, stretching and shearing. Stresses generated in the part during forming are greater than the yield stress but lesser than the ultimate stress. Jackson and Allwood [6] experimentally investigated the mechanism of ISF for copper sheet through forming a cone geometry and explained that the deformation mechanism is a combination of stretching and shearing with greatest strain component being shear in the tool direction. Global bending occurs near the clamp and tool causes local bending around the contact point. Deformation is nearly in plain strain condition for axisymmetric parts. Circumferential strain is rather small but increases with increase in curvature. Biaxial stretching condition occurs at corners in plane-symmetric parts

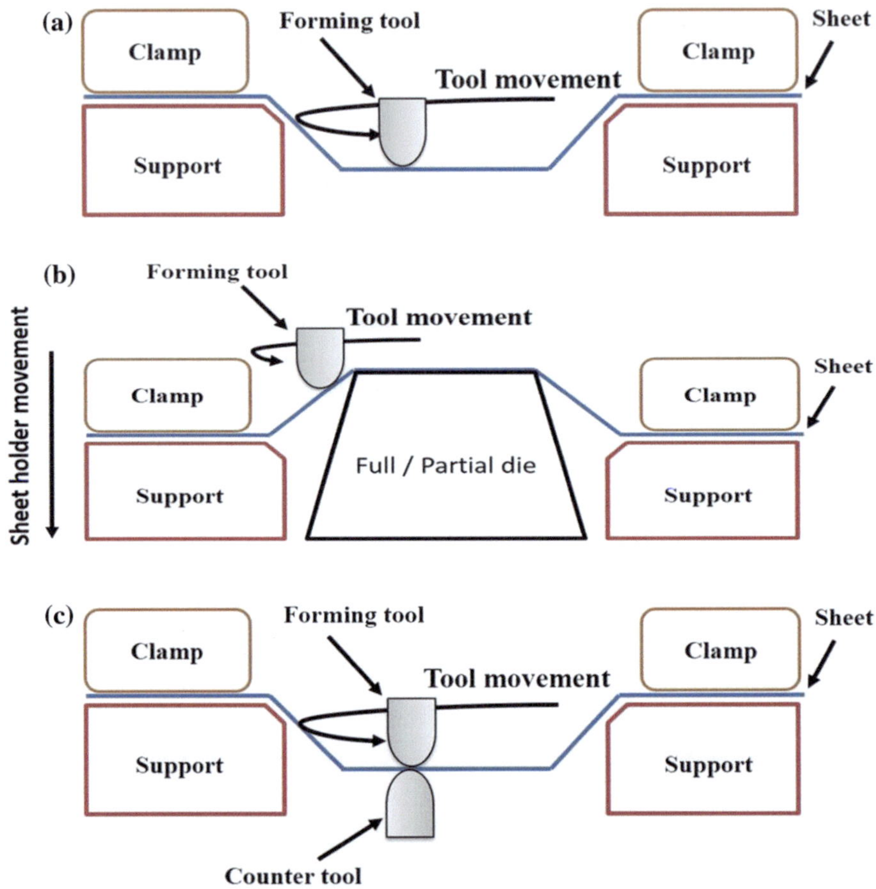

Fig. 2 Types of ISF; **a** single-point incremental forming (SPIF), **b** two-point incremental forming (TPIF), **c** double-sided incremental forming (DSIF) or duplex incremental sheet forming

and causes fracture in some cases [7]. Sheet deforms under through-thickness shear in tool movement direction as discussed by Liu et al. [8]. Thinning and cracking are the main limitations of ISF. The change in thickness can be measured as discussed by Hirt et al. [9] using sine law as $t = t_0 * \sin \alpha$, which is primarily used in conventional spinning process [10]. In this equation, t is the final thickness, t_0 is the initial sheet thickness and α is the forming wall angle. Forming Limit Diagram (FLD) was used by Silva et al. [11] to investigate the failure of the sheet in ISF. Many authors reported that the classical FLD is incapable to predict the failure correctly [12]. Ambrogio et al. [13] developed a failure criteria based on the forming forces by taking forming force gradient as a function of the sheet thickness. Malhotra et al. [14] studied the deformation mechanism, forming forces and fracture based on the damage accumulation and fracture initiation theory for ductile material proposed by Xue [15]. The role of voids in ductile failure was first modelled by McClintock to analyse the evolution of cylindrical voids in ductile elastoplastic material [16]. The evolution of spherical voids in elastic-perfectly plastic material was studied by Rice and Tracey [17]. The yield criteria for rigid plastic porous materials were first proposed by Gurson [18] and improved by Tvergaard and Needleman to add microvoids interactions and microvoid coalescence. This form of Gurson model is popularly referred to as GTN model. The GTN model is micromechanical damage model based on the mechanism of voids and describes the nucleation, growth and coalescence of voids in the material as the damage proceeds [19]. Li et al. [20] simulated the ISF process to model damage as a function of void volume fraction in the material using Gurson–Tvergaard–Needleman (GTN) model with Hill anisotropy. A comparative study is done by Cao et al. [21] at three ductile damage approaches, namely coupled phenomenological, uncoupled phenomenological and micromechanical model for fracture prediction in cold forming process. Kiran and Khandelwal compared four computation ductile damage models based on void mechanism [22]. Bjorklund [23] carried out the review on modelling of failure for the thin sheet of high-strength steel. Focus has been given on phenomenological models and categorized in four categories, namely stress-based, strain-based, combined stress–strain-based and damage-based models. Sahu et al. [24] simulated the ISF with phenomenological Johnson–Cook (J-C) material model, i.e. stress-based phenomenological model, but only plasticity is modelled and failure was not simulated. Although extensive research on finite element analysis of ISF has been conducted over the past decade, J-C plasticity model with constant equivalent plastic strain criteria is rarely discussed in the literature. Therefore, this work investigates the effectiveness of the constant equivalent plastic strain criterion for accurate prediction of fracture in ISF through FE analysis.

3.1 Plasticity Model

The equation which defines the flow stress in terms of strain, strain rate, temperature and other stress state variables like triaxiality and lode parameter is called the constitutive equation. Within the elastic region, Hook's law is sufficient to calculate the flow stress, while in plastic region nonlinear equation is required due to nonlinearity of the material. In the present work, Johnson–Cook (J-C) model [25] had been used to calculate the flow stress in the plastic region.

The typical constitutive equation of Johnson–Cook model is given below [25]:

$$\sigma = \left(a + b * \varepsilon_p^n\right)\left(1 + c * \ln\frac{\dot{\varepsilon}}{\dot{\varepsilon}_0}\right)\left(1 - T^m\right) \tag{1}$$

where σ is the flow stress, ε_p is the plastic strain, a is the yield stress, b is the hardening modulus, n is the hardening exponent, c is the strain rate coefficient, $\dot{\varepsilon}$ is the strain rate, $\dot{\varepsilon}_0$ is the reference strain rate, m is the temperature exponent and

$T = (t - 298)/(t_{\text{melt}} - 298)$, where t is the material temperature (K) and t_{melt} is the melting temperature (K), while room temperature (t_{room}) is assumed at 298 K [26].

The J-C relation includes strain hardening component, strain rate component and thermal softening component. The first part of equation defines the strain hardening effect, while the second and third define the strain rate and thermal softening effect, respectively. This constitutive model is suitable for the materials subjected to large strains, high strain rate and high temperature.

3.2 Fracture Criterion

To model the failure in ISF, constant equivalent plastic strain criteria are used as a fracture criterion. When the equivalent plastic strain ($\acute{\varepsilon}$) exceeds the value of critical strain ($\acute{\varepsilon}_f$), the fracture of the material occurs.

The equivalent plastic strain $\acute{\varepsilon}$ is defined as

$$\acute{\varepsilon} = \sqrt{\frac{2}{3}\left(\varepsilon_1^2 + \varepsilon_2^2 + \varepsilon_3^2\right)} \tag{2}$$

where ε_1, ε_2 and ε_3 are the principle strains and the fracture condition for material fracture is given by

$$\acute{\varepsilon} \geq \acute{\varepsilon}_f \tag{3}$$

where $\acute{\varepsilon}_f$ is the strain at fracture in uniaxial tensile test.

The constant equivalent plastic strain criteria are based on the value of plastic strain at failure. When equivalent plastic strain reaches the value of plastic strain at fracture, material fails. Thus, critical value of equivalent stress (σ_{ec}) at this condition can be calculated as

$$\sigma_{\text{ec}} = E\acute{\varepsilon}_f + a = \sigma_f \tag{4}$$

This equation reveals that the equivalent plastic strain failure criterion is independent of both stress triaxiality and lode angle. Hence, the single circle contour with radius of $\sqrt{\frac{2}{3}}\sigma_f$ on π-plane or deviatoric plane in the principle stress space represents the failure surface of the material. Here, σ_f is the failure stress that corresponds to plastic strain at failure ($\acute{\varepsilon}_f$) under uniaxial loading, a is yield stress and E is the Young's modulus.

Although the criterion used here is a bit inaccurate as compared to Xue-Wierzbicki (X-W) and Wilkins, as discussed by Wierzbicki et al. [27], even Gurson–Tvergaard–Needleman (GTN) model, a porosity-based material model predicts fracture with much better accuracy. However, all such models are tedious to calibrate and require exhaustive experimentation for parameter estimation.

On the other hand, the criterion deployed here provides results of acceptable accuracy in much less time as compared to others and can be calibrated easily by uniaxial tensile test.

The deformation behaviour of any material depends upon the mechanical properties of that material. The correct acquisition of material properties leads to an accurate simulation of material deformation. In the case of Incremental Sheet Forming (ISF), owing to significance of vast material nonlinearity, material modelling and parameter acquisition is of paramount importance for simulating ISF with reasonable accuracy.

3.3 *Stress–Strain Analysis*

A comprehensive knowledge of stress and strain evolution during deformation is a prerequisite for modelling the behaviour of a given material. In order to acquire such knowledge without going through rigors and intricacies of experimental modelling, a uniaxial tensile test is used to acquire the parameters. Stress and strain generated during uniaxial tensile test are considered for the estimation of material model parameters. A typical stress–strain curve representative for non-ferrous ductile materials is shown in Fig. 3.

The complete stress–strain curve can be divided into two distinct regions, i.e. elastic region and plastic region. Both regions are explained in detail in the coming sections.

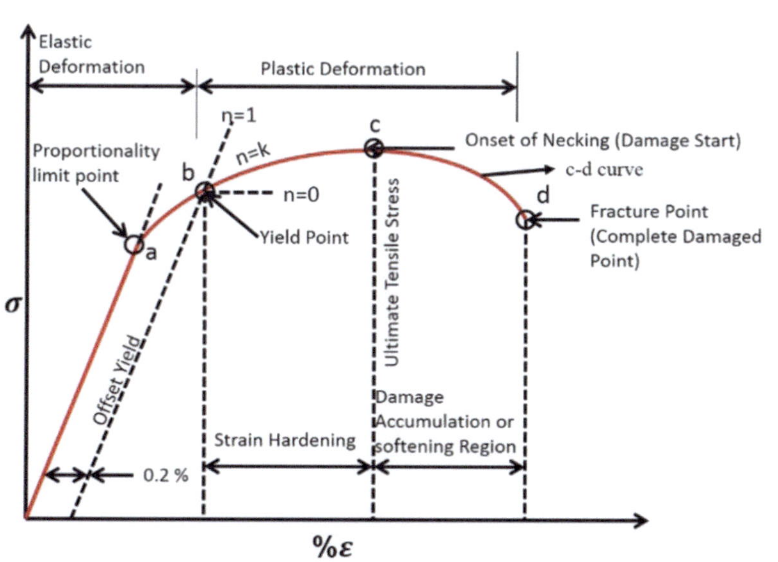

Fig. 3 Stress–strain profile of non-ferrous material

3.3.1 Elastic Region

Initially, the material behaves linearly and exhibits elastic behaviour. The region of stress–strain curve exhibiting elastic behaviour is called elastic region. The point up to which stress is directly proportional to the strain is called the proportionality limit and is shown by point '*a*' in Fig. 3. In the elastic region, Hook's law governs the stress evolution with increases in strain. Hook's law is shown below:

$$\sigma \propto \varepsilon \tag{5}$$

After using a proportionality constant,

$$\sigma = E\varepsilon \tag{6}$$

Here, σ represents stress, ε represents stain and E is the proportionality constant and also known as modulus of elasticity of the material or as the stiffness of material in units of force per unit area. The material has the ability to regain its shape and size while unloading, when deformation solely takes place in this region. This property is called elasticity of a material.

3.3.2 Plastic Region

The part of the stress–strain curve other than that belonging to the elastic region is known as plastic region. Initiation of plastic deformation is depicted by the yield point '*b*' in Fig. 3. In most non-ferrous materials like aluminium, it is not possible to find yield point based on geometry of stress–strain curve. In these materials, the concept of offset yield is used to define the yield stress. The value of this is commonly set at 0.2% plastic strain. This yield point divides the stress–strain curves in elastic and plastic regions. After the yield point, strain hardening of the material starts. The region *b–c* of the curve represents hardening of the material. Point '*c*' also shows the ultimate stress point or strength of the material. The nonlinearity of the material in hardening region is defined by the hardening exponent n. It varies from 0 to 1 indicating ideally plastic or non-hardening material and linear elastic material, respectively. In the real world, the value of n lies always between 0 and 1. At the last, in plastic region, damage accumulation (region *c–d*) occurs. It starts with onset of necking in tensile test at ultimate stress point.

3.4 Parameter Acquisition

In the present work, aluminium sheet is considered with 0.91 mm thickness. All the material model parameters are obtained from the uniaxial tensile tests with E8 (ASTM Standard) dog bone sheet type specimen, as shown in Fig. 4. The stress–strain curve

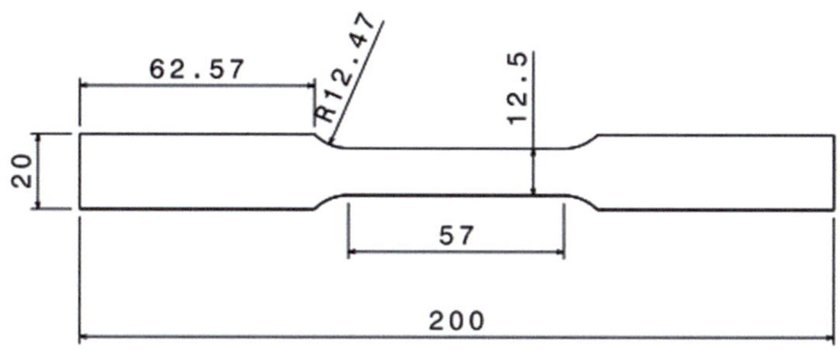

Fig. 4 Dog bone E8 sheet type specimen

for Al1050-H14 (thickness = 0.91 mm) obtained from a uniaxial tensile test is given in Fig. 5.

The moduli of elasticity, density and Poisson's ratios for the materials are taken as per the material handbook [28]. The other three parameters required to complete plasticity model parameters, i.e. initial yield stress, hardening modulus or hardening coefficient and hardening exponent, are calculated as discussed in Fig. 6.

At the first stage, the force versus displacement data is exported as an excel file from the tensile test. Then, engineering stress, engineering strain, true stress, true strain and plastic strain are calculated by the following formulae:

Engineering stress:

$$\sigma_e = \frac{F}{A} \tag{7}$$

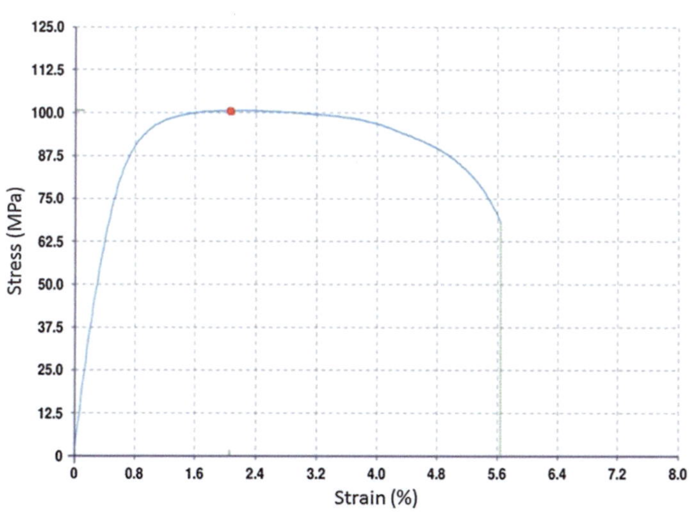

Fig. 5 Experimentally obtained stress–strain curve for Al1050H14 (0.91 mm)

Fig. 6 Process to estimate plasticity parameters

Engineering strain:

$$\varepsilon_e = \frac{\delta L}{L} \tag{8}$$

True stress:

$$\sigma_t = \sigma_e(1 + \varepsilon_e) \tag{9}$$

True strain:

$$\varepsilon_t = \ln(1 + \varepsilon_e) \tag{10}$$

Plastic strain:

$$\varepsilon_p = \left(\varepsilon_t - \frac{\sigma_t}{E}\right) \tag{11}$$

The mechanical properties of the material were identified by a uniaxial tensile test and are listed in Table 1.

Table 1 Material model parameters

Parameters	Value
Young's modulus (E)	69,000 MPa
Initial yield stress (a)	87.54 MPa
Strain hardening coefficient (b)	143.9 MPa
Strain hardening exponent (n)	0.08529
Density (ρ)	2.7×10^3 kg/m^3
Poisson's ratio (υ)	0.33
Ultimate tensile stress (σ_{max})	103 MPa
Max. plastic strain ($\%\varepsilon_{max}$)	5.38

4 Finite Element Modelling

In the present work, the entire preprocessing was done in HyperMesh, a preprocessing module of HyperWorks. The ISF process is defined under preprocessing as follows.

4.1 Geometric Modelling

At first, the CAD models of the hemispherical tool with 6 mm diameter and sheets with 0.91 mm thickness were modelled in CatiaV5 and saved as STEP file. The STEP file is preferred, as it does not lead to any geometrical error during exchange of CAD data. These STEP files were used for the FE analysis. The CAD models of the tool and sheet are shown in Fig. 7.

4.2 Tool Path Generation

In the present work, frustum of a cone was considered as test geometry for the FE analysis. The CAD model of the test geometry was developed in CatiaV5. After that, CAD model was imported into in-house developed code to generate the tool path with G-codes. This tool path was used to form the truncated cone experimentally. The *X*-, *Y*- and *Z*-coordinates of the tool path were plotted in MATLAB R2014a as shown in Fig. 8.

The *X* versus time, *Y* versus time and *Z* versus time data were plotted as a tool path for the simulation in HyperMesh as shown in Fig. 9.

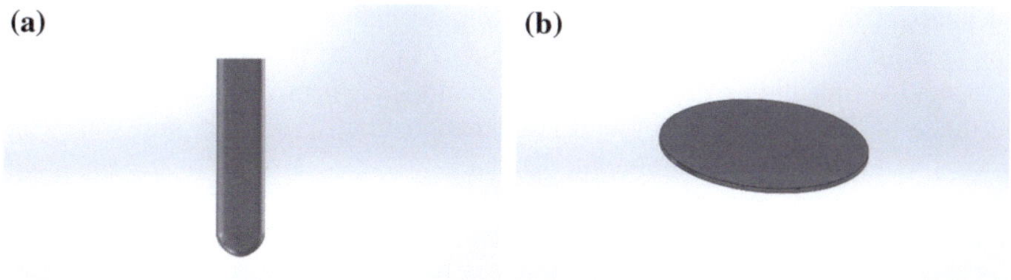

Fig. 7 **a** CAD model of tool, **b** CAD model of circular blank

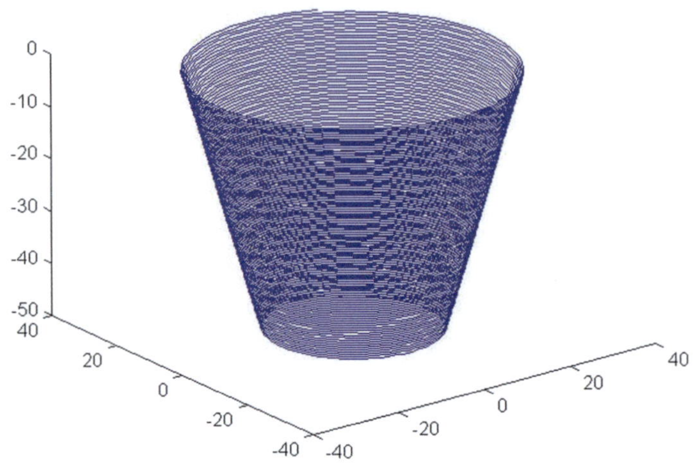

Fig. 8 Tool path trajectory

4.3 Discretization and Element Formulation

After importing CAD models, geometries were cleaned with an auto cleanup tool to eliminate the geometrical error, if any, and then the solid models were discretized. In this approach, four-node quadrilateral reduced integration shell element (1 point quadrature element) was used with Improved Flanagan Belytschko (IFB) shell formulations. In this shell element formulation, anti-hourglass forces were used to eliminate the zero-energy mode of the element created due to the reduced integration scheme. IFB is most suitable shell element formulation when the material goes under elastoplastic deformation. Mid-surface of the sheet was extracted to mesh it with shell element. Seven integration points were considered through the thickness of the sheet [29]. Adaptive meshing technique was used to generate the fine mesh in the deformable zone and coarse mesh in the rest of the geometry. The computation time highly depends on the size of the element. Therefore, to maintain the trade-off between accuracy and computation time, 2 and 0.5 mm were considered as maximum and minimum element sizes, respectively, for the fine mesh zone. A number of elements in sheet and tool were 6631 and 2579, respectively. A number of nodes in tool and sheet were 6759 and 2580, respectively. The meshing of tool and sheet blank is shown in Fig. 10.

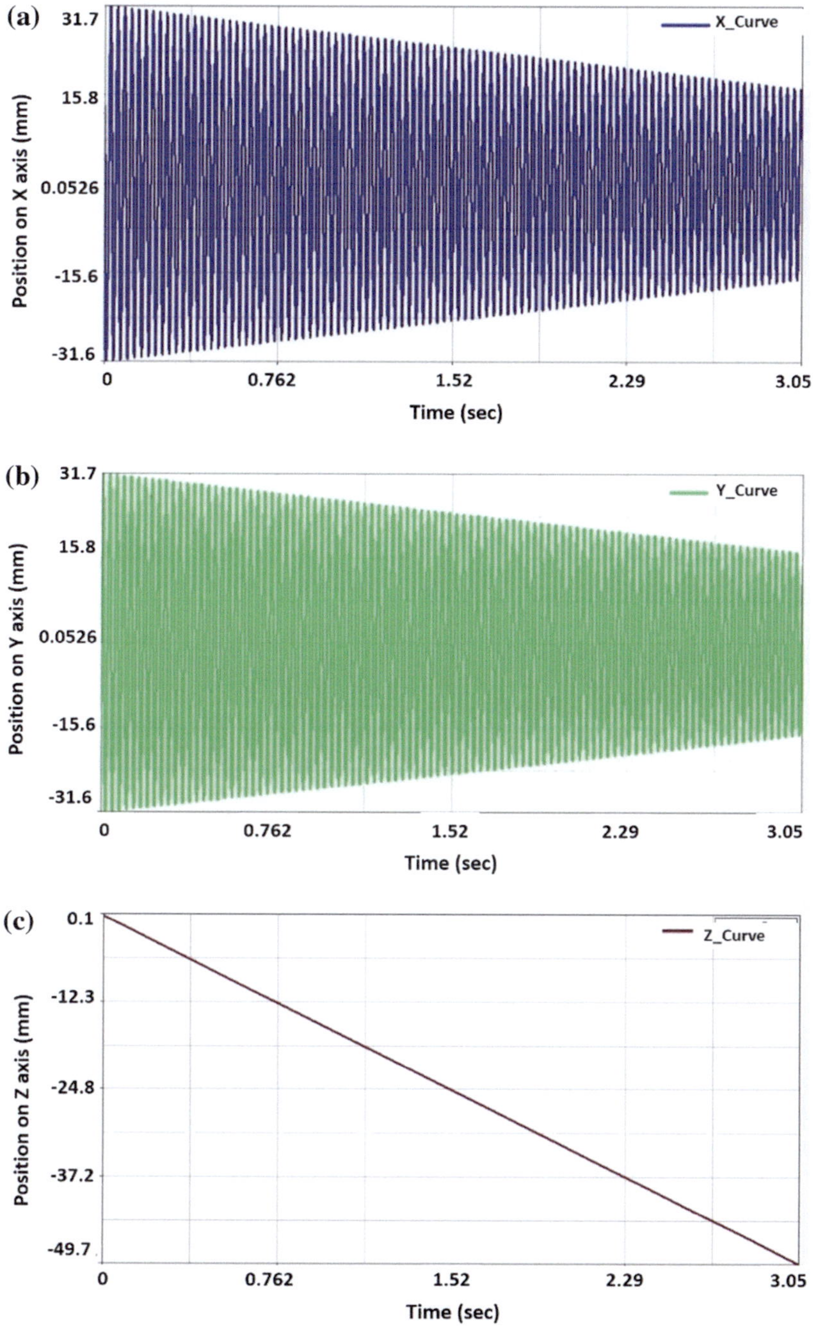

Fig. 9 Time plot in HyperMesh: **a** X versus time, **b** Y versus time, **c** Z versus time

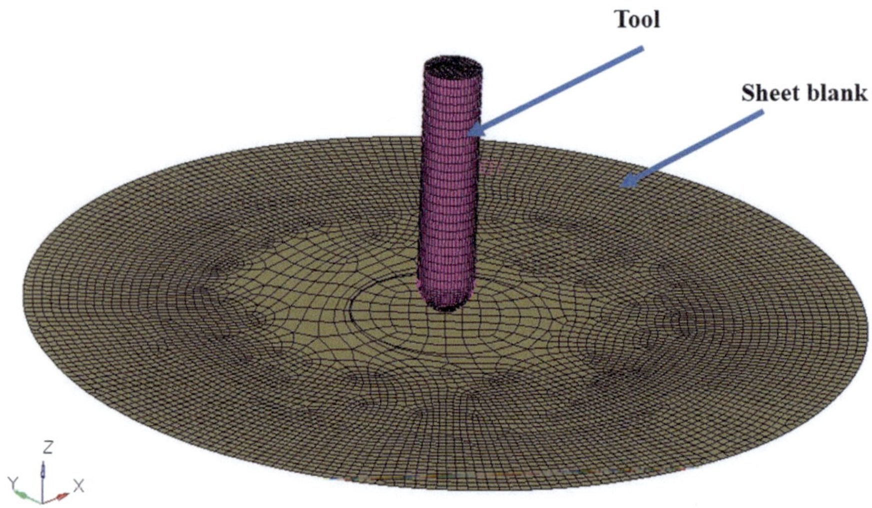

Fig. 10 Meshing of tool and sheet blank

4.4 Material Modelling

The material model, which was used in FE analysis, is described in detail in Sect. 3. The constant equivalent plastic strain criteria with Johnson–Cook plasticity are listed as MAT/LAW2 [26] in material models library of HyperMesh 12.0. The tool was modelled as a non-deformable body with MAT/LAW1 [26].

4.5 Boundary Conditions and Interface

The *X*, *Y* and *Z* translation, as well as rotation motion, means all six degrees of freedom of the nodes of the outer periphery of the sheet were restricted to fix the sheet. The tool was only constrained with rotations in *X*- and *Y*-directions. To define contact, master-and-slave algorithm was used as detailed in the previous FE modelling approach. Contact modelling of tool sheet interface in HyperMesh 12.0 is shown in Fig. 11. The apex of the pyramid (white colour pyramid for sheet and green colour pyramid for tool) of each element indicates the normal direction for sheet and tool contact.

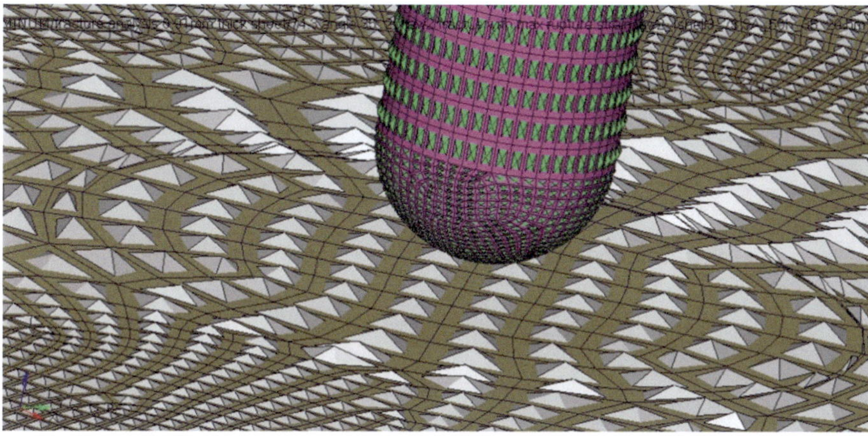

Fig. 11 Contact modelling between tool and sheet blank

4.6 *Loading Condition*

In HyperMesh 12.0, load collector was used to define the loading conditions. A master node of the tool was created to assign the tool path with imposed displacement load collector (IMPDISP). For the rotation in the *Z*-direction, 1000 rpm was given to the tool with initial velocity load collector (INIVEL). The boundary condition load collector (BCS) was used to fix the blank and to move the tool [26].

4.7 *Solver*

In the present work, Radioss solver of HyperWorks 12.0 was used. ISF is a highly nonlinear process, and thus an explicit time integration scheme was adapted because explicit scheme handles nonlinearity (contact, material nonlinearity and geometric nonlinearity) with relative ease as compared to the implicit scheme. The Lagrangian formulation was used for large strain formulation. An iterative algorithm option of the solver configuration was used to compute the plastically admissible stress, when the stress state is in the plastic region.

4.8 *Post-processing*

For post-processing, Hyperview and Hypergraph were used. The 3D animation files generated by the solver were viewed in Hyperview. The 3D and 2D plots of the output parameters were generated by Hyperview and Hypergraph, respectively.

To generate the time-dependent output of plastic strain in Hypergraph file or time history file, time history control cards were added as an output block in HyperMesh file. The output parameters were written as local variables in the output block.

5 Experimental Validation

All the experiments were performed on Okuma 3-axis CNC vertical milling machine to validate the numerical prediction of fracture in incrementally formed truncated cones. Figure 12 shows the setup, which was developed and experiments were performed as a validation exercise.

The circular blank was clamped on its outer periphery with the help of circular clamping fixture. The forming tool was made up of hardened stainless steel with hemispherical end. The diameters of the sheet blank and tooltip were 88 and 6 mm, respectively. For lubrication purpose, hydraulic lubricating oil of grade 68 was used to decrease the friction and to improve the surface finish and formability of the part. Spiral tool path was preferred over constant *Z* tool path as it results in better formability and surface finish. Process parameters, which are used in the experiments, are given in Table 2.

Two truncated cones with different wall angles were formed as test case geometries to validate the proposed work. The wall angles of the geometries considered were slightly greater than 73°, as for 0.95 mm Al1050H14 sheets the critical angle for ISF is 73°, i.e. any geometry with the wall angle greater than this will fracture instantly as the forming commences. An instant fracture will make measurement of fracture depth incomprehensible as the depth formed would be too small to measure. Figure 13 shows the fractured part.

Fig. 12 Experimental setup

Table 2 Process parameters

Parameters	Value
Tool rotation (rpm)	700
Feed rate (mm/min)	1000
Incremental depth (mm)	0.5
Tool path	Spiral
Tool type	Hemispherical end

Fig. 13 Fractured (left) and formed (right) parts

6 Results and Discussions

In ISF, material failure is a vital factor due to the uncertainty of fracture in the components during forming. Prior understanding about the fracture depth assists in identifying the safe forming limit. In the present model, fracture depth prediction is based on the critical value of plastic strain, which is the plastic strain of the material at fracture point in uniaxial tensile condition. When any of the elements exceeds this value, that element will vanish (in FEA) and this deletion of the element is considered as the fracture in the material. In FEA, the time step at which the element gets deleted at first is noted down and simultaneously, the position of the tool in Cartesian coordinate system is checked at the same time step in the tool path. The *Z*-coordinate of the real-time position of the tool is considered as the fracture depth in that case. The complete description of the fracture depth prediction is given only for the truncated cone of wall angle 73.3°. The same procedure is followed to predict the fracture in truncated cone of wall angle 75.3°. Figure 14 shows the vanished elements in the finite element model of truncated cone of wall angle 73.3°.

The 3D animation files are generated at a difference of 0.01 time step. At 0.8700 time step, no element deletion is observed in the animation while, at 0.8800 time step, two elements having element Ids 333 and 311 vanished. The output file created by the solver is checked to get the exact time step at which the elements had vanished. The time step at which these elements got deleted were 0.8704 and 0.8722, respectively, as shown in Fig. 15 with red box along with the *Z*-coordinate

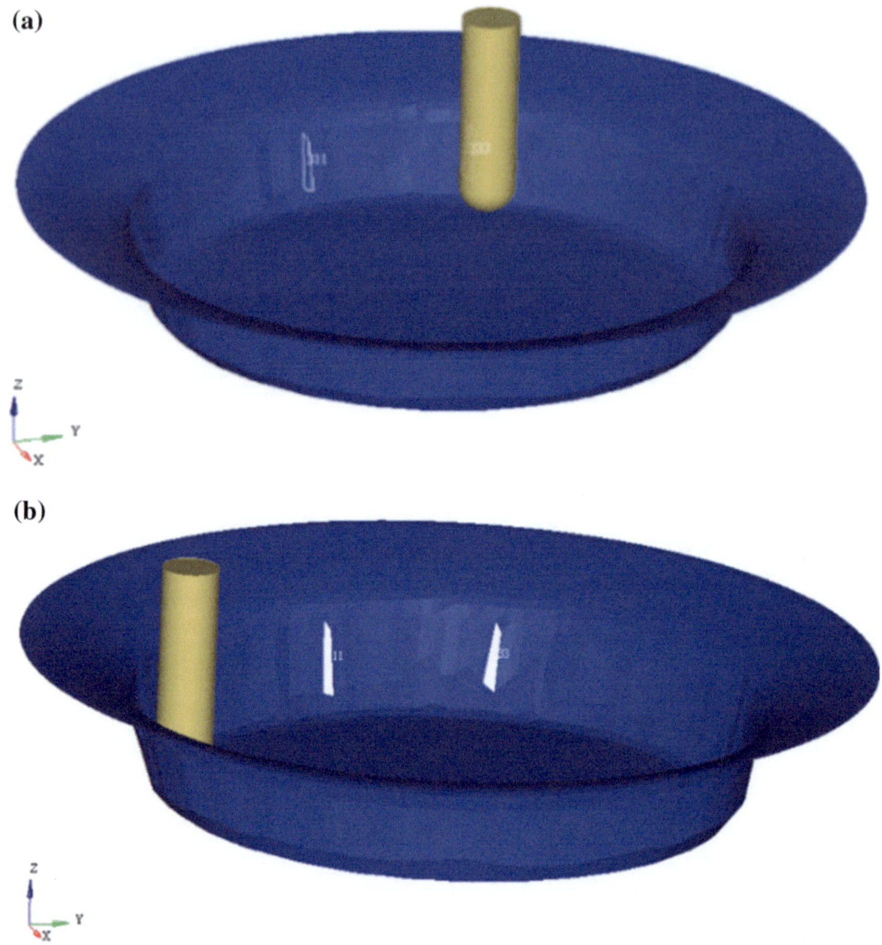

Fig. 14 **a** No element deletion at time step 0.87, **b** element with Ids 311 and 333 deleted at time step 0.88

of the tool at that time step. Thus, the fracture depth obtained by FEA for forming of truncated cone with wall angle of 73.3° is 14.3 mm.

The plastic strain is also plotted in Fig. 16, which clearly depicts the evolution of maximum plastic strain with respect to time in both the ruptured elements. It was observed that the maximum plastic strain had reached the critical value of equivalent plastic strain (5.38) near 0.85 s time.

Table 3 shows the comparison of fracture depth prediction between experimental results and simulation results.

As equivalent plastic strain is evaluated on the basis of the increments in the principle plastic strain, this criterion takes care of the history of the deformation (in terms of plastic strain increment) upon which the hardening depends. The proposed

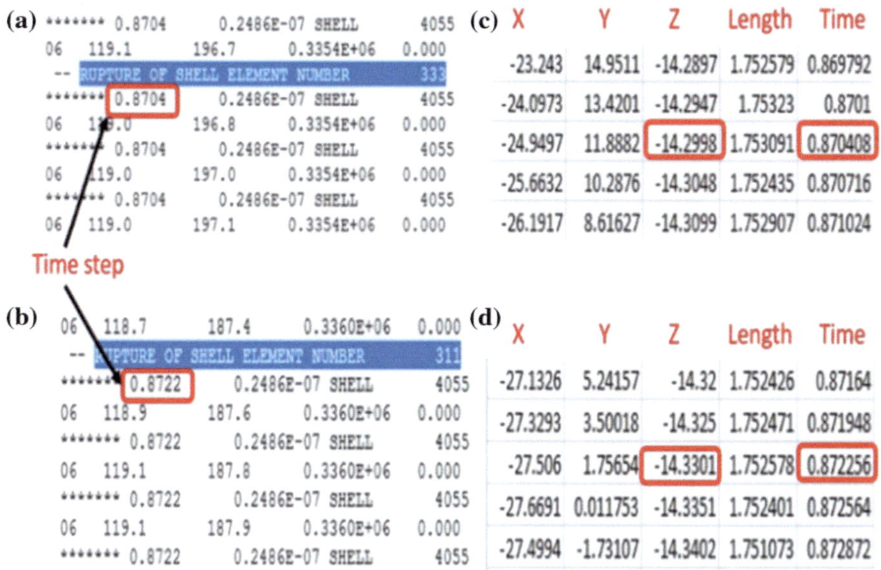

Fig. 15 **a**, **b** Rupture of element in simulation output file; **c**, **d** corresponding *Z*-coordinate and time in tool path

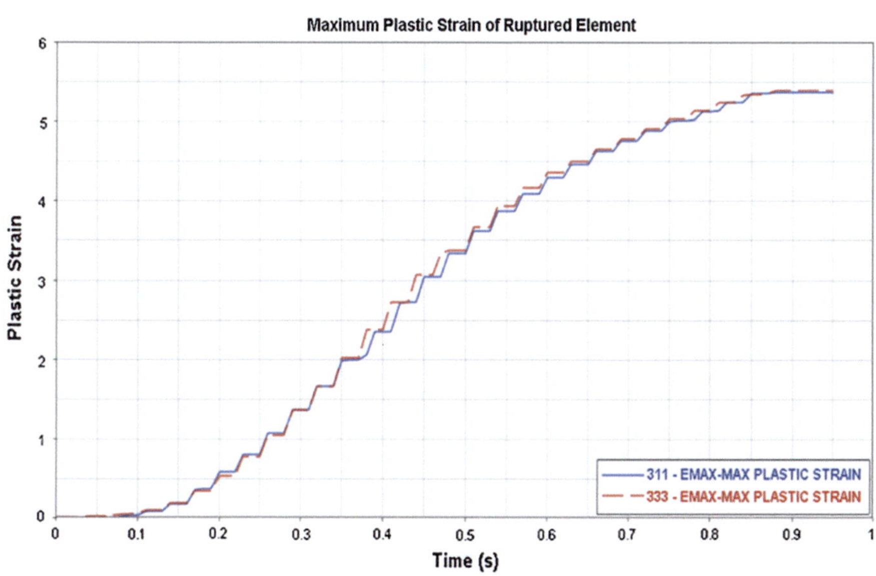

Fig. 16 Plastic strain versus time plot of ruptured elements

Table 3 Comparison of fracture depth prediction

Geometry (truncated cone)		Simulation fracture depth (mm)	Experiment fracture depth (mm)
Wall angle	Height (mm)		
73.3°	50	14.3	15
75.3°	50	11.76	12

model employs this concept to predict the failure of the material. This fracture model being very simple and easy to implement, as it requires only one parameter to estimate the fracture, helps in quick understanding the failure characteristics of the material in the case of ISF.

7 Conclusion

In the present work, numerical investigation has been carried out to predict the fracture depth in incrementally formed truncated cones with different wall angles. The Johnson–Cook plasticity model along with the constant equivalent plastic strain criteria is modelled to simulate the elastoplastic deformation and failure of material during forming in ISF. Flow stress in the plastic region along with the failure of the component is successfully computed and simulated. A Lagrangian solver with explicit time integration scheme and iterative configuration is used to handle contact, material nonlinearity and geometric nonlinearity and compute results. An extensive experimental campaign was undertaken to evaluate the accuracy of the simulation. After comparing the simulation results with experimental outcomes, it was found that the present material model predicts the fracture depth with acceptable accuracy. The associated error is only approximately 5%.

References

1. Emmens, W. C., Sebastiani, G., & van den Boogaard, A. H. (2010). The technology of incremental sheet forming—A brief review of the history. *Journal of Materials Processing Technology, 210*(8), 981–997.
2. Leszak, E. (1967). *Apparatus and process for incremental dieless forming*. United States Patent, Patent No. 3342051.
3. Iseki, H. (2001). An approximate deformation analysis and FEM analysis for the incremental bulging of sheet metal using a spherical roller. *Journal of Materials Processing Technology, 111*(1–3), 150–154.
4. Hagan, E., & Jeswiet, J. (2003). A review of conventional and modern single-point sheet metal forming methods. *Proceedings of the Institution of Mechanical Engineers, Part B: Journal of Engineering Manufacture, 217*(2), 213–225.

5. Jeswiet, J., Micari, F., Hirt, G., Bramley, A., Duflou, J., & Allwood, J. (2005). Asymmetric single point incremental forming of sheet metal. *CIRP Annals—Manufacturing Technology, 54*(2), 88–114.
6. Jackson, K., & Allwood, J. (2009). The mechanics of incremental sheet forming. *Journal of Materials Processing Technology, 209*(3), 1158–1174.
7. Fang, Y., Lu, B., Chen, J., Xu, D. K., & Ou, H. (2014). Analytical and experimental investigations on deformation mechanism and fracture behavior in single point incremental forming. *Journal of Materials Processing Technology, 214*(8), 1503–1515.
8. Liu, Z., Daniel, W. J. T., Li, Y., Liu, S., & Meehan, P. A. (2014). Multi-pass deformation design for incremental sheet forming: Analytical modeling, finite element analysis and experimental validation. *Journal of Materials Processing Technology, 214*(3), 620–634.
9. Hirt, G., Junk, S., & Witulski, N. (2002). Incremental sheet forming: Quality evaluation and process simulation. In *Proceeding of the 7th ICTP Conference* (pp. 925–930).
10. Reagan, J., & Smith, E. (1991). *Metal spinning*. Bradley, IL: Lindsay Publication.
11. Silva, M. B., Nielsen, P. S., Bay, N., & Martins, P. A. F. (2011). Failure mechanisms in single-point incremental forming of metals. *International Journal of Advanced Manufacturing Technology, 56*(9–12), 893–903.
12. Henrard, C. (2008). *Numerical simulations of the single point incremental forming process*. Ph.D., thesis, University of Liège, Belgium.
13. Ambrogio, G., Filice, L., & Micari, F. (2006). A force measuring based strategy for failure prevention in incremental forming. *Journal of Materials Processing Technology, 177*(1–3), 413–416.
14. Malhotra, R., Xue, L., Belytschko, T., & Cao, J. (2012). Mechanics of fracture in single point incremental forming. *Journal of Materials Processing Technology, 212*(7), 1573–1590.
15. Xue, L. (2007). Damage accumulation and fracture initiation in uncracked ductile solids subject to triaxial loading. *International Journal of Solids and Structures, 44*(16), 5163–5181.
16. McClintock, F. A. (1968). A criterion for ductile fracture by the growth of holes. *Journal of Applied Mechanics, 35*(2), 363–371.
17. Rice, J. R., & Tracey, D. M. (1969). On the ductile enlargement of voids in triaxial stress fields∗. *Journal of the Mechanics and Physics of Solids, 17*(3), 201–217.
18. Gurson, A. L. (1977). Continuum theory of ductile rupture by void nucleation and growth: Part I—Yield criteria and flow rules for porous ductile media. *Journal of Engineering Materials and Technology, 99*(76), 2–15.
19. Tvergaard, V., & Needleman, A. (1984). Analysis of the cup-cone round tensile fracture. *Acta Metallurgica, 32*(I), 157–169.
20. Li, J., Li, S., Xie, Z., & Wang, W. (2015). Numerical simulation of incremental sheet forming based on GTN damage model. *The International Journal of Advanced Manufacturing Technology, 81,* 9–12.
21. Cao, T. S., Bobadilla, C., Montmitonnet, P., & Bouchard, P. O. (2015). A comparative study of three ductile damage approaches for fracture prediction in cold forming processes. *Journal of Materials Processing Technology, 216,* 385–404.
22. Kiran, R., & Khandelwal, K. (2014). Computational models for ductile fracture prediction in structural engineering applications. *Procedia Materials Science, 3,* 1947–1955.
23. Bjorklund, O. (2008). *Modelling of failure*. Master's thesis, Department of Management and Engineering, Linköping University, Sweden.
24. Sahu, G. N. (2015). *Development of finite element analysis methodology for incremental sheet forming process*. Master's thesis, PDPM Indian Institute of Information Technology, Design and Manufacturing, Jabalpur.
25. Johnson, G. R., & Cook, W. H. (1983). A constitutive model and data for metals subjected to large strains, high strain rates and high temperatures. In *7th International Symposium on Ballistics* (pp. 541–547).
26. Altair. (2013). *Altair RADIOSS 12.0.*

27. Wierzbicki, T., Bao, Y., Lee, Y. W., & Bai, Y. (2005). Calibration and evaluation of seven fracture models. *International Journal of Mechanical Sciences, 47*(4–5 SPEC. ISS.), 719–743.
28. Minford, J. D. (1993). *Handbook of aluminium bonding technology and data*. CRC Press.
29. Sahu, G. N., Saxena, S., Jain, P. K., Roy, J. J., Samal, M. K., & Tandon, P. (2015 November). Shell element formulation based finite element modeling, analysis and experimental validation of incremental sheet forming process. In *International Mechanical Engineering Congress and Exposition, American Society of Mechanical Engineers (ASME)* (pp. V02AT02A034–V02AT02A034).

Finite Element Analysis of V-Shape Incremental Equal Channel Angular Pressing

Pintu Kumar and Sudhansu Sekhar Panda

1 Introduction

The lightweight high-strength material in existing technology is a product of alloying or heat treatment process. The current technology required improvement of the specific strength of these metals without using any costly additive as well as applying energy consuming heat treatment processes. Reducing the size of grains from micro- to nano-level when coarse-grained (CG) subjected to SPD processes and consolidation of nanoparticles are method of achieving the entire requirements [1].

SPD is a part of metal forming processes which contemplated as pre-eminent method to refines the size of grain in general engineering materials up to nano-level and induces higher plastic strain [2]. Unfortunately, the method to produce UFG through SPD processes is not matched by equally splendid engineering attempt to explore the method at higher scale due to which the industrial utilization still under consideration and limited to production of a few smaller components [3]. Although many SPD processes have been developed, most of them fail to produce continuous and required properties of the billet as demanded by the manufacturing industries [3, 4]. Very successful early implementations of SPD method is an ECAP as developed by Segal in 1970 [5, 6]. ECAP is most popular SPD method in which square billet as well as cylindrical billet pushed from constant profile inlet channel to exit channel turned at different abrupt angles as shown in Fig. 1.

P. Kumar · S. S. Panda (✉)
Indian Institute of Technology Patna, Patna, India
e-mail: sspanda@iitp.ac.in

P. Kumar
e-mail: pintu.pme13@iitp.ac.in

U. S. Dixit and R. Kant (eds.), *Simulations for Design and Manufacturing*,
Lecture Notes on Multidisciplinary Industrial Engineering,
https://doi.org/10.1007/978-981-10-8518-5_7

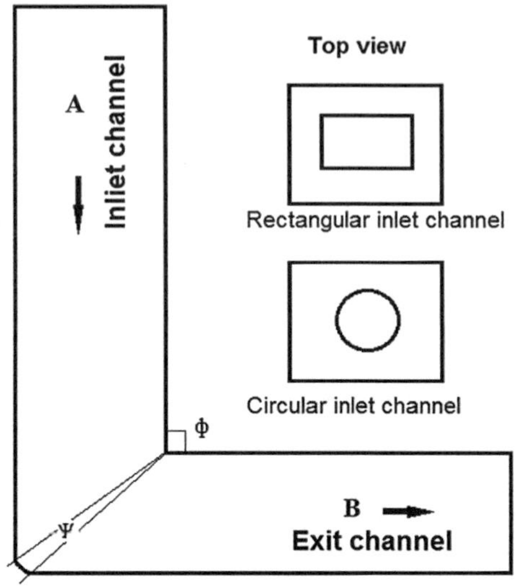

Fig. 1 Schematic of L-shape ECAP channel with rectangular and circular inlet section

Simple machine kinematic and less deformation load makes ECAP as most popular process for laboratory purpose but is limited to industrial applications. Poor material utilization at the end effects and material losses between each pass are major disadvantages due to which productivity reduces. However, among all possible routes of rotation, route C rotation is popular for low wastage of material [7]. Implementation of a two-turn incremental die channel follows route C rotation in which materials effective strain get double in a single press and therefore improves the productivity [8–11]. Route A rotation is another way to minimize the wastage of material before moving to next pass or by using the multi-bend channel. Route A can also be implemented in three-turn, V-shape ECAP channel where plane orientation and strain are aligned towards the direction of extrusion. Hence, the present aim focused to observe the improvement of grain refinement (effective strain) using V-shape ECAP in a single pass through finite element simulation.

2 ECAP Die Design

In L-shape ECAP die design, the channel is bend at an abrupt angle known as channel angle (φ) (lies between 0° and 180°) and corner angle (ψ) (varies from 0° to 20°) as shown in Fig. 1 [5, 12, 13]. '*A*' and '*B*' are defined as inlet and exit length of one-turn ECAP channel. The angle at which both channels are intersecting is known as channel angle (ϕ) and arc make an angle by outer surface of both channels is known as corner angle (ψ) [5]. A reciprocating punch is used to compress the billet. Billet feed inside the die inlet channel with minor clearance between channel and billet. By application of continuous punch load, billet deformed inside the inlet channel moves forward with its plane shearing where consecutive plane shear overlap and results uniform strain distribution along the billet [14, 15]. L-shape ECAP is still an open area for continuous processing of long billet. Relation used for maximum effective strain (ε) after each turn is given in Eq. (1) [14]. For one-turn die, maximum effective strain distribution in L-shape ECAP channel at 90° channel angles and 0° corner angles is 1.15.

$$\varepsilon = \frac{N}{\sqrt{3}}\left\{2\cot\left(\frac{\phi+\psi}{2}\right) + \psi\operatorname{cosec}\left(\frac{\phi+\psi}{2}\right)\right\} \quad (1)$$

Depending upon the type of materials, effective strain distribution in L-shape channel increases with increase in number of turns (N). Numerous operations involved to achieve large plastic strain when billet undergoes different routes of rotation such as route A, B_A, B_C and C as shown in Fig. 2a–d.

Method of reusing extruded material is done by these four different routes of rotation along its longitudinal directions (X) as shown in Fig. 3 [15]. In route A, the billet is repressed without rotating in any direction. Figure 2a shows plane observed along the direction of extrusion (X) is repressed inside an inlet channel without rotation of shear plane. In route B_A, billet is rotated at 90° along longitudinal direction either in a clockwise or in anticlockwise, and rotation for next pass continued with following previous rotation as shown in Fig. 2b. For route B_c as shown in Fig. 2c, billet is rotated at 90° either in a clockwise or in anticlockwise and rotation for next pass continued conflicting the previous rotation. Grain refinement of route B_c processed billet is much better as compare to other routes of rotation. In route C, any of the direction is valid when the sample is rotated at 180°

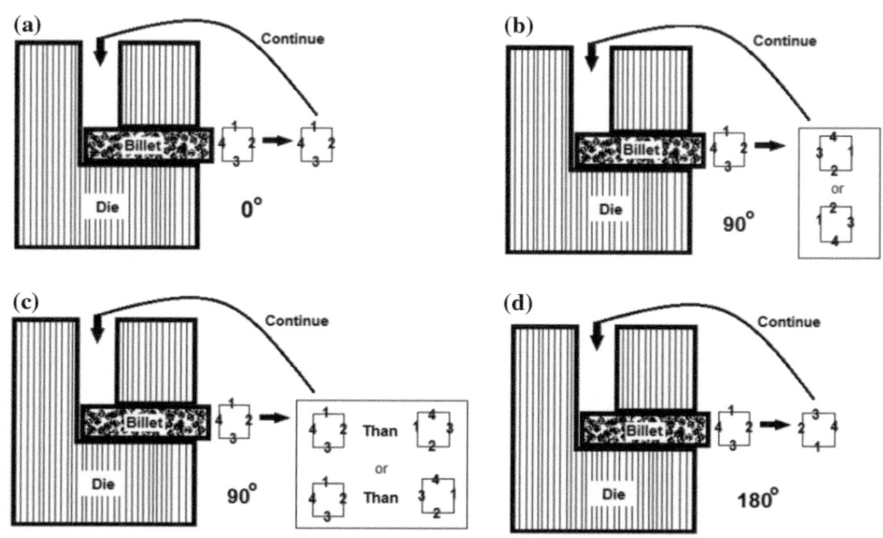

Fig. 2 Schematic for different route of rotation **a** A, **b** B_A, **c** B_C and **d** C

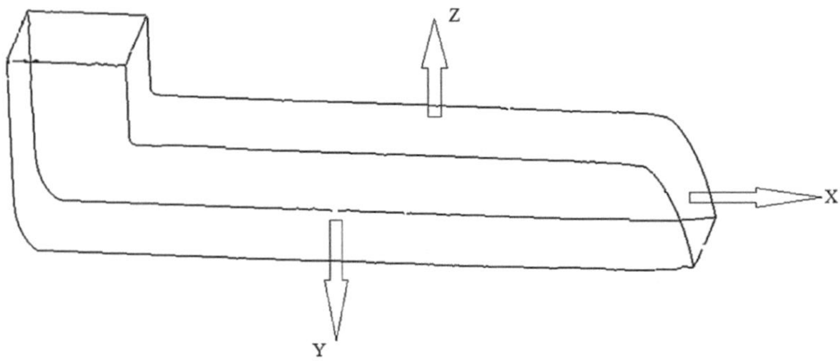

Fig. 3 Direction of processed ECAP billet

for each pass. It is noticed that the billet following route C rotation in L-shape ECAP takes more time with minor loss of material in each number of pass.

Route *C* rotation through S-shape ECAP requires higher deformation load but improves the material mechanical properties as shown in Fig. 4. Requirement of higher deformation load is reduced with increasing the channel angle for a fixed channel section. In designing the S-shape ECAP, inlet channel and exit channel are offset to each other. Fillet given at each die turn is termed as inner die radius and outer die radius. Requirement of higher deformation load is also reduced using three-turn, V-shape ECAP channel as shown in Fig. 5. Three number of turn in V-shape ECAP is possible with two offset lengths. Minimum offset between inlet and exit channel for square and rectangular channel will be greater than summation of half of width of first channel, half of width of second channel and required wall thickness between two channels. Similarly, for circular channel section, it will be greater than summation of radius of first channel, radius of second channel and required wall thickness between two channels. Channel angle at each turn may be similar or can be changed based on the type of analysis. Separation of punch load eliminates friction between die and billet during feeding stroke, this enables billet to reuse it for next pass with or without following different route of rotation.

3 FE Model

Earlier work [11] presented for S-shape two-turn ECAP channel to the elastic/plastic flow of material is validated with commercial FEA-based DEFORM 3D software and it is further extended to three-turn, V-shape ECAP channel. Working procedure of FEA-based DEFORM 3D software is shown in Appendix 1. The design of die with a punch is assumed to be rigid throughout the deformation. The cross section of a square channel is 8×8 mm^2. Length of inlet channel and exit channel in S-shape ECAP is 56 and 40 mm long while both channels are 56 mm long for V-shape ECAP. In S-shape ECAP, channel is offset by 16 mm whereas same distance is maintained between each die turn of three-turn, V-shape

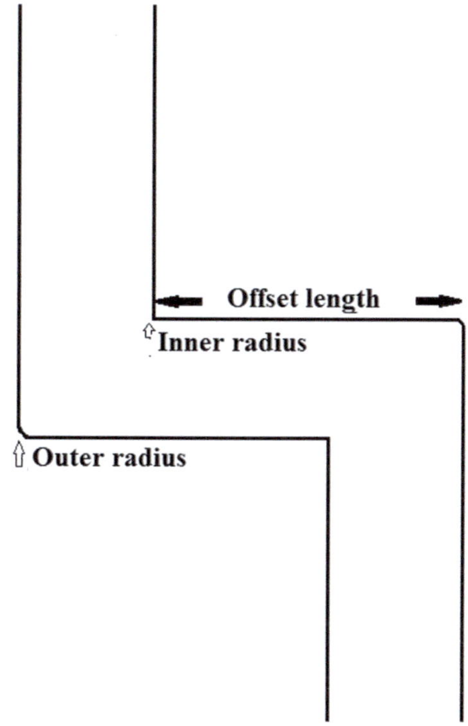

Fig. 4 Two-turn ECAP channel

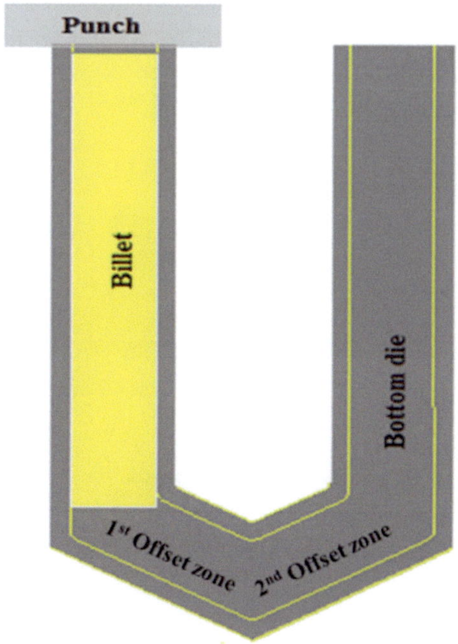

Fig. 5 V-shape ECAP channel

ECAP. The channel angle for each turn of S-shape ECAP is 90° and for three-turn V-shape ECAP is at 120°. The outer radius and inner radius of channel for both the case are 1 and 1.5 mm, respectively. Constant mesh ratio of $56 \times 8 \times 8\ \text{mm}^3$ long billet is divided into 10,000 tetrahedral mesh elements and used to detect the flow behaviour of billet during simulation without approaching to re-meshing criteria. Punch speed of 1 mm/s is given to deform the billet. The flow rule for pure aluminium with elastic–plastic, isotropic, Huber/Mises material is considered during simulation as given in Eq. (2). For S-shape ECAP authors [11] have assumed that coefficient of friction (μ) between die and billet is 0.05 and same is taken for V-shape ECAP.

$$\sigma = 159(0.02 + \varepsilon)^{0.27} \tag{2}$$

3.1 FEA Validation of S-Shape ECAP

Figure 4 shows the movement of billet when it passes through S-shape incremental ECAP channel. The process illustrates the flow patterns of mesh element with varying effective strain. Maximum effective strain at the end of first turn is found to be 1.5 mm/mm and in second turn it reaches to 2.8 mm/mm for 16 mm offset length. Variation of this effective strain can also be seen by plotting the contours line over transverse direction of billet. Two different sections have been used by author [11] to see the variation in effective strain as shown in Fig. 6. Strain variations along the transverse direction of a billet selected in offset zone and exit channel with Sects. 1 and 2 are shown in Fig. 7. A comparison of strain variation along the transverse direction as proposed by [11] is compared with our present results and shown in Fig. 7.

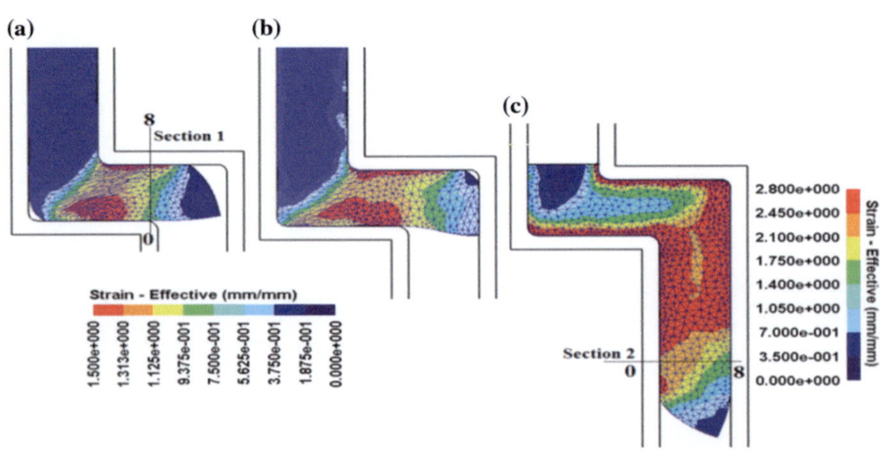

Fig. 6 Effective strain along longitudinal direction in S-shape ECAP

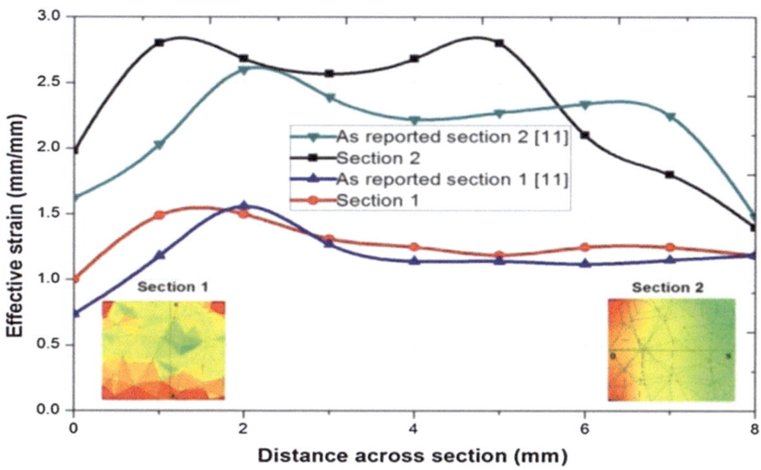

Fig. 7 Effective strain in transverse section of S-shape ECAP channel

Length of plateau through which highest effective strains are traced out is equal to width of billet (0–8 mm). Average effective strain for Sects. 1 and 2 are 1.15 and 2.3 mm/mm where both sides of billet have smaller effective strain as reported in Fig. 6 [11]. In the present work, a similar pattern of strain along plateau is observed, where average effective strain for Sects. 1 and 2 are 1.26 and 2.31 mm/mm which is at higher side and close deviation up to 8%. Similarly, maximum effective strain observed in the present case is 1.5 and 2.8 mm/mm in Sects. 1 and 2, respectively.

Figure 6a–c show that corner gap at first die turn is not occupied by the deformed billet when billet enters into offset zone but it is gradually filled up when billet enters into second die turn resulting in the increase of deformation load from 16 to 37 kN. In comparing the observed punch load, it is found that maximum load in present case matches with the work reported by Rosochowski et al. [11] as shown in Fig. 8. From Fig. 8, it is also found that punch load in the present case is shifted by 1.5 mm as compared to reported work [11] at the start of deformation but similar pattern punch load is observed throughout the stroke length. The maximum stress by the punch is the ratio of maximum load to the cross section of billet and same is transmitted to bottom die. Hence, the maximum stress sustained by bottom die is 600 MPa and same is matching with reported work [11].

3.2 *FEA of V-Shape ECAP*

As observed from the previous section that adopted governing equations as well as boundary conditions evaluate the results which more closely agree with the reported work. Hence, the similar situation is used in a different die with an intention to achieve similar output with less deformation load.

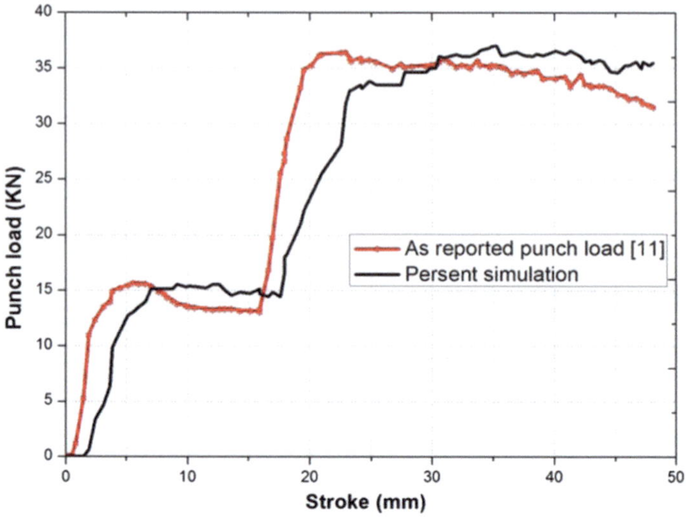

Fig. 8 Punch load versus stroke length in two-turn, S-shape ECAP channel

Figure 9a–i shows the flow of billet through V-shape incremental ECAP channel. Process illustrates the flow patterns of mesh element with varying effective strain. Maximum effective strain distribution at the end of first turn, second turn and third turn are observed as 0.856, 1.624 and 2.5 mm/mm, respectively. Figure 9b–i shows that corner gap at first die turn is not occupied by the deformed billet when billet enters into first offset zone but it is gradually filled up when billet enters into second offset zone resulting in the increase of deformation load. Region near to the tip of the billet is neither sheared nor deformed and hence strain in that region is almost zero as observed from Fig. 9i. It is also observed that material flow near the tip adheres to outer die wall in first as well as second offset zone, but when billet enters into the exit channel, material flow direction abruptly changes and adheres to the inner die wall.

Variation in this effective strain can also be seen by plotting the contours line normal to the direction of movement of billet. Three different sections have been used to see these variations as shown in Fig. 10. Selection of section in each turn of V-shape ECAP channel is same as S-shape ECAP channel. Sections 1, 2 and 3 are selected in first offset zone, second offset zone and in exit channel of die. Length of plateau through which highest effective strains are traced out is equal to width of billet. Maximum effective strain of 0.852 mm/mm is observed in first offset zone which gets doubled to 1.651 mm/mm in second offset zone and tripled to 2.503 mm/mm in the exit channel of die. Average effective strain for all these three-turns over length of plateau are 0.656, 1.314 and 1.969 mm/mm, respectively. It is also observed that both sides of the billet in contact with die wall have lower effective strain. Maximum effective strain for S-shape die is skewed towards left side whereas in present case; it is skewed towards the right side. This happened because of sheared plane are aligned in one direction. The maximum stress by the punch is the ratio of

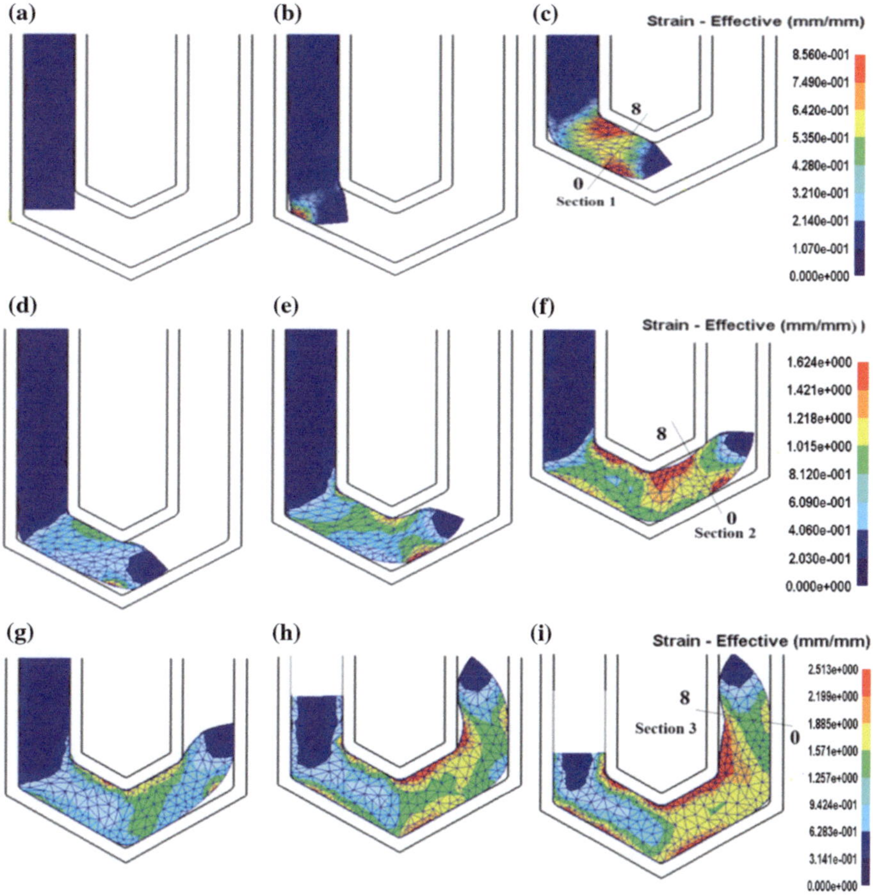

Fig. 9 Development of meshed billet and effective strain in V-shape ECAP (a-i)

maximum load to the cross section of billet and same is transmitted to the bottom die. Hence, the maximum stress sustained by the bottom die is 438 MPa which is much lower as compared to S-shape die. So, V-shape ECAP channel is best suitable to produce highly grain refined material at lower die stress.

4 Result and Discussion

4.1 Effective Strain

Distribution of effective strain of each processed materials with change in frictional coefficient between the billet and die channel is shown in Fig. 11a–j. Following observations are observed during the simulation through V-shape incremental ECAP channel.

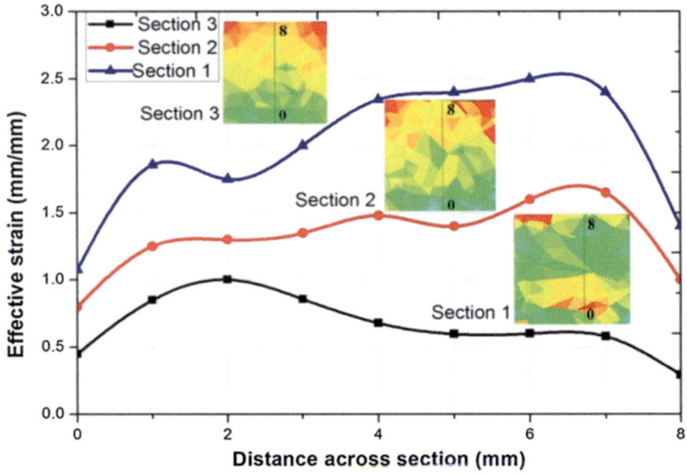

Fig. 10 Effective strain along the width of billet distribution in V-shape ECAP channel

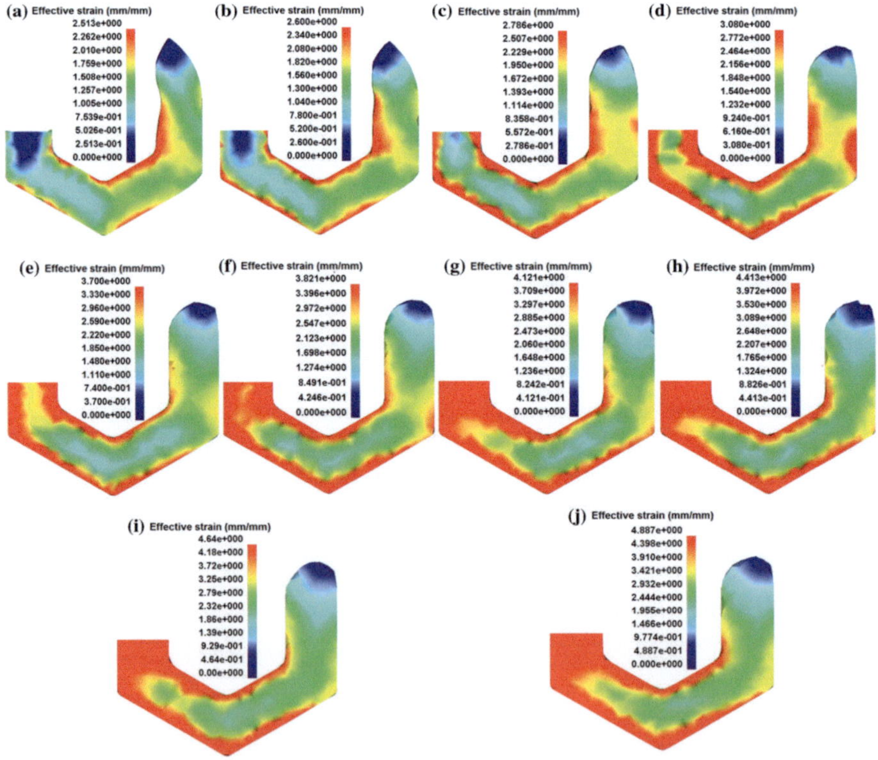

Fig. 11 Flow of effective strain with friction coefficient of **a** 0.05, **b** 0.1, **c** 0.15, **d** 0.2, **e** 0.25, **f** 0.3, **g** 0.35, **h** 0.4, **i** 0.45 and **j** 0.5

Throughout the simulation, it is observed that non-sheared part of billet (tip) has very low effective plastic strain due to compression as well as friction. A plane of billet continuously sheared in each die turns thus making it strain hardened. The first die turn, second die turn and third die turn amounts of effective strain observed are 0.58, 1.62 and 2.5 mm/mm, respectively, when the coefficient of friction is 0.05 as shown in Fig. 9. Strain developed in case of V-shape ECAP channel is lower as compared to S-shape ECAP channel. This happened due to route a rotation of billet in V-shape ECAP channel. In addition to deformation of billet, effects of friction coefficients are also shown in Fig. 11a–j. Distribution of effective strain is high for higher coefficient of friction due to grain dislocation. Increase in effective strain for higher coefficient of friction (0.5) is 4.8 mm/mm. During deformation of billet, it is observed that increasing coefficient of friction increases contact surfaces between the billet and die channel which restrict the free flow of material resulting in higher effective strain observed at punch and billet contact surfaces.

4.2 Equivalent Stress

Distribution of equivalent stress of each processed material with change in frictional coefficient between the billet and die channel is shown in Fig. 12. During deformation, it has been observed that undeformed tip of billet has low equivalent stress resulted due to compressive load when billet enters into each die turns. Planes of billet continuously sheared resulting in higher shearing stress developed in first die, second die turn and in third die turn. Rise of equivalent stress for each die turns are 155, 190 and 249 MPa, respectively, when coefficient of friction between the billet and die channel is 0.05. In three-turn V-shape ECAP die, billet after primary yielding at the first turn undergoes continuous loading till exit channel resulting in increase of equivalent stress along extrusion direction. In addition to deformation of billet, effects of friction coefficient are also shown in Fig. 12a–j. With increase in effective plastic strain, materials become strain harden and result in equivalent stress increase. Distribution of equivalent stress increases due to increase in coefficient of friction between a billet and die channel. Amount of equivalent stress observed at 0.5 coefficient of friction is 177, 265 and 354 MPa when billet crosses first, second and third turns, respectively, as shown in Fig. 12j. In the deformation of billet, it is found that increasing the coefficient of friction increases the contact surfaces between a billet and die channel which restricts the free flow of material resulting in higher equivalent stress observed at punch and billet contact area.

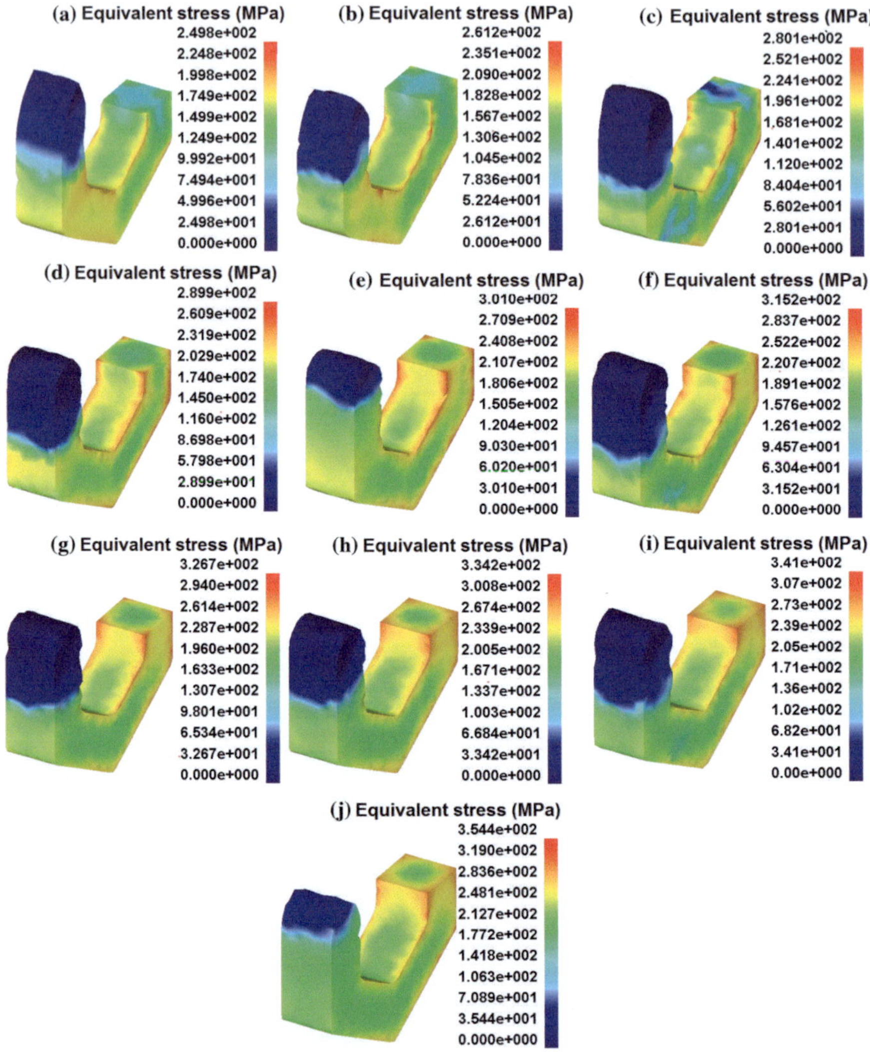

Fig. 12 Flow of effective strain with friction coefficient of **a** 0.05, **b** 0.1, **c** 0.15, **d** 0.2, **e** 0.25, **f** 0.3, **g** 0.35, **h** 0.4, **i** 0.45 and **j** 0.5

4.3 Punch Load

The billet undergoes deformation through V-shape ECAP; hence, load required by punch to deform the billet is shown in Fig. 13. The following observations are observed through V-shape incremental ECAP channel.

For inlet channel, punch load sharply increases with increase in punch stroke; thereafter, it becomes constant till it reaches to second die turn. Load required to cross the first die turn at coefficient of friction 0.05 is 6.5 kN which increased to 16 kN in second die turn but when the billet passes through the third die turn it gradually increases up to 27.5 kN as shown in Fig. 13. Continuously applied load causes grain dislocation of a sheared plane, due to which sharp loading changed into gradually loading as shown in Fig. 13. For a single press, in comparison to S-shape channel, three-turn V-shape ECAP required 27% lower punch load. Maximum 87 kN punch load is experienced to deform the billet when the coefficient of friction reached 0.5. Magnitude of punch load is higher at higher coefficient of friction. During deformation of billet, it is found that increase in contact surfaces of a billet and die channel with increasing coefficient of friction results in punch load increase as shown in Fig. 13.

4.4 *Loss of Volume*

ECAP is most popular SPD technique to enhance the mechanical properties of a billet (effective strain) without change in initial dimension as well as volume of the sample. Through simulation, it is observed that contact surface between a billet and the die walls plays an important role in determining the volume loss of billet. From Fig. 14, it is found that at low coefficient of friction, contact surface between a billet

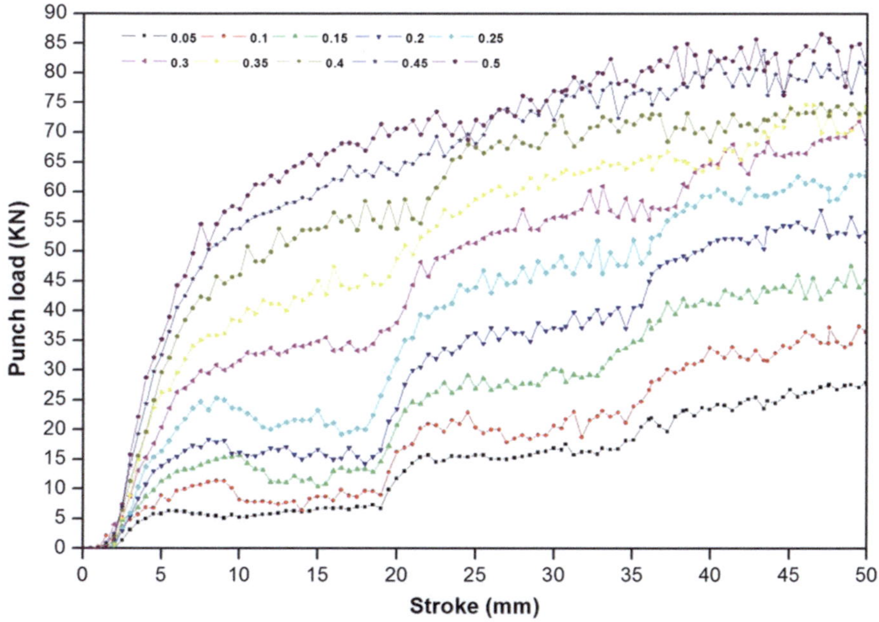

Fig. 13 Load experienced by punch at coefficients of friction: 0.05, 0.1, 0.15, 0.2, 0.25, 0.3, 0.35, 0.4, 0.45 and 0.5

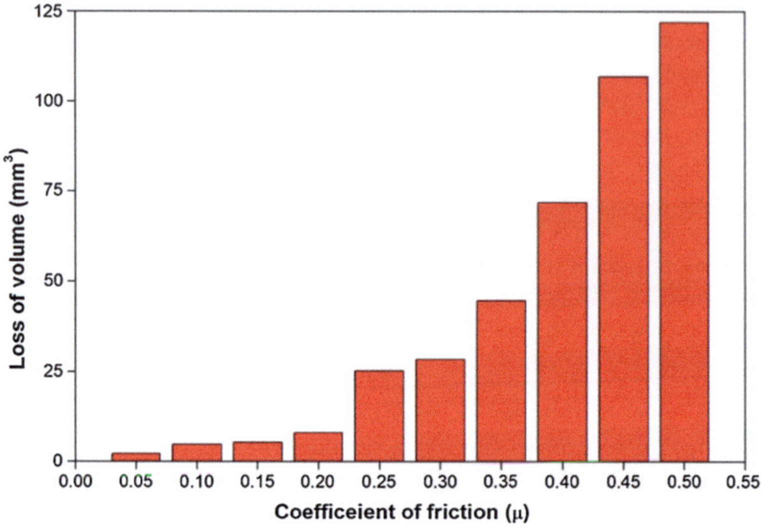

Fig. 14 Loss of volume at coefficients of friction 0.05, 0.1, 0.15, 0.2, 0.25, 0.3, 0.35, 0.4, 0.45 and 0.5

and die wall is very small due to which material flows easily inside the channel resulting in minimum volume loss. On the other hand, at higher coefficient of friction, the material flow inside the channel is restricted resulting in maximum volume losses.

5 Conclusions

The present work evaluates the load and effective strain for S-shape die and finds a close agreement of results with reported work. Hence, a similar situation is used in a V-shape die with an intention to achieve similar output with less deformation load. Method and design concept derived from results will provide a platform for industrial applications.

For S-shape ECAP

- Top surface of compressed billet in the offset zone has higher elemental density.
- Element orientations as well as its distribution in die inlet and exit channel are same.
- Distribution of effective plastic strain is more uniform when entered into an exit channel.
- Punch load sharply increases in die inlet channel, at first turn and at second turn.
- Punch load gradually increases in die offset channel and exit channel.

For V-shape ECAP

- Lower effective strain and equivalent stress are on tip of billet.
- Elemental orientation and its distribution are elongated along the direction of extrusion.
- In comparison to offset channel, developed equivalent strain is higher along the die corner.
- Effective strain is high for higher coefficient of friction.
- Back pressure increases with increasing coefficient of friction.
- Load required to deform the billet is comparatively 27% lower than load required for two-turn ECAP processes at constant coefficient of friction (0.05).
- Punch load sharply increases at inlet channel, first turn, second turn and third turn.
- Punch load gradually increases in first offset channel and second offset channel.
- Loss of volume is higher with increasing coefficient of friction.

Appendix 1

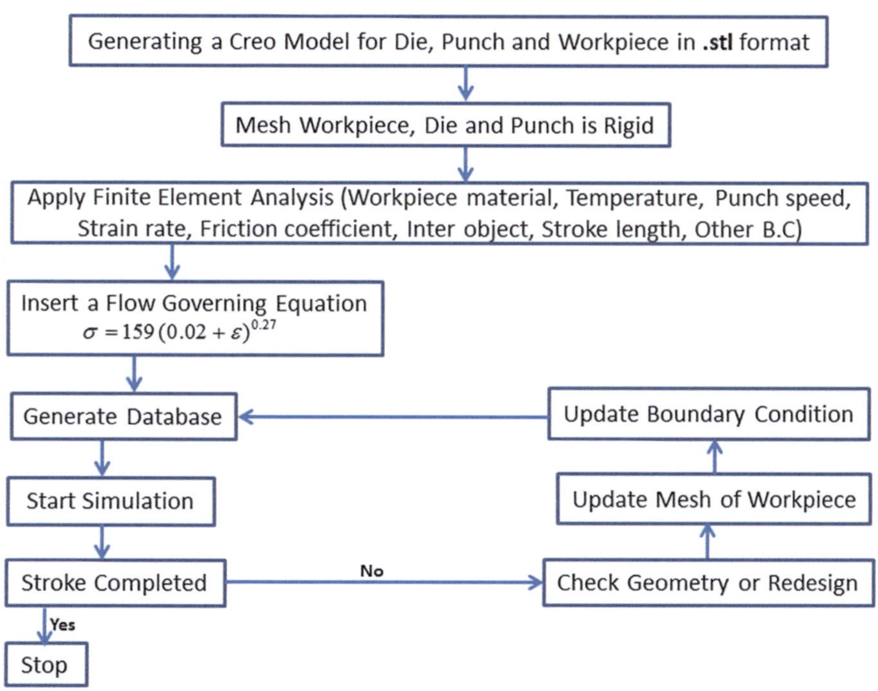

References

1. Yoon, S. C., Nghiep, D. M., Hong, S. I., Horita, Z., & Kim, H. S. (2007). Achieving both powder consolidation and grain refinement for bulk nanostructured materials by equal-channel angular pressing. *Key Engineering Materials, 345,* 173–176.
2. Valiev, R. Z., Korznikov, A. V., & Mulyukov, R. (1993). Structure and properties of ultrafine-grained materials produced by severe plastic deformation. *Materials Science and Engineering A, 168*(2), 141–148.
3. Azushima, A., Kopp, R., Korhonen, A., Yang, D. Y., Micari, F., Lahoti, G. D., et al. (2008). Severe plastic deformation (SPD) processes for metals. *CIRP Annals-Manufacturing Technology, 57*(2), 716–735.
4. Kumar, P., & Panda, S. S. (2014). A review on micro-extruded microstructure from ultra-fine grained and as cast. In *5th International & 26th All India Manufacturing Technology, Design and Research Conference (AIMTDR 2014)*, December 12th–14th, 2014, IIT Guwahati (pp. 1–6).
5. Segal, V. M. (1995). Materials processing by simple shear. *Materials Science and Engineering A, 197,* 157–164.
6. Segal, V. M. (1999). Equal channel angular extrusion: from macromechanics to structure formation. *Materials Science and Engineering A, 271*(1), 322–333.
7. Barber, R. E., Dudo, T., Yasskin, P. B., & Hartwig, K. T. (2004). Product yield of ECAE processed material. In *Ultrafine Grained Materials III, TMS Annual Meeting, North Carolina, U.S.A.* (pp. 667–672).
8. Kim, H. S. (2002). Finite element analysis of deformation behaviour of metals during equal channel multi-angular pressing. *Materials Science and Engineering A, 328,* 317–323.
9. Kumar, P., & Panda, S. S. (2015). Deformation validation and property analysis of pure Aluminium during two turn equal channel angular pressing. In *International Conference on Precision, Meso, Micro and Nano Engineering*, December 10–12, 2015 (pp. 1–5).
10. Aour, B., Zairi, F., Gloaguen, J. M., Nait-Abdelaziz, M., & Lefebvre, J. M. (2009). Finite element analysis of plastic strain distribution in multi-pass ECAE process of high density polyethylene. *Journal of Manufacturing Science and Engineering, 131*(3), 1–11.
11. Rosochowski, A., & Olejnik, L. (2002). Numerical and physical modelling of plastic deformation in 2-turn equal channel angular extrusion. *Journal of Materials Processing Technology, 125,* 309–316.
12. Langdon, T. G., Furukawa, M., Nemoto, M., & Horita, Z. (2000). Using equal-channel angular pressing for refining grain size. *JOM Journal of the Minerals Metals and Materials Society, 52*(4), 30–33.
13. Horita, Z., Fujinami, T., Nemoto, N., & Langdon, T. G. (2000). Equal channel angular pressing of commercial aluminum alloys: Grain refinement, thermal stability and tensile properties. *Metallurgical and Materials Transactions A: Physical Metallurgy and Materials Science, 31,* 691–701.
14. Segal, V. M. (1995). Materials processing by simple shear. *Materials Science and Engineering A, 197*(2), 157–164.
15. Furukawa, M., Iwahashi, Y., Horita, Z., Nemoto, M., & Langdon, T. G. (1998). The shearing characteristics associated with equal-channel angular pressing. *Materials Science and Engineering A, 257,* 328–332.

Dynamic Analysis of Robotic Manipulator for Incremental Sheet Metal Forming

Swagatika Mohanty, Srinivasa Prakash Regalla and Y. V. Daseswara Rao

1 Introduction

Robotic manipulator has been widely used in manufacturing industries for flexible manufacturing capabilities. Stamping involves die and punch to get the desired shape. So stamping can be advantageous for production part. As stamping is a part dependent process, to reduce the initial production cost of the product in proto and new product introduction phase, for past few years incremental sheet metal forming (ISF) has been extensively used. Generally, ISF is executed using computer numerical control (CNC) machine or any robotic manipulator. The ISF process mainly involves a forming tool, a blank, a backing plate a fixture and a blank holder. Here the blank is firmly fixed clamped on the fixture and backing plate. Tool path for the forming tool can be programmed and can be controlled by a CNC milling machine. In this process, it is easier to form complex shaped parts using a suitable forming tool path strategy. The tool incrementally deforms the material till the desired shape is achieved.

ISF can be of two types, single-point incremental forming (SPIF) and two-point incremental forming (TPIF). In single-point incremental forming, single tool is utilised to form the sheet incrementally to get the desired shape. TPIF process involves either two tools, or a partial die and single forming tool. But to make the

---

S. Mohanty (✉) · S. P. Regalla · Y. V. Daseswara Rao
Department of Mechanical Engineering, Birla Institute of Technology and Science, Pilani Hyderabad Campus, Hyderabad 500078, India
e-mail: swagatika.bits@gmail.com

S. P. Regalla
e-mail: regalla@hyderabad.bits-pilani.ac.in

Y. V. Daseswara Rao
e-mail: yvdrao@hyderabad.bits-pilani.ac.in

U. S. Dixit and R. Kant (eds.), *Simulations for Design and Manufacturing*, Lecture Notes on Multidisciplinary Industrial Engineering,
https://doi.org/10.1007/978-981-10-8518-5_8

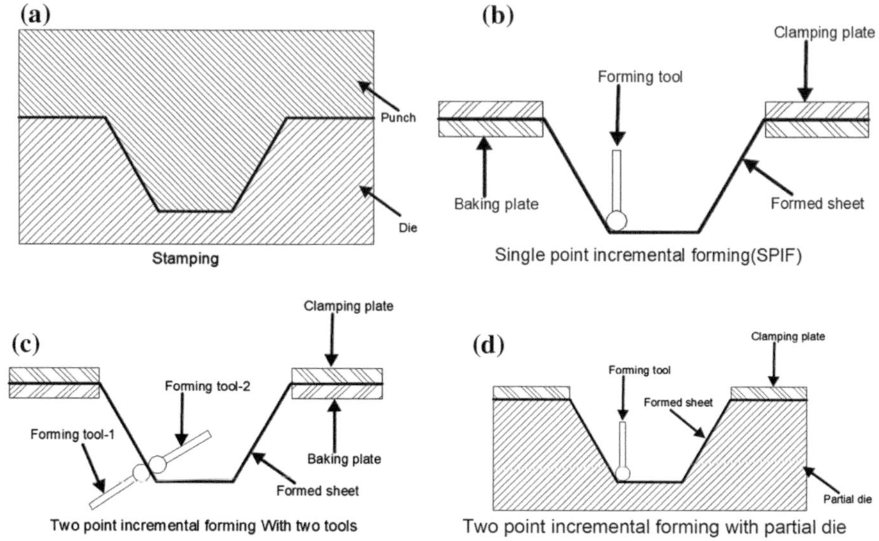

Fig. 1 Stamping (**a**), single-point incremental forming (**b**), two-point incremental forming by two tools (**c**) and two-point incremental forming with tool and partial die (**d**)

process part independent, partial die has been eliminated. Figure 1 shows the schematic diagram of stamping, TPIF and SPIF.

Mode of deformation in ISF can be due to shearing [1], while some believe that it is due to stretching [2]. The accuracy of the part formed in ISF is inferior to that in stamping due to a significant amount of springback. Non-uniform thickness distribution, wall angle limitation are the major disadvantages in ISF. Steeper wall angles are also difficult to form using conventional ISF though the recently proposed multistage forming has begun to yield some results. There are various process parameters which affect accuracy, surface quality and formability in ISF [3]. To make ISF process more flexible in terms of uniform thickness distribution, robotic manipulators have been used in previous research. Forming tool was attached to multi-degree-of-freedom end effector of the robot to form complex parts. In ISF tool can move in single direction but when the tool is controlled by a robotic manipulator, it can move in multiple directions which helps in forming complex parts.

A four-degree-of-freedom robot with C-frame support equipped by a linear actuator has been used for two-point incremental forming for achieving uniform thickness distribution [4]. 5-axis robots, anthropomorphic robots have also been used for forming complex parts [5, 6]. But tool path planning in robots has been a difficult task in roboforming. DELMIA software was used for designing, simulating tool path for forming tool attached to a KUKA robot in incremental forming [7]. While using robotic manipulator, mechanical specifications such as maximum load carrying capacity, velocity and acceleration of the moving links and joints have to be measured. When robot is used for sheet metal forming, dynamic analysis plays a major role. Methods to obtain dynamic behaviour of the robot and motion equation mainly done by either of the two methods, Newton–Euler (NE) and Euler–Lagrange (EL) formulation. The NE formulation is the most efficient method and based on free body diagram of each body part [8]. Dynamic analysis of any multibody robotic manipulator requires kinematic analysis. Optimal dynamic balancing was formulated by minimization of the RMS (root mean square) value of the input torque of two-degrees-of-freedom (2-DOF) serial manipulator and results were simulated using ADAMS software [9].

In a 2-DOF manipulator, the dynamic parameters were estimated and validated through simulation [10]. Neuro-fuzzy controller can be used for position control proposed for two-degrees-of-freedom (2-DOF) serial manipulator [11]. Minimisation of joint reaction force by optimal selection of angular rotation and by attaching counterweights to the link in a 2-DOF manipulator was done to improve the dynamic performance of the manipulator [12]. By using optimal redistribution of link masses, driving torques of the 2-DOF serial manipulator could be minimised [13]. Feature-based computer-aided design and computer-aided engineering (CAD-CAE) methodology was used to do static and dynamic analysis of parallel kinematic machine [14]. Inverse kinematics (position, velocity, acceleration) and rigid body dynamics using virtual work concept and Jacobean matrices for three rotational degrees of parallel manipulator were done in decoupled way [15]. Dynamic analysis of parallel manipulators with flexible platform was done using inverse dynamics [16].

In the present work, the tool has been attached to three axes CNC milling machine and part has been manipulated using a two-degree-of-freedom manipulator. The manipulator design involves prior analysis to maintain the stiffness as well as for long-run use.

Figure 2 shows the block diagram giving the steps involved in designing a manipulator. The first step is to do trajectory analysis. In the ISF, spiral tool path gives more formability, better surface quality. Therefore, in the present work the tool path followed is spiral tool path, which can be achieved by combining, 3-axis CNC milling machine and a manipulator. Position analysis is an important part of the robot analysis, as this will identify the joint angle movement and link translation.

As the trajectory to be followed is known, link length, position and joint angle can be determined by inverse kinematics. It is also necessary to conduct the velocity and acceleration analysis of each link and joint of the manipulator. To find out the joint torque and behaviour of the manipulator when external force is applied on the effector, dynamic analysis is necessary. Kinematic and dynamic analysis of the manipulator involves various steps. A block diagram of the procedure is depicted in Fig. 2.

Fig. 2 Steps involved in kinematic and dynamic analysis

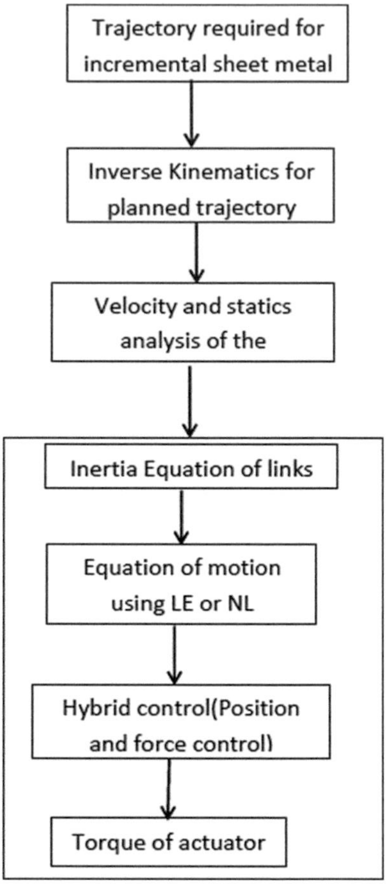

2 Trajectory Planning

In ISF, tool path has an influence on various parameters like force and roughness. In dynamic analysis, the external force needs to be given as input to obtain forces in different links and joints. To minimise the force experienced by tool due to sheet deformation and to maintain surface quality, tool path needs to be properly selected. Out of all path generation techniques used previously, the profile tool path and spiral tool path (Helical tool path) have given satisfactory results till date [17]. Spiral tool path was found to be better in strain and thickness distribution as well as maintaining good surface finish and forming force [18].

Continuous circular motion of the end effector in 2-DOF manipulator and vertical movement of the tool attached to 3-axis CNC milling machine, when integrated give spiral tool path. Joint and link for additional 2-DOF has to be selected in such a way that end effector's output path should be circular. Revolute and prismatic joint were selected for trajectory analysis. Roboanalyzer [19] has been used to

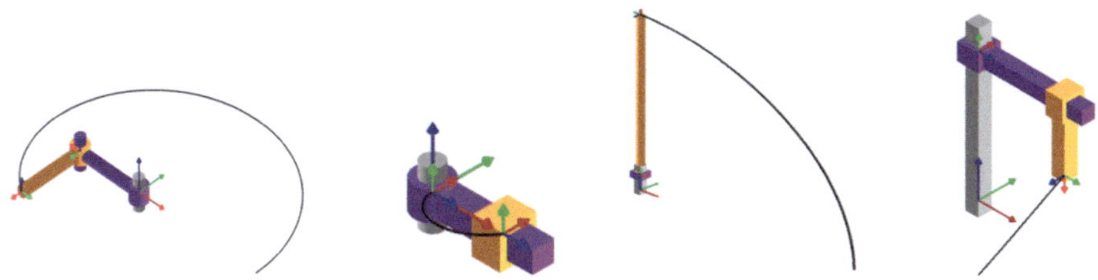

Fig. 3 Workspace for four most popular robotic configurations determined using Roboanalyzer software [19]

check the output of different combinations of joints. Figure 3 shows the roboanalyzer analysis output, in which RR combinations give a circular trajectory but workspace is more in this case. RR combinations with second joint axis perpendicular to first joint axis will give circular end effector path with lesser workspace.

3 Kinematic Analysis

Kinematic analysis involves the study of position, velocity and acceleration. The main aim of this section is to find out the optimum velocity and position of the link and joints to follow a specific trajectory. If the path to be followed by the end effector and number of joints and links to be used in manipulator are known to us, position orientation and velocity of the links and joints can be easily calculated using kinematic analysis. This can be analysed in two ways, one using inverse kinematics and other using forward kinematics. In forward kinematics, the joint angle in revolute joints and displacement of the link in prismatic joints are prescribed and using these parameters, end effector configuration is to be obtained. But in incremental sheet metal forming tool path trajectory is known and joint and link configuration need to be calculated to obtain the desired trajectory. For this, inverse kinematics has to be done. The forward kinematics problem helps in solving the end effector pose when complete set of joint variables are known. But the inverse kinematics problem helps in finding joint variables for known end effector pose. In incremental sheet metal forming, tool path trajectory is shown in Fig. 4. The manipulator to be used in incremental forming is a serial manipulator whose inverse kinematics and tool path planning have been done [20]. Both the kinematic analysis requires frame transformation of the model.

3.1 Frame Assignment of Reference Frame

Frame assignment of the link is the first step for developing a mathematical model for robot. In frame assignment, following basic rules need to be followed:

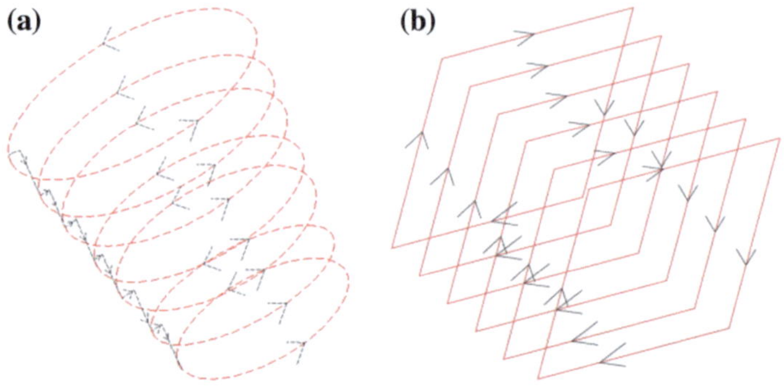

Fig. 4 Tool paths: tool path obtained for circular cone (**a**), and tool path obtained for square cup (**b**)

(a) Z vector is always on the joint axis direction.
(b) The X vector of the frame lies along the common perpendicular to Z_{i-1} and Z_i, where i indicates the ith joint.

Frame assignment for 2-DOF manipulator is shown in Fig. 5.

In a serial robot of N number of links, it has $N + 1$ number of serially mounted rigid bodies connected by N number of joints. These are numbered from 0 to N. The first rigid body is fixed to base and the end effector is connected to the Nth rigid body. Rigid bodies are also called as links and are actuated by motors (hydraulic/pneumatic/electrical) via transmission mechanism. Joints are numbered from 1 to N and the ith joint connects $(i - 1)$th and ith links. Position and orientation of the end effector can be controlled fully. Configuration of the manipulator can be described by joint angle or angular position vector. Figure 5 shows the frame transformation of the manipulator. In this case, first link is fixed to the base. So it

Fig. 5 Frame assignment

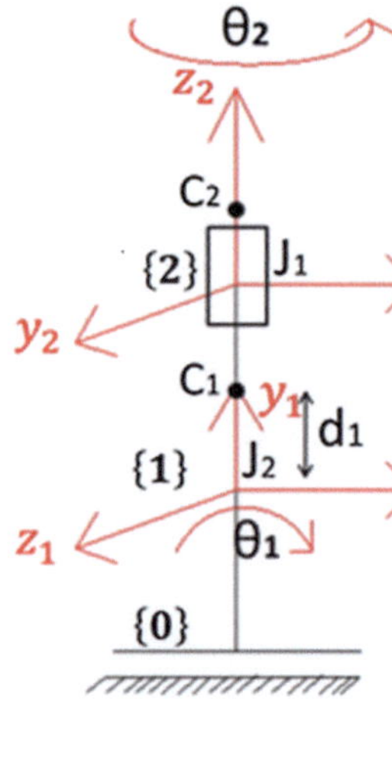

Table 1 Denavit–Hartenberg parameters

Link	d	a	α	θ	Type
1	0	0	0	0	Fixed
2	L_2	L_2	0	θ_1	Revolute
3	L_1	L_1	90	θ_2	Revolute

has been assigned to frame {0}. Second link has been connected to fixed first link by revolute joint and is assigned as frame {1}. First revolute joint is denoted as $J_{1.}$ End effector or the sheet metal holder is connected to first link by revolute joint $J_{2.}$ End effector is assigned as frame {2} as shown in Fig. 5. θ_1 is the angular rotation by J_1 or rotation of second link about first link. The mathematical description of the robotic manipulator usually represented as Denavit and Hartenberg (DH) parameters. The DH parameters include link offset (d), link length (a), link angle (α) and link twist (θ). The θ is a variable when joint is revolute and constant when joint is translational. When joint is revolute the parameter d is a constant whereas it is a variable when joint is prismatic. Here, d is the translation distance along z axis, θ is the rotation angle around z axis, a is the translational distance along x axis and α is the rotation angle around x axis. For above-mentioned frame, DH parameters are as shown in Table 1.

3.2 *Linear and Angular Velocities*

Position and orientation of the links and joints can be obtained by inverse kinematics. Linear velocity and angular velocity can be defined as time derivative of position vector and rotation matrix of rigid body. Linear velocity of an arbitrary point on rigid link undergoing translation and rotation with respect to base frame has to be calculated first. Using this linear velocity, both linear and angular velocity of the links and joints can be calculated as follows. Before calculating linear and angular velocities, homogeneous transformation matrices for all joints need to be obtained.

The transformation matrices of frames 1 and 2 with respect to reference frame are

T_j^i is transformation matrix of frame i with respect to j

$$T_0^1 = \begin{bmatrix} \cos\theta_1 & 0 & \sin\theta_1 \\ 0 & 1 & 0 \\ -\sin\theta_1 & 0 & \cos\theta_1 \end{bmatrix} \tag{1}$$

$$T_1^2 = \begin{bmatrix} \cos\theta_2 & -\sin\theta_2 & 0 \\ \sin\theta_2 & \cos\theta_2 & 0 \\ 0 & 0 & 1 \end{bmatrix} \tag{2}$$

$$T_0^2 = T_1^2 \times T_0^1$$
$$T_0^2 = \begin{bmatrix} \cos\theta_2 & -\sin\theta_2 & 0 \\ \sin\theta_2 & \cos\theta_2 & 0 \\ 0 & 0 & 1 \end{bmatrix} \begin{bmatrix} \cos\theta_1 & 0 & \sin\theta_1 \\ 0 & 1 & 0 \\ -\sin\theta_1 & 0 & \cos\theta_1 \end{bmatrix} \quad (3)$$
$$T_0^2 = \begin{bmatrix} \cos\theta_1\cos\theta_2 & -\cos\theta_1\sin\theta_2 & \sin\theta_1 \\ \sin\theta_2 & \cos\theta_2 & 0 \\ -\sin\theta_1\cos\theta_2 & -\sin\theta_1\sin\theta_2 & \cos\theta_1 \end{bmatrix}$$

Using Eqs. (1), (2) and (3), the linear velocity can be calculated as follows:

$$V_i^i = {}_{i-1}^{i}T(V_{i-1}^{i-1} + \omega_{i-1}^{i-1} \times O_i^{i-1})$$

where V_j^i is velocity of link i in frame j.

Angular velocity can be calculated as follows:

$$\omega_i^i = {}_{i-1}^{i}T\,\omega_{i-1}^{i-1} + \dot{\theta}_i(0 \quad 0 \quad 1)^T$$

For fixed frame {0},
$i = 0$

$$V_0^0 = 0 \quad (4)$$

$$\omega_0^0 = 0 \quad (5)$$

For frame {1},
$i = 1$

$$V_1^1 = {}_0^1T(V_0^0 + \omega_0^0 \times O_1^0) = 0 \quad (6)$$

$$\omega_1^1 = {}_{i-1}^{i}T\omega_{i-1}^{i-1} + \dot{\theta}_i(0 \quad 0 \quad 1)^T = \begin{pmatrix} 0 & 0 & \dot{\theta}_1 \end{pmatrix}^T \quad (7)$$

For frame {2},
$i = 2$

$$V_2^2 = {}_1^2T(V_1^1 + \omega_1^1 \times O_2^1) = \begin{pmatrix} -\sin\theta_2 l_1\dot{\theta}_1 \\ \cos\theta_2 l_1\dot{\theta}_1 \\ 0 \end{pmatrix} \quad (8)$$

$$\omega_2^2 = {}_1^2T\omega_1^1 + \dot{\theta}_2(0 \quad 0 \quad 1)^T = \begin{pmatrix} 0 & 0 & \dot{\theta}_1 + \dot{\theta}_2 \end{pmatrix}^T \quad (9)$$

Linear and angular velocities of the end effector are given by

$$\omega_E^E = \begin{pmatrix} 0 & 0 & \dot{\theta}_1 + \dot{\theta}_2 \end{pmatrix}^T$$

$$V_E^E = \begin{pmatrix} -\sin\theta_2 l_1 \dot{\theta}_1 \\ \cos\theta_2 l_1 \dot{\theta}_1 \\ 0 \end{pmatrix}$$

The linear velocity of the end effector in the fixed coordinate system can be calculated by multiplying the rotation matrix with V_E^E. Angular velocity will be same as there is no joint connection in the end effector.

ω_j^i is angular velocity of link j with respect to frame i.

$$\omega_E^0 = \begin{pmatrix} 0 & 0 & \dot{\theta}_1 + \dot{\theta}_2 \end{pmatrix}^T \tag{10}$$

$$V_E^0 = \begin{bmatrix} \cos\theta_1 \cos\theta_2 & -\cos\theta_1 \sin\theta_2 & \sin\theta_1 \\ \sin\theta_2 & \cos\theta_2 & 0 \\ -\sin\theta_1 \cos\theta_2 & -\sin\theta_1 \sin\theta_2 & \cos\theta_1 \end{bmatrix} \times \begin{pmatrix} -\sin\theta_2 l_1 \dot{\theta}_1 \\ (\cos\theta_2 l_1 + l_2)\dot{\theta}_1 \\ 0 \end{pmatrix}$$

$$= \begin{bmatrix} -\cos\theta_2 \sin\theta_2 \dot{\theta}_1 (\cos\theta_1 l_1 + l_2) \\ \dot{\theta}_1 \left[l_1 (\cos\theta_2 - \sin^2\theta_2) + \cos\theta_2 l_2 \right] \\ -\sin\theta_1 \sin\theta_2 l_2 \dot{\theta}_1 \end{bmatrix} \tag{11}$$

$$V_E^0 = \begin{bmatrix} -\cos\theta_2 \sin\theta_2 \dot{\theta}_1 (\cos\theta_1 l_1 + l_2) \\ \dot{\theta}_1 \left[l_1 (\cos\theta_2 - \sin^2\theta_2) + \cos\theta_2 l_2 \right] \\ -\sin\theta_1 \sin\theta_2 l_2 \dot{\theta}_1 \end{bmatrix}$$

The velocity and acceleration of the end effector by quasi-static analysis have been measured in CREO parametric, and the results are shown in Figs. 6 and 7.

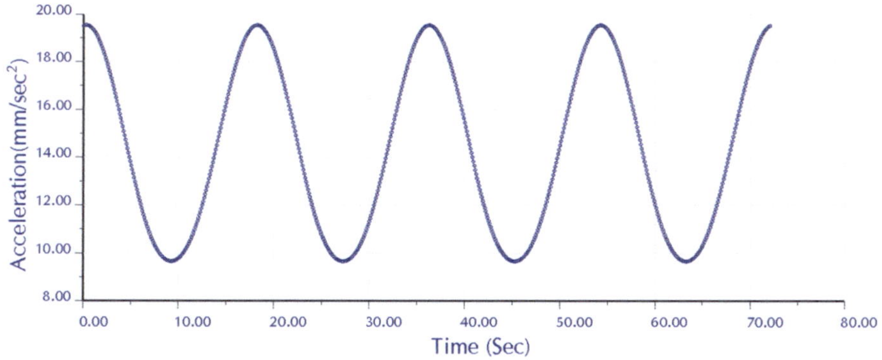

Fig. 6 Acceleration analysis

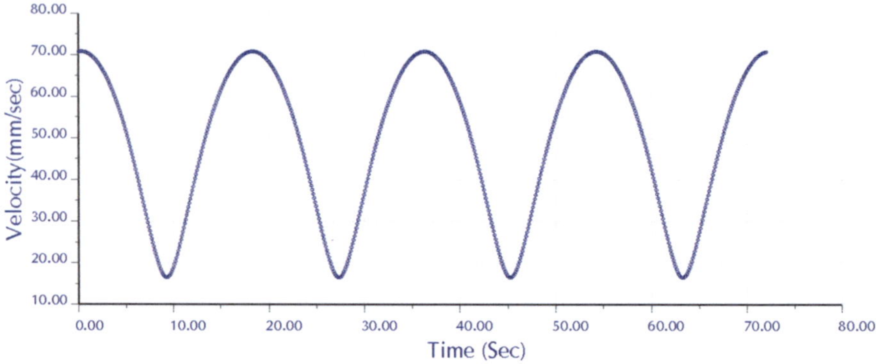

Fig. 7 Velocity analysis

4 Dynamic Analysis

The objective of this section is to analyse the behaviour of the link and joint variables when end effector undergoes external forces. When the manipulator is intended for high-speed application or high torque application or when external force application is high, dynamic analysis plays an important role. Dynamic analysis results will give maximum torque value experienced by each link and joint as well as dynamic stress due to application of external force. There are several ways to derive dynamic equation of motion of a manipulator. Equation of motion can be generated using the NE formulation, the LE formulation, D'Alembert's principle and Hamilton's principle. Kane's formulation has also been used by few researchers. Kane's formulation can be applied to any frame which can be represented in Newtonian's frame in terms of generalised coordinates. Two additional parameters are involved in this formulation, *partial angular velocity* and *partial velocity*.

NE and LE formulation are two widely used formulations for dynamic analysis. NE formulation requires linear and angular momentum to be estimated.

Here manipulator is intended for high forming load. So results of dynamic analysis provide input for the required actuator torque as well as the dynamic stress that will occur in the link members. This work represents dynamic analysis of a serial spatial manipulator using LE formulation.

4.1 Euler–Lagrange (EL) Formulation

In the EL formulation of the present work, the kinetic energy and the potential energy of each link of the manipulator is computed first. By applying Lagrange's equation to the kinematic model, the dynamic equations are derived. In this model, a massive forming force is assumed and gravitational and inertia effects are considered.

The dynamic equations are derived for this 2-DOF manipulator using Lagrange's formulation. In this formulation, the input joint angles are chosen as per the consistent generalised coordinate system. The Lagrangian, which is a function of generalised coordinates depends on potential energy as well as kinetic energy of the manipulator. Therefore, position and velocity analysis as done in kinematic analysis are required in this analysis. This approach is advantageous as it removes internal constraint forces from the equation. Here Lagrange multipliers can also be used to inspect stress in the manipulator link induced by the external forces. The dynamic equation of motion of our 2-DOF manipulator can be derived using the EL formulation as follows.

The robotic manipulator's generalised coordinate is

$$\theta = [\theta_1, \theta_2]^T$$

Equations of motion of the manipulator according to EL formulation are as follows:

$$\frac{\mathrm{d}}{\mathrm{d}t}\left(\frac{\partial L}{\partial \dot{p}_i}\right) - \frac{\partial L}{\partial p} = \tau_i$$

where L *is* Lagrangian, p_i is the generalised coordinate, τ_i is generalised force. Here $L = T - U$, T is kinetic energy and U is potential energy.

$$T = \frac{1}{2}\left(m_i \vec{\dot{c}_i}^T \vec{\dot{c}_i} + \vec{\omega_i}^T I_i \vec{\omega_i}\right)$$

where $\dot{c}$ is cartesian velocity of mass centre.

ω = Angular Velocity of the link.

$$\vec{\omega_i} = \overrightarrow{\omega_{i-1}} + \dot{\theta}_i \vec{e_i}$$

$$\vec{\dot{c}_i} = \overrightarrow{\dot{c}_{i-1}} + \omega_{i-1} \times r_{i-1} + \omega_i \times d_i$$

Link-1

$$\vec{c_1} = \begin{bmatrix} \frac{l_1}{2}\cos\theta_1 \\ \frac{l_1}{2}\sin\theta_1 \\ 0 \end{bmatrix} \quad \vec{\dot{c}_1} = \begin{bmatrix} -\frac{l_1}{2}\sin\theta_1\dot{\theta}_1 \\ \frac{l_1}{2}\cos\theta_1\dot{\theta}_1 \\ 0 \end{bmatrix}$$

$$\left(\vec{\dot{c}_1}^T \vec{\dot{c}_1}\right) = \frac{l_1^2}{4} sin^2\theta_1\dot{\theta}_1^2 + \frac{l_1^2}{4} cos^2\theta_1\dot{\theta}_1^2 + \frac{l_1^2}{4}\dot{\theta}_1^2 = \frac{l_1^2}{4}\dot{\theta}_1^2$$

Kinetic Energy due to linear motion = $\frac{1}{2} m_1 \frac{l_1^2}{4}\dot{\theta}_1^2$

Kinetic Energy due to rotational motion = $\frac{1}{2}\overrightarrow{\omega_i}^T I_i \overrightarrow{\omega_i}$

$$\overrightarrow{\omega_1} = \dot{\theta}_1 \overrightarrow{e_1}$$

$$\overrightarrow{\omega_1} = \dot{\theta}_1 \begin{bmatrix} 0 \\ 0 \\ 1 \end{bmatrix} = \begin{bmatrix} 0 \\ 0 \\ \dot{\theta}_1 \end{bmatrix}$$

$$I = \begin{bmatrix} I_{xx} & I_{xy} & I_{xz} \\ I_{yx} & I_{yy} & I_{yz} \\ I_{zx} & I_{zy} & I_{zz} \end{bmatrix}$$

$$I\overrightarrow{\omega_1} = \begin{bmatrix} I_{xx} & I_{xy} & I_{xz} \\ I_{yx} & I_{yy} & I_{yz} \\ I_{zx} & I_{zy} & I_{zz} \end{bmatrix} \begin{bmatrix} 0 \\ 0 \\ \dot{\theta}_1 \end{bmatrix} = \begin{bmatrix} I_{xz}\dot{\theta}_1 \\ I_{yz}\dot{\theta}_1 \\ I_{zz}\dot{\theta}_1 \end{bmatrix}$$

$$\frac{1}{2}\overrightarrow{\omega_1}^T I\overrightarrow{\omega_1} = \frac{1}{2}\begin{bmatrix} 0 & 0 & \dot{\theta}_1 \end{bmatrix} \begin{bmatrix} I_{xz}\dot{\theta}_1 \\ I_{yz}\dot{\theta}_1 \\ I_{zz}\dot{\theta}_1 \end{bmatrix} = \frac{1}{2} I_{zz}\dot{\theta}_1^2 = \frac{1}{2}\frac{m_1 l_1^2}{12}\dot{\theta}_1^2$$

$$T_1 = \frac{1}{2} m_1 \frac{l_1^2}{4}\dot{\theta}_1^2 + \frac{1}{2}\frac{m_1 l_1^2}{12}\dot{\theta}_1^2 = \frac{m_1 l_1^2}{6}\dot{\theta}_1^2 \tag{12}$$

m_1 = mass of the link-1
U = Potential Energy

$$U_1 = -m_1(-g)\left(\frac{l_1}{2} - \frac{l_1}{2}\cos\theta\right)$$

$$U_1 = m_1 g \frac{l_1}{2}(1 - \cos\theta_1) \tag{13}$$

From Eqs. 12 and 13, L_1 can be calculated as mentioned below

$$L_1 = T_1 - U_1 = \frac{m_1 l_1^2}{6}\dot{\theta}_1^2 - m_1 g \frac{l_1}{2}(1 - \cos\theta_1) \tag{14}$$

Link-2

$$\dot{\overrightarrow{c_2}} = \dot{\overrightarrow{c_1}} + \vec{\omega}_1 \times r_1 + \vec{\omega}_2 \times d_2 = e_1 \times (l_1 + d_2)\dot{\theta}_1 + e_2 d_2 \dot{\theta}_2$$

$$\vec{\omega}_2 = \vec{\omega}_1 + \dot{\theta}_2 e_2 = \begin{bmatrix} 0 \\ 0 \\ \dot{\theta}_1 \end{bmatrix} + \dot{\theta}_2 \begin{bmatrix} \sin\theta_2 \\ 0 \\ \cos\theta_2 \end{bmatrix}$$

$$\vec{\omega}_2 = \begin{bmatrix} (\sin\theta_2)\,\dot{\theta}_2 \\ 0 \\ (\cos\theta_2)\,\dot{\theta}_2 + \dot{\theta}_1 \end{bmatrix}$$

$$\vec{\dot{c_2}} = \begin{bmatrix} 0 \\ 0 \\ 1 \end{bmatrix} \times \left(l_1 + \frac{l_2}{2}\right)\dot{\theta}_1 + \dot{\theta}_2 \frac{l_2}{2} \begin{bmatrix} \sin\theta_2 \\ 0 \\ \cos\theta_2 \end{bmatrix}$$

$$\vec{\dot{c_2}} = \begin{bmatrix} (\sin\theta_2)\dot{\theta}_2 \frac{l_2}{2} \\ 0 \\ (l_1 + \frac{l_2}{2})\dot{\theta}_1 + (\cos\theta_2)\dot{\theta}_2 \frac{l_2}{2} \end{bmatrix}$$

$$\text{Kinetc Energy due to linear motion} = \frac{1}{2} m_2 \vec{\dot{c_2}}^T \vec{\dot{c_2}}$$

$$= \frac{1}{2} m_2 \left[(\sin\theta_2)\dot{\theta}_2 \tfrac{l_2}{2} \quad 0 \quad (L_1 + \tfrac{l_2}{2})\dot{\theta}_1 + (\cos\theta_2)\dot{\theta}_2 \frac{l_2}{2} \right]$$

$$\times \begin{bmatrix} (\sin\theta_2)\dot{\theta}_2 \frac{l_2}{2} \\ 0 \\ (l_1 + \frac{l_2}{2})\dot{\theta}_1 + (\cos\theta_2)\dot{\theta}_2 \frac{l_2}{2} \end{bmatrix}$$

$$= \frac{1}{2} m_2 \left[\frac{l_2^2}{4} \left(\dot{\theta}_1^2 + \dot{\theta}_2^2 + \dot{\theta}_1 \dot{\theta}_2 \cos\theta_2 \right) + l_1^2 \dot{\theta}_1^2 + l_1 l_2 \dot{\theta}_1 \left(\dot{\theta}_1 + \frac{\dot{\theta}_2}{4} \cos\theta_2 \right) \right]$$

m^2 = mass of the link-2

Kinetic Energy due to Rotational Motion = $\frac{1}{2} \overrightarrow{\omega_2}^T I_2 \overrightarrow{\omega_2}$

$$I_2 \vec{\omega}_2 = \begin{bmatrix} I_{xx} & I_{xy} & I_{xz} \\ I_{yx} & I_{yy} & I_{yz} \\ I_{zx} & I_{zy} & I_{zz} \end{bmatrix} \begin{bmatrix} (\sin\theta_2)\dot{\theta}_2 \\ 0 \\ (\cos\theta_2)\dot{\theta}_2 + \dot{\theta}_1 \end{bmatrix}$$

$$= \begin{bmatrix} I_{xx}(\sin\theta_2)\dot{\theta}_2 + I_{xz}\left[(\cos\theta_2)\dot{\theta}_2 + \dot{\theta}_1\right] \\ I_{yx}(\sin\theta_2)\dot{\theta}_2 + I_{yz}\left[(\cos\theta_2)\dot{\theta}_2 + \dot{\theta}_1\right] \\ I_{zx}(\sin\theta_2)\dot{\theta}_2 + I_{zz}\left[(\cos\theta_2)\dot{\theta}_2 + \dot{\theta}_1\right] \end{bmatrix}$$

$$\frac{1}{2} \overrightarrow{\omega_2}^T I_2 \overrightarrow{\omega_2} = \frac{1}{2} \left[(\sin\theta_2)\dot{\theta}_2 \quad 0 \quad (\cos\theta_2)\dot{\theta}_2 + \dot{\theta}_1 \right]$$

$$\begin{bmatrix} I_{xx}(\sin\theta_2)\dot{\theta}_2 + I_{xz}\left[(\cos\theta_2)\dot{\theta}_2 + \dot{\theta}_1\right] \\ I_{yx}(\sin\theta_2)\dot{\theta}_2 + I_{yz}\left[(\cos\theta_2)\dot{\theta}_2 + \dot{\theta}_1\right] \\ I_{zx}(\sin\theta_2)\dot{\theta}_2 + I_{zz}\left[(\cos\theta_2)\dot{\theta}_2 + \dot{\theta}_1\right] \end{bmatrix}$$

Kinetic Energy due to rotational motion

$$= \frac{1}{2}\left[I_{xx}(\sin^2\theta_2)\dot{\theta}_2^2 + 2I_{zx}\sin\theta_2\dot{\theta}_2\left[(\cos\theta_2)\dot{\theta}_2 + \dot{\theta}_1\right] + I_{zz}\left[(\cos\theta_2)\dot{\theta}_2 + \dot{\theta}_1\right]^2\right]$$

$$T_2 = \frac{1}{2}m_2\left[\frac{l_2^2}{4}\left(\dot{\theta}_1^2 + \dot{\theta}_2^2 + \dot{\theta}_1\dot{\theta}_2\cos\theta_2\right) + l_1^2\dot{\theta}_1^2 + l_1 l_2\dot{\theta}_1\left(\dot{\theta}_1 + \frac{\dot{\theta}_2}{4}\cos\theta_2\right)\right] + \frac{1}{2}\left[I_{xx}(\sin^2\theta_2)\dot{\theta}_2^2 + 2I_{zx}\sin\theta_2\dot{\theta}_2\left[(\cos\theta_2)\dot{\theta}_2 + \dot{\theta}_1\right] + I_{zz}\left[(\cos\theta_2)\dot{\theta}_2 + \dot{\theta}_1\right]^2\right] \tag{15}$$

$$U_2 = -m_2(-g)\left(\frac{l_1 + l_2}{2} - \frac{l_1 + l_2}{2}\cos\theta_1\right) \tag{16}$$

Using Eqs. 15 and 16, L_2 can be calculated as follows:

$$L_2 = T_2 - U_2 = \frac{1}{2}m_2\left[\frac{l_2^2}{4}\left(\dot{\theta}_1^2 + \dot{\theta}_1^2 + \dot{\theta}_1\dot{\theta}_2\cos\theta_2\right) + l_1^2\dot{\theta}_1^2 + l_1 l_2\dot{\theta}_1\left(\dot{\theta}_1 + \frac{\dot{\theta}_2}{4}\cos\theta_2\right)\right] + \frac{1}{2}\left[I_{xx}(\sin^2\theta_2)\dot{\theta}_2^2 + 2I_{zx}\sin\theta_2\dot{\theta}_2\left[(\cos\theta_2)\dot{\theta}_2 + \dot{\theta}_1\right] + I_{zz}\left[(\cos\theta_2)\dot{\theta}_2 + \dot{\theta}_1\right]^2\right] - m_2(-g)\left(\frac{l_1 + l_2}{2} - \frac{l_1 + l_2}{2}\cos\theta_1\right) \tag{17}$$

Lagrangian Equation for Manipulator

Using Eqs. 17 and 14, τ_1 and τ_2 can be calculated as mentioned below.

$$\frac{\mathrm{d}}{\mathrm{d}t}\left(\frac{\partial L_1}{\partial\dot{\theta}_1}\right) - \frac{\partial L_1}{\partial\theta_1} = \tau_1, \frac{\mathrm{d}}{\mathrm{d}t}\left(\frac{\partial L_2}{\partial\dot{\theta}_2}\right) - \frac{\partial L_2}{\partial\theta_2} = \tau_2$$

$$\tau_1 = \frac{1}{3}m_1 l_1^2\ddot{\theta}_1 + \frac{1}{2}m_1 g l_1\sin\theta_1 \tag{18}$$

$$\tau_2 = \frac{m_2}{4}l_2^2\ddot{\theta}_2 + I_{zz}\ddot{\theta}_2\cos\theta_2 - \frac{m_2}{4}\left[l_2^2\dot{\theta}_2\ddot{\theta}_2 + (l_2 + l_1)\times\frac{l_2\dot{\theta}_1}{2}\left(\ddot{\theta}_2\cos\theta_2 - \dot{\theta}_2\sin\theta_2\right)\right] - \frac{I_{zz}}{2}\left(2\left(\ddot{\theta}_2\cos\theta_2 - \dot{\theta}_2\sin\theta_2\right)\left(\dot{\theta}_2\cos\theta_2 + \dot{\theta}_1\right)\right) \tag{19}$$

Equations (18) and (19) show the Lagrangian equation for manipulator, which are difficult to solve analytically hence numerical method of solution was resorted to using COMSOL multibody dynamic analysis module.

5 Multibody Dynamic Analysis

Dynamic analysis of the mechanism shows us how our design will respond to dynamic forces (friction/gravity). Before creating physical prototype, dynamic analysis can be conducted in design phase. Kinematic and dynamic simulation of a rescue robot has been done using MATLAB and ADAMS previously [21]. Here in our manipulator, links are connected by two joints which restrict their relative motion. Multibody dynamic analysis shows dynamic behaviour of a manipulator or links and joints, when an external force is applied. Displacement and velocity analysis of 2-DOF manipulator has been done in COMSOL multiphysics. In COMSOL, the EL equation is the governing equation for dynamic analysis. Maximum force has been calculated by conducting ISF experiments without using

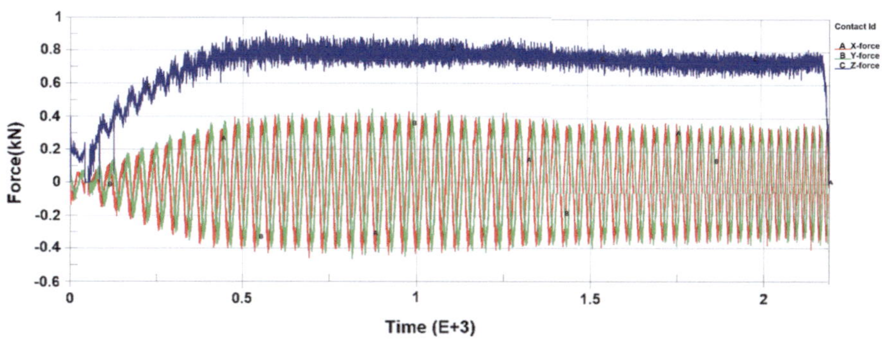

Fig. 8 Experimental force value

Table 2 Parameters used for simulation

Parameter name	Value	Description
F	1000 N	Total force
Angular	25 rad/s	Angular velocity (for both joints)

Table 3 Material properties

Name	Value	Unit
Density	7850	kg/m^3
Young's modulus	200e9	Pa
Poisson's ratio	0.33	–

manipulator. Force plot has been shown in Fig. 8. Here vertical *Z* force is higher as compared to *X* and *Y* forces. Therefore, while carrying out the simulation, only the *Z* force has been considered. *Z* force was found to be less than 1000 N in experiments; hence, in the simulation, the maximum load has been taken to be 1000 N. The sheet metal is made of Aluminium alloy AA-1100 and the angular velocity was chosen as 200 rad/s, as mentioned in Table 2. Material properties used in the simulation are shown in Table 3. Simulation results are given in Figs. 9 and 10.

The manipulator was modelled in CREO as an assembly of three links and then was imported into COMSOL. In COMSOL, the appropriate settings were made to declare the manipulator as assembly so that the interfaces of individual links are recognised. Structural properties of the manipulator material are selected as shown in Table 3.

The COMSOL Multiphysics simulation gave the displacement at end effector to be maximum, which is 0.54 m at 1.0 s and when the end effector is at its maximum position. Velocity is also maximum on the end effector. Displacement at intermediate stages is plotted in Fig. 9, which shows that as the end effector moves up, due to application of external force, the displacement reduces.

Figure 10 shows the velocity at intermediate time steps.

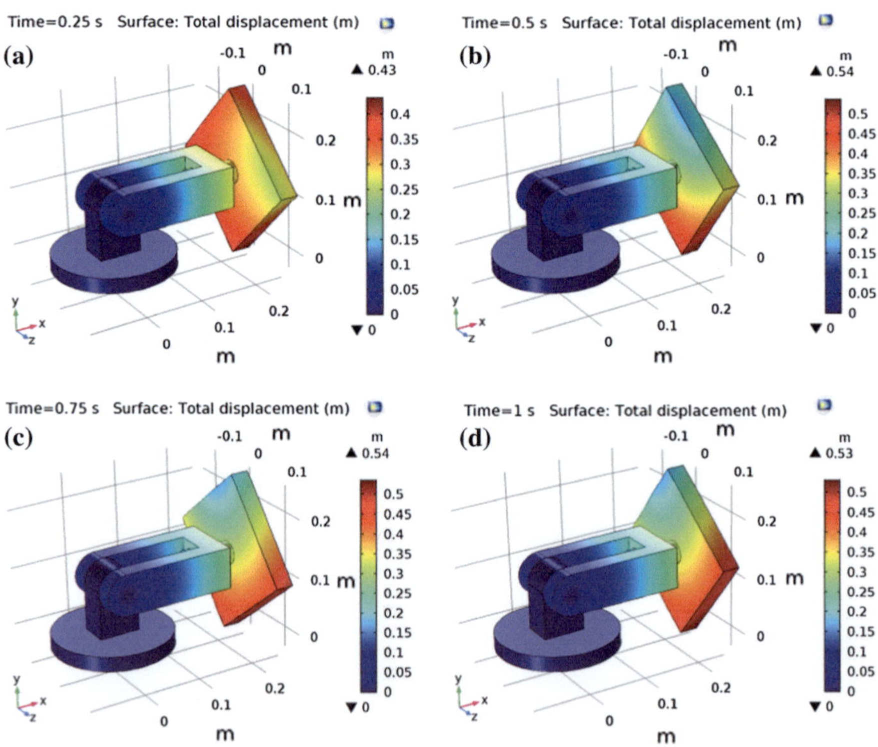

Fig. 9 Displacement results from multibody dynamic analysis, at 0.25 s (**a**), 0.5 s (**b**), 0.75 s (**c**) and 1.0 s (**d**)

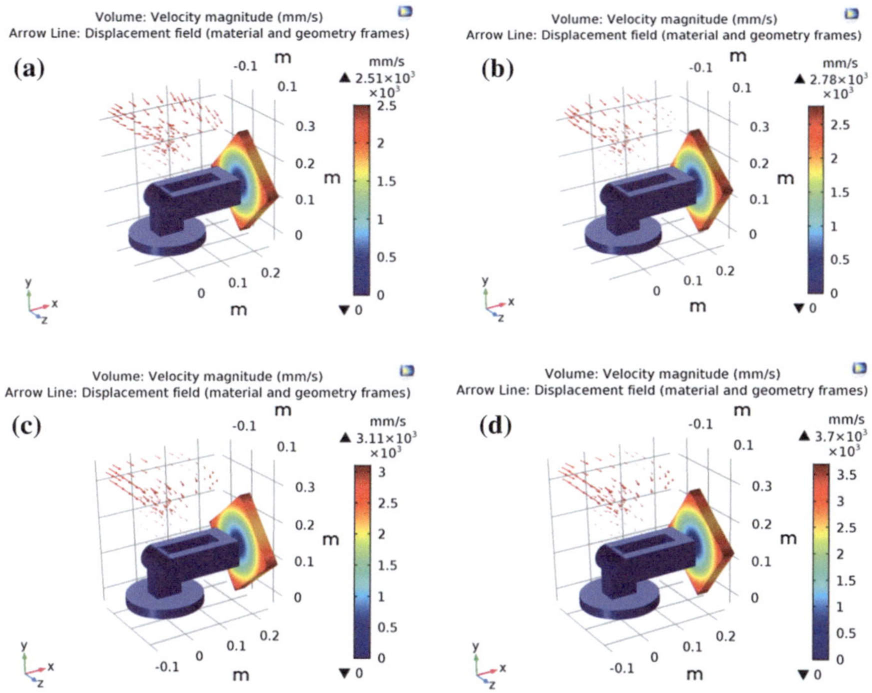

Fig. 10 Velocity results from multibody dynamic analysis, at 0.25 s (**a**), 0.5 s (**b**), 0.75 s (**c**) and 1.0 s (**d**)

6 Conclusions

This contribution investigates the dynamic analysis of a robotic manipulator aimed to be used for incremental sheet metal forming. Analysis of the dynamic behaviour of 2-DOF manipulator to be used for incremental sheet metal forming has been carried out using two methods, first the EL formulation and then using finite element multibody dynamic simulation using COMSOL Multiphysics. The link lengths of the manipulator are chosen as generalised coordinates and the velocity has been expressed in terms of the generalised coordinates. Linear and angular velocity of each joint and link have been calculated. The influence of tool force on the manipulator joint torques has been analysed. Multibody dynamic analysis has been done by considering Aluminium alloy as the forming material, for which the average forming was estimated from separate ISF experiments. Under the action of this force and the inertia force, the dynamic analysis helped estimate the displacement, velocity and accelerations for the range of rotational degrees-of-freedom of the manipulator to be used in the incremental forming, helping in the design of the manipulator link material, dimensions that would withstand both the static and dynamic loads and would not impair the workpiece geometric and surface accuracy.

References

1. Allwood, J. M., Shouler, D. R., & Tekkaya, A. E. (2007). The increased forming limits of incremental sheet forming processes. *Key Engineering Materials, 344,* 621–628.
2. Silva, M. B., Skjoedt, M., Martins, P. A. F., & Bay, N. (2008). Revisiting the fundamentals of single point incremental forming by means of membrane analysis. *International Journal of Machine Tools and Manufacture, 48*(1), 73–83.
3. Shakir, G., Ou, H., & McCartney, G. (2016). Review on the influence of process parameters in incremental sheet forming. *International Journal of Advanced Manufacturing Technology, 87,* 1–21.
4. Imre, P. (2014). A novel, single-robot based two sided incremental sheet forming system. In *Proceedings of 41st International Symposium on Robotics, ISR/Robotik 2014* (pp. 547–553). VDE.
5. Lamminen, L. (2005). Incremental sheet forming with an industrial robot—Forming limits and their effect on component design. *Advanced Materials Research, 6–8,* 457–464.
6. Sasso, M., Callegari, M., & Amodio, D. (2008). Incremental forming: An integrated robotized cell for production and quality control. *Meccanica, 43,* 153–163.
7. Chera, I., Bologa, O., Racz, G., & Breaz, R. (2013). Robot-Forming—An incremental forming process using an industrial robot by means of DELMIA software package. *Applied Mechanics and Materials, 371,* 416–420.
8. Mata, V., Provenzano, S., Valero, F., & Cuadrado, J. (2002). Serial-robot dynamics algorithms for moderately large numbers of joints. *Mechanism and Machine Theory, 37*(8), 739–755.
9. Arakelian, V., Baron, J. P. L., & Mottu, P. (2011). Torque minimization of the 2-DOF serial manipulators based on minimum energy consideration and optimum mass redistribution. *Mechatronics, 21,* 310–314.
10. Biradar, R., & Kiran, M. B. (2012). The dynamics of fixed base and free-floating robotic manipulator. *International Journal of Engineering Research & Technology, 1*(5), 1–9.
11. Tavoosi, J., Jokandan, A. S., & Daneshwar, M. A. (2012). A new method for position control of a 2-DOF robot arm using neuro—Fuzzy controller. *Indian Journal of Science and Technology, 5*(3), 2253–2257.
12. Salinic, S., Boskovic, M., & Bulatovic, R. (2015). Minimization of dynamic joint reaction forces of the 2-DOF serial manipulators based on interpolating polynomials and counterweights. *Theoretical and Applied Mechanics, 42*(4), 249–260.
13. Gupta, V., Chaudhary, H., & Saha, S. K. (2015). Dynamics and actuating torque optimization of planar robots. *Journal of Mechanical Science and Technology, 29,* 2699–2704.
14. Ma, Y., Niu, W., Luo, Z., Yin, F., & Huang, T. (2016). Static and dynamic performance evaluation of a 3-DOF spindle head using CAD–CAE integration methodology. *Robotics and Computer-Integrated Manufacturing, 41,* 1–12.
15. Yongjie, Z., Qiu, K., & Shuangxi, W. Z. Z. (2015). Inverse kinematics and rigid-body dynamics for a three rotational degrees of freedom parallel manipulator. *Robotics and Computer Integrated Manufacturing, 31,* 40–50.
16. Long, P., Khalil, W., & Martinet, P. (2014). Dynamic modelling of parallel robots with flexible platforms. *Mechanism and Machine Theory, 81,* 21–35.
17. Skjoedt, M., Hancock, M. H., & Bay, N. (2007). Creating 3D spiral tool paths for single point incremental forming. *Key Engineering Materials, 344,* 583–590.
18. Blaga, A., Bologa, O., Oleksik, V., & Breaz, R. (2011). Influence of tool path on main strains, thickness reduction and forces in single point incremental forming process. *Proceedings in Manufacturing Systems, 6*(4), 191–196.

19. http://www.roboanalyzer.com/.
20. Mohanty, S., Regalla, S. P., & Daseswara Rao, Y. V. (2017). Tool path planning for manufacturing of asymmetric parts by incremental sheet metal forming by means of robotic manipulator. *Materials Today: Proceedings, 4*(2), 811–819.
21. Xu, Z., & Gao, J. Y. (2014). The kinematics and dynamics analysis of rescue Robot based on Adams and Matlab simulation. *Advanced Materials Research, 945–949,* 1400–1407.

FEM Analysis of Different Materials Based on Explicit Dynamics ANSYS in Electrochemical Discharge Machine

Pravin Pawar, Raj Ballav and Amaresh Kumar

1 Introduction

The micro level products have a great importance as they are used in various fields like mechanical, medical, fluidics, etc., but the manufacturing of these products is very difficult task to the industries because of difficulties in cutting material at micro levels. Hence, these types of materials are machined by using the nonconventional machining process which gives high accuracy and precision. The advantage of this process is that it does not produce chips during machining conditions. The different nonconventional processes are based on removal of material by chemical action, thermal action, electrochemical/chemical dissolution, impact of water, and abrasion by abrasive particles. The electrochemical discharge machining (ECDM) is one of the nonconventional processes which specifically preferred to fabricate the material at micro scale. The electrically conductive and nonconductive materials can be machined by this hybrid nonconventional machining process, i.e., ECDM. The electrochemical discharge machining is also named as electrochemical spark machining and spark assisted chemical engraving [1–3]. It combines the electrochemical machining process (ECM) and electro-discharge machining process (EDM) which is used to machine the nonconducting materials having good machining quality [4]. The Classification of ECDM process is shown in Fig. 1. The ECDM process first introduced by Kurafuji and Suda [5], they successfully drilled hole on glass. Crichton and McGeough [6] described the arc discharge

P. Pawar (✉) · R. Ballav · A. Kumar
Department of Manufacturing Engineering, National Institute of Technology Jamshedpur, Jamshedpur 831014, Jharkhand, India
e-mail: pravin.1900@gmail.com

R. Ballav
e-mail: rballav.prod@nitjsr.ac.in

A. Kumar
e-mail: akumar.prod@nitjsr.ac.in

U. S. Dixit and R. Kant (eds.), *Simulations for Design and Manufacturing*, Lecture Notes on Multidisciplinary Industrial Engineering,
https://doi.org/10.1007/978-981-10-8518-5_9

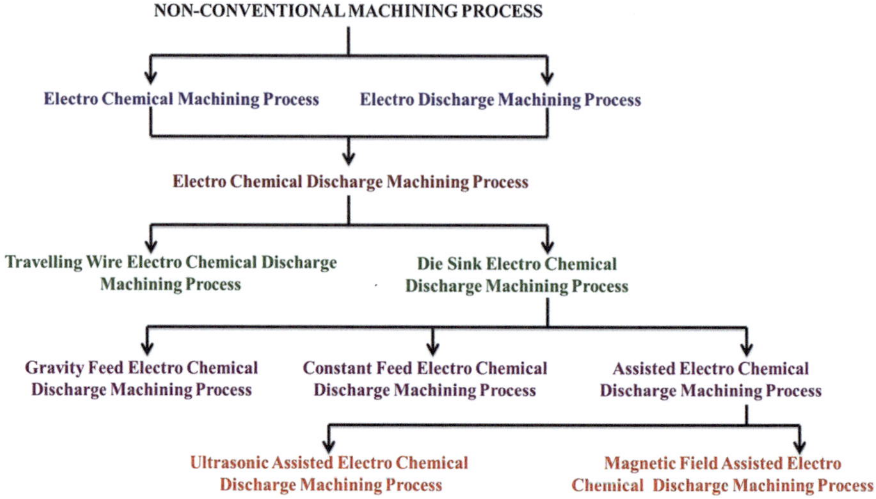

Fig. 1 Classification of ECDM process

phenomenon due to which the electrical discharge between the cathode tool and the electrolyte interface was occurred. In this process, material removal takes place due to the combined effect of anodic dissolution and electrical sparks. The chain of micro-explosions creates on the workpiece surface layer because of electrochemical reaction and electrical sparks which cause melting and vaporization process on work surface [7]. Wuthrich and Fascio [8] reviewed first ECDM phenomenon with special reference to the electrochemical point of view. According to them, ECDM is a complex process. The key factors of this process are change in wettability of the electrode, hydrodynamic instabilities, local Joule heating, combination of wettability and hydrodynamic effects, and bubble coalescence as a percolation. The limitations of this process are efficiency, accuracy, and machining quality [8]. Similarly, Paul and Hiremath [9] explored the ECDM process based on theoretical, design, and development of experimental setup, experimental investigations. Goud et al. [10] explained this process with special focus to the material removal mechanism. According to Pawar et al. [11], most of the researchers used NaOH as an electrolyte, tungsten carbide material as cathode tool, and graphite material as anode tool for the ECDM process.

This chapter presents the finite element analysis of ECDM machine structure by using explicit dynamics ANSYS is carried out. The explicit dynamics are powerful and efficient tools which analyze the model in short duration with high energy dynamics events, i.e., crash and impact, blast, ballistics, drop testing, metal forming [12]. The materials such as Al6061T6, magnesium alloy, nickel alloy, silicon carbide, and titanium alloy are used for the present investigation as they have wide range of applications in different fields such as aircraft, aerospace components, automotive, nuclear applications, gas and steam turbine power plants, chemical and petrochemical industries, electrical and electronics technologies, biomedical field, jet engines, airframes, industrial applications, computer, and armor applications [13]. The force and displacement applied to the cathode tool electrode on different workpiece material to obtain the stress, strain, and deformation observations. The stress is the force applied to the workpiece area and the strain is nothing but the change in length per unit length. The deformation is change in shape of the body caused by an applied force. Hence, the deformation is proportional to the stress. For this study, only physical model of ECDM is considered.

1.1 Basic Mechanism of ECDM

Figure 2 shows the schematic diagram for basic mechanism of ECDM, it is also called as electrochemical spark machining (ECSM). The mechanism of ECDM is still not very clear. In this process, the electrolyte container is made up of acrylic material. The machining gap between cathode tip and workpiece is maintained for stable electrochemical reaction and discharge mechanism. The cathode and anode are dipped into an electrolyte chemical and DC power supply is provided between them. When the applied voltage is higher than the critical voltage, sparks are created from the tool tip. The dissolution of metal ions takes place in anode electrode; hence, it is larger than cathode. The constant DC supply voltage is applied between cathode and anode; it results in electrolysis due to which the hydrogen gas bubbles are formed at the cathode electrode and oxygen bubbles are produced at the

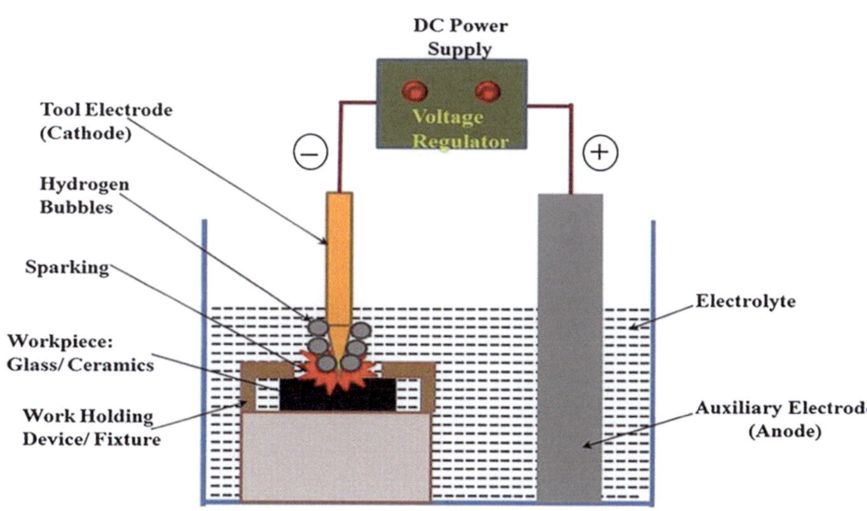

Fig. 2 Schematic diagram of ECDM

anode electrode. The increase in supply voltage increases the current density due to which the higher numbers of bubbles are formed. The bubbles layer generates around the electrodes which cause the bubbles coalescence into a gas film around the tool electrode. The light emission can be seen during electrical discharge so that it is recognized as electrochemical discharges. The discharge takes place at the tool electrolyte interface because of high gradient developed across the hydrogen bubbles. The high electric field is created at the electrolyte and cathode tip which produces an arc discharge in the gas layers covering the tip. The electrons moved to the workpieces when cathode tip is near to it. The continuous flow of electrons increases the temperature of workpieces [4, 14–17]. The current between the cathode and anode has valuable evidence on key process parameters [18].

The tool-glass bond is produced during drilling process due to the chemical reaction [19]. The stable and dense gas film obtained at high voltage or voltage beyond critical voltage is called as transition voltage due to which electrochemical discharge action takes place [20, 21]. The tool electrode wettability changes the gas film thickness which causes higher machining stability [22, 23]. This process ensues because of etching and machining which can take place at different temperatures [24]. The hardness and density of the machined zones get reduced during the ECDM process [25]. The different cathode and anode reactions are occurred when a suitable potential is reached between the interelectrode gap of the machining zone [26]. In this process, mainly two types of reaction occur at the cathode and anode which are given below.

(1) The two types of reaction occur at the cathode are (a) plating of metal ions and (b) evolution of hydrogen gas.

  (a) The reaction for metal plating is

$$M^+ + e^- \rightarrow M,$$

  where M denotes anode material

  (b) The reaction for hydrogen evolution is

$$2H^+ + 2e^- \rightarrow H_2 \uparrow \text{ (in acidic electrolytic solution)}$$

$$2H_2O + 2e^- \rightarrow 2(OH)^- + H_2 \uparrow \text{ (in alkaline solution)}$$

(2) The two types of reaction occur at the anode are (a) dissolution of metal ions in the electrolytic solution and (b) evolution of oxygen gas at the auxiliary electrode surface.

  (a) The anodic dissolution reaction is

$$M \rightarrow M^+ + e^- \text{(in acidic electrolyte solution)}$$

(b) The oxygen evolution reaction is

$$2H_2 \rightarrow O \rightarrow O_2 \uparrow + 4H^+ + 4e^- \text{ (in acidic electrolyte reaction)}$$

$$4(OH)^- \rightarrow 2H_2O + O_2 \uparrow + 4e^- \text{ (in alkaline electrolyte reaction)}$$

1.2 Theoretical Modeling in ECDM

The researchers have developed theoretical modeling for ECDM process which is based on analytical, mathematical, and finite element method modeling technique. Basak and Ghosh [27] developed a theoretical model for the discharge phenomenon ensued at the tool tip. They evaluated the critical voltage and current required to initiate discharge between the electrode and the electrolyte. The developed model is shown in Eq. 1.

$$V_c = J_c A_c [R_{1c} + R_{2c} + R_{3c}] \tag{1}$$

where V_c is the critical voltage, J_c is the current density, A_c is the equivalent surface area of the electrode, and R_{1c}, R_{2c}, R_{3c} are the critical values of interface resistance between the tool and the electrolyte.

Liu et al. [28] developed a model shown in Eq. 2 stated that the electric field acting on hydrogen bubble in ECDM process. They concluded that this model can predict the position of the maximum field strength at the surface of the bubbles along with the critical breakdown voltage for spark initiation and the spark action is in the form of an arc.

$$E_0 = \frac{161}{260} \text{Max } E_{in} \tag{2}$$

where Max E_{in} is the dielectric strength of hydrogen, and E_0 is the original electric field strength. Fascio et al. [29] studied two models for spark assisted chemical engraving (SACE). The first model is based on percolation theory which predicts voltage and current and second model to evaluate the spark's characteristics, i.e., amplitude and duration. Kamaraj et al. [30] developed a mathematical model (Eq. 3) to predict the overcut of machined surface in the ECDM process. The model estimates that the overcut observed with a variation of less than 15% and the tool diameter shows the minor effect on overcut at micron scale.

$$\frac{1}{\frac{\{T(r,t)-t_0\} \times 4\pi K}{\left(V \times I_m - I_m^2 \times R\right)} + \frac{1}{\pi}\sqrt{\frac{pc}{KC_f t}}} \tag{3}$$

where R is the interelectrode resistance of the electrolyte, c is the heat capacity, K is the thermal conductivity of the work material, V is the machining voltage, I_m is the mean machining current, t is the total machining time, r is the overcut, p is the density, and C_f is the ratio of mean machining current to peak current.

Bhondwe et al. [31] developed a thermal model for the analysis of material removal rate (MRR) during electrochemical spark machining which is shown in Eq. 4. Also, they achieved the temperature distribution within the influence zone of a single spark by using finite element method.

$$\mathrm{MRR_m} = \mathrm{MRR_v} \times \rho \tag{4}$$

where $\mathrm{MRR_m}$ is the mass material removal rate, $\mathrm{MRR_v}$ is the volume of material removal rate, and ρ is the density of the workpiece material.

Wei et al. [32] obtained a finite element model for machining depth in ECDM drilling of discharge regime which is shown in Eq. 5.

$$Z_\infty \left\{ 1 - \exp\left[\frac{2tr_0^2 z_0 A_1 u^2 \exp(-B_1/u)}{z_\infty (4dl + d^2)} \right] \right\} \tag{5}$$

where Z_∞ is the machinable depth, t is the time, r_o and z_o are the intercepts of the isotherm and coordinate axes, u is the applied voltage, d is the tool diameter, and l is the tool immersion depth.

Skrabalak et al. [33] developed a simplified model to estimate the current of electrochemical dissolution and electro-discharge machining in ECDM process. In this model, they adapting fuzzy logic controller which provides an advantage to decrease the number of surface roughness and microcracks.

1.3 Progress in ECDM Process Parameters

The advancement in technology is altered vastly; hence, to improve the performance of ECDM process, many innovative methods with different process parameters are applied by the researchers for getting better results. Coteata et al. (2016) applied a rotating tool and a passivating electrolyte mechanism for electrochemical discharge drilling [34]. The rotational tool with controlled feed can improve the process performance [35]. Similarly, the rotational tool increases the circularity of the machined hole [36], and reducing the tool electrode wears [37]. The electrolyte flow technique gives better results of the material removal rate [38]. The super high pressure interior flushing improves the machining efficiency and surface quality [39]. However, the powder-mixed electrolyte improves the machining quality and surface roughness [40]. The increasing voltage improves the material removal rate. The tool tip shape is the key reason for controlling the spark generation in ECDM [41]. An abrasive cutting tool improves the material removal and increased machined depth [42]. Likewise, rotary abrasive electrode at higher supply voltages enhances the volume of material removal and gives better quality of holes [43]. The stainless steel electrode shows zero tool wear whereas tungsten shows the highest tool wear [44]. The side insulated tool immersed in the electrolyte generates more stable spark discharges related to non-insulated tool electrodes [45]. Whereas tapered tool electrodes can enhance the consistency of spark generation and suppress the generation of minor discharges [46]. The spherical tool electrode gives enhanced shape accuracy for the micro through hole [47]. The electro-chemical spark machining (ECSM) with reverse polarity cuts the quartz plate at a faster rate. The material removal in electrochemical spark machining with reverse polarity is due to the melting, vaporization, and chemical reaction on quartz [48]. Mediliyegedara et al. [49] proposed control system architecture which helps to investigate the optimum parameters of ECDM process which improves the surface finish and metal removal rate. The ECDM process used to fabricate the micro-holes on Pyrex glass with 8.8 aspect ratio [50], cutting and drilling holes in composites material [51], piezoelectric ceramics and carbon fiber epoxy composites material [15], micromachining of e-glass–fiber–epoxy composite material [52], particulate reinforced metal matrix composite material [28], holes and channels on borosilicate and soda-lime glass [53], alumina and borosilicate glass [42], Al_2O_3 [54], blind holes and through holes in ceramic material [16], silicon nitride [55], and zirconium oxide [56]. Didar et al. [57] produced 2D micromachining on glass with constant speed. They obtained five types of micro-channels, i.e., well-defined linear micro-channel edges and a smooth channel surface, jagged outline contours with a smooth micro-channel surface, heat-affected edges with a smooth micro-channel surface, heat-affected edges with an unsmooth micro-channel surface and thermal cracks [57]. Whereas, Zheng et al. [17] obtained 3D machining on Pyrex glass. The complex 3D microstructures are done layer-by-layer ECDM micro milling without any mask [17]. Also, 3D micro-pattering is obtained in an electrolyte using electro-discharge assisted etching at the tool tip [58]. Likewise, Cao et al. [59] manufactured a 3D microstructure and various microstructures less than 100 μm in size. Lee et al. [60] developed reversible interconnections for glass-based micro-fluidic systems by using ECDM to machine the three-dimensional vias in glass substrates. The ECDM process is used to fabricate small and shallow dies on materials [61]. The scanning electron microscope plays a key role in analyzing all output results obtained from ECDM process [62].

1.4 Development in ECDM Machine Setup

The ECDM machine set up get changes due to the advancement in machine design which helps to achieve better results. The feeding mechanism is the key part of this setup. Basically, it is two types, i.e., gravity-feed mechanisms and constant feed mechanism. Figure 3 shows the three-dimensional model of ECDM setup. Bhattacharyya et al. [26] developed the machining set up by adding the modular mechatronic feature in electrochemical discharge machine. Sahasrabudhe et al. [54] have established the gravity-feed ECDM set up for machining of non-conductive alumina ceramic. The setup has the fixture which is used to hold the

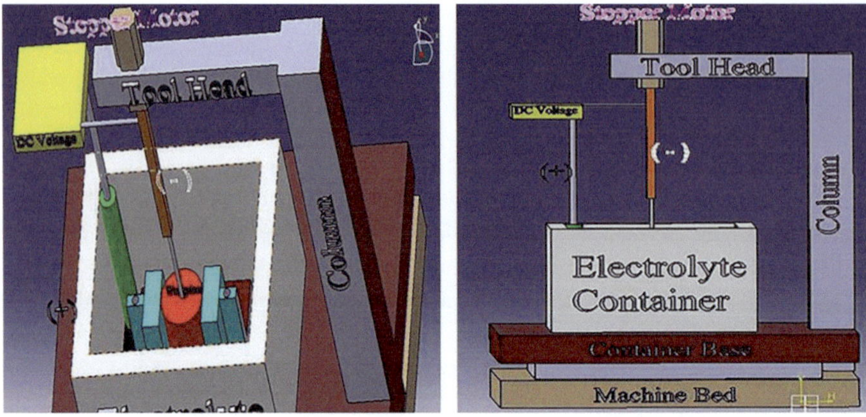

(a) Top View (b) Side View

Fig. 3 Three-dimensional model of ECDM

workpiece and the tool holder to guide the cathode tool, whereas the anode electrode was placed at the bottom side of the fixture. A self-alignment of the ECDM system used to maintain the coordinate of the tool tip [16].

The gravity-feed drilling is most generally useful for micro-hole drilling of glass material [63]. The four types of micro-holes are obtained in gravity-feed drilling, i.e., well-defined cylindrical holes with smooth surface, holes with heat-affected zone, jagged outline holes, and holes with thermal cracks [64]. The electrodes made up of material having high thermal conductivity give faster drilling in the discharge regime during gravity-feed drilling [65]. However, Mallick et al. [66] applied spring feed mechanism job to allow the movement in upward direction. Whereas, Paul et al. [67] done feeding through a servo motor actuator with high precision. Coteata et al. [43, 68] used slotter ram mechanism for the alternative linear motion of the electrode tool to attain the small diameter holes. Abou Ziki and Wuthrich [69] have applied constant feed force from 0 to 5 N for stainless steel cathode tool electrodes with sizes of 250 and 500 μm in diameter. Han et al. [70] developed ultrasonic electrolyte vibration assisted ECDM which provides an adequate electrolyte flow. It gives stability to the gas film formation and maintains the consistent spark generation resulted in improved surface quality of the hole [70]. Similarly, Rusli and Furutani [71] found that the ultrasonic vibration changes the discharge behavior and enhance the electrolyte circulation which resulted in decreased removal rate and better surface integrity. However, adequate tool vibration possibly improves the mean material removal rate in gravity-feed drilling [72]. Likewise, vibration amplitude uniform current signal shows the considerable effect on the material removal rate [73]. Whereas, according to Razfar et al. [74], the ECDM process does not influence by vibration when a rotating micro-drill is used as the cathode electrode. Cheng et al. [75] developed magnetic field assisted ECDM they found that the magnetic field enhances electrolyte circulation in the micro-hole which gives higher machining efficiency. Similarly, Hajian et al. [76] found that the effect of magnetic field on material removal rate and surface quality depends on electrolyte concentration. Chak and Rao [77] applied trepanning method in ECDM machine setup which gives the possibility of large size holes drilling with efficient and economical way. Schopf et al. [78] developed new technology for trueing and dressing of metal bonded diamond tools which is useful for high precision grinding of superhard cutting materials. Kulkarni et al. [79] developed a novel automated 5-DOF electrochemical spark micromachining machine which is useful for the surface treatment, micromachining, surface modification, and deposition.

1.5 Applications of ECDM

The ECDM process has a wide range of applications in the machining of non-conductive ceramic materials as well as composite materials are as follows [26].

- To obtain the different channel surface textures and micro-texturing of glass.
- Used for the micro-texturing in microfluidic devices such as single cell capturing device, digital micro-mirror device, and mini-bioreactors.
- Employed in microelectromechanical systems.
- Used in applications of micro-factories such as probes, sensors, and automated factories.
- In the biomedical field for equipment, i.e., electron microscope accessories [4, 80, 81].
- Micro-fabrication of miniaturizing machine tools for micromachining such as miniature gears for ophthalmic industry, valves and mixing devices, microscale fuel cells, and microscale pumps.
- Micro-fabrication of arrays of holes in SU-8 material.
- Formation of interconnections.
- Fabrication of micro-filters required for the micro-EDM process, micro-channels on silicon for heat sinks, surface modification.
- Utilized for the micro-seam welding of copper plates and foils.
- Applicable in heat treatment.
- Formation of Si-metal contacts to form an array of Schottky diodes.
- Used to the repairing of copper tracks on printed circuit boards.
- Utilized in the formation of nanoparticle and hydrogen gas.
- Employed in 3-D prototypes of metal and nonmetal as well as for the building of prototypes [79, 82].

2 FEM Analysis of Different Materials Based on Explicit Dynamics ANSYS in ECDM

The finite element simulation process is classified as preprocessing, solution, and post-processing [30]. In preprocessing, the three-dimensional model is built in ANSYS workbench then explicit dynamics analysis conditions are applied to the tool electrode, i.e., 1 N force and 1 mm displacement toward the workpiece. In solution, the strain, stress, and total deformation were automatically solved and recorded by ANSYS solver. In the post-processing, the strain, stress, and total deformation are calculated and results are generated. The electrochemical discharge machine is a hybrid chemical and discharge process.

The stress wave propagation and impact or rapidly changing time-dependent loads are involved in explicit dynamics analysis due to which the dynamic response of a structure can be determined. This type of analysis can also be advantageous to model mechanical phenomena that are highly nonlinear. The nonlinearities may ensue from the materials, from the contact and from the geometric deformation. The events with time scales of less than 1 s are efficiently simulated with this type of analysis. The major advantage of the explicit method is that there is no requirement for solving simultaneous equations, which mean that no global matrix inversion is required, which save the computational timing. The time integration routines are necessary for numerical analysis to solve for accelerations, velocities, and displacements of the dynamic problem, characterized by the system equation. The explicit integration method is stable only if the time step size is smaller than the critical time step size for the finite element system being simulated. The finite element method simulates physical phenomena by converting a continuum into a discrete domain, i.e., nodes and elements [83–88].

2.1 Basic Formulations for Explicit Dynamic FEM

The equation of dynamic equilibrium governing the response of the FEM system which is derived from the virtual displacement principle and it is written in matrix form, shown in Eq. 6 [84–86].

$$[M]\{\ddot{u}_n\} + [C]\{\dot{u}_n\} + [K]\{u_n\} = \{F_n^a\} \tag{6}$$

where $[M]$ is the global mass, $[C]$ is damping, and $[K]$ is stiffness matrices, $\{u\}$ is global nodal degree-of-freedom, and $\{F_n^a\}$ is load vectors. They are integrated by their individual matrix of the FEA model. The consistent mass matrix M, damping matrix C, and the stiffness matrix K are as follows [85]:

$$[M] = \int \rho[N]^{\mathrm{T}}[N]\mathrm{d}V \tag{6a}$$

$$[C] = \int c[N]^{\mathrm{T}}[N]\mathrm{d}V \tag{6b}$$

$$[K] = \int [B]^{\mathrm{T}}[E][B]\mathrm{d}V \tag{6c}$$

$$\{F_n^a\} = \int [N]^{\mathrm{T}}[f]\mathrm{d}V + \int [N]^{\mathrm{T}}\{t\}\mathrm{d}s + \sum_{i=1}^{n}\{p\}_i \tag{6d}$$

where N is the finite element interpolation function matrix, ρ is the material density, B is the strain–displacement matrix, $[E]$ is a stress–strain matrix, $[f]$ is the body force, s is the surface, $\{t\}$ is the surface traction, and $\{p\}$ is the concentrated load. These integrals are taken over the element domain d, i.e., the element volume V.

The explicit analysis method makes use of the central difference method to calculate the state of the next increment. The method of central difference is an explicit algorithm, but it has a stable condition that at the time of solving a problem by using this method the size of time step must be less than a threshold determined by the nature of the problem-solving equation. The central difference is used to replace the acceleration, velocity derivative central difference, and the expressions for the nodal accelerations (Eq. 7) and velocities (Eq. 8). Equation 6 shows motion equation for the body is integrated using the explicit central difference integration rule with u displacement degree-of-freedom with derivatives. The central difference integration operator is explicit in which the values from the previous increment, i.e., $\ddot{u}_n$ and $\dot{u}_n$ can be made kinematic state advanced [84, 86, 88].

$$\ddot{u}_n = \frac{1}{\Delta t^2}(u_{n-1} - 2u_n + u_{n+1}) \tag{7}$$

$$\dot{u}_n = \frac{1}{2\Delta t}(u_{n+1} - u_{n-1}) \tag{8}$$

Substituting (7) and (8) into (6) to obtain an expression to solve for the unknown displacements $\{u_{n+1}\}$, where n is the current step, n + 1 is the next step, and n − 1 is the previous step in time. Δt is the increment between time steps.

$$\left(\frac{1}{\Delta t^2}M + \frac{1}{2\Delta t}C\right)u_{n+1} = F_a^n - \left(K - \frac{2}{\Delta t^2}M\right)u_n - \left(\frac{1}{\Delta t^2}M - \frac{1}{2\Delta t}C\right)u_{n-1} \tag{9}$$

From Eq. 9, it can be said that u_{n+1} (displacement at $t + \Delta t$) is determined by both u_n (displacement at t) and u_{n-1} (displacement at $t - \Delta t$). Therefore, before starting the procedure, the displacement at time $t = -1$, u_{-1} is determined using Eqs. 7 and 8, where $i = 0$.

$$\left(u_{-1} = u_0 - \Delta t \dot{u}_0 + \frac{\Delta t^2}{2} \ddot{u}_0\right) \tag{10}$$

After solving u_{-1}, the state of motion $\ddot{u}_{n+1}$, $\dot{u}_{n+1}$, and u_{n+1} can be fully solved. According to the magnitude of Δt, the method of central difference has selective convergence. However, the convergence and accuracy both are linearly proportional to the square of Δt. Nodal displacements can be attained at $n + 1$ which is a time step by Eq. 9 then the area of deformation is updated from which a new magnitude of the time increment to guarantee convergence can be finalized. After the calculation of time increment in every element, the global time increment magnitude can be defined by applying Eq. 11.

$$\Delta t_{n+1} = \alpha \min\{\Delta t_1 + \Delta t_2 \ldots \Delta t_n\} \tag{11}$$

where N is the total element number and Δt_n is the time increment of the nth element. The safe constant α is often selected to be less than 0.9. The critical time increment is determined by

$$\Delta t_c = L_s/C \tag{12}$$

where L_s is the characteristic length, which is the given element area divided by the largest edge. The propagation speed C is determined by Eq. 13.

$$C = \sqrt{\frac{E_t}{\rho}} \tag{13}$$

where E_t is the tangent modulus and ρ is material density. The response of the system changes over time.

In the structure of the electrochemical discharge machine, the effect of force and displacement is applied specifically from cathode tool to the workpiece. The ECDM setup is built in ANSYS workbench. However, the analysis was done in explicit dynamics ANSYS. The design of machine parameters is based on reference to Abou Ziki and Wuthrich [69]. In this model, the tool electrode diameter is 2 mm; 1 N force, 1 mm displacement is applied from stainless steel tool material to Al6061T6, magnesium alloy, nickel alloy, titanium alloy, and silicon carbide workpiece material.

2.2 *Methodology and Assumptions*

The following are the assumptions taken for the analysis:

(a) Loading and boundary conditions were applied to rectangular plate workpiece material, which is shown in Fig. 4c.
(b) 1 N force and 1 mm displacement is applied to the 47 × 38 × 10 mm rectangular plate with supporting 10 mm diameter four round bars of 20 mm height. 2 mm diameter stainless steel material is taken as the ECDM cathode tool electrode which is identical for all materials analysis.
(c) The rectangular plates of different materials, i.e., Al6061T6, magnesium alloy, nickel alloy, titanium alloy, and silicon carbide are used for the analysis.
(d) The material properties of different materials were shown in Table 1.
(e) Calculation of maximum equivalent von Mises stresses, equivalent elastic strain, and total deformation of each rectangular plate is done and then the obtained results are compared with one another.

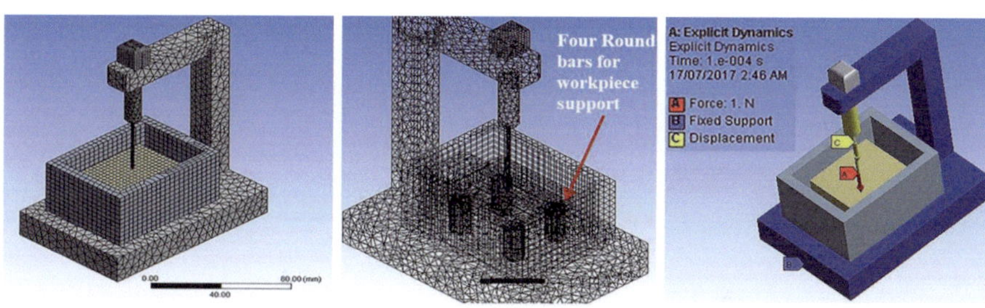

(a) Meshed Model (b) Wireframe Model (c) Boundary Condition

Fig. 4 **a** Meshed model of ECDM machine, **b** wireframe model of ECDM, and **c** boundary condition of ECDM machine

Table 1 Properties of different materials

Material	Density (kg/m^3)	Yield strength (MPa)	Tensile ultimate strength (MPa)	Young's modulus (MPa)	Poisson's ratio	Bulk modulus (MPa)	Shear modulus (MPa)
Stainless steel	7750	207	586	193,000	0.31	169,300	73,664
Structural steel	7800	250	460	210,000	0.3	166,670	76,923
Al6061-T6	2703	240	310	68,900	0.33	68,000	27,600
Magnesium alloy	1800	193	255	45,000	0.35	50,000	16,667
Nickel alloy	8810	170	345	190,000	0.30	177,300	68,800
Titanium alloy	4420	930	1070	105,000	0.36	114,290	35,294
Silicon carbide	3215	1395	1625	450,000	0.37	176,000	193,500

(f) Figure 4a shows the meshed model of ECDM machine which has 11,010 nodes and 18,233 elements.
(g) In this analysis, the tool electrode is considered as a rigid body while a workpiece is a flexible body and the supporting four round bars and base plate are again considered as a rigid body.
(h) The materials other than the tool and workpiece, i.e., base plate, vertical column, horizontal beam, tool holder, four supporting bars, are considered as structural steel material.

2.3 Simulation Method of ECDM Model

The model of ECDM is developed using ANSYS 14.5 software. Figure 5 indicates the exported ECDM model viewed by CATIA software. In the simulation method for obtaining strain, stress, and total deformation of different materials in ECDM,

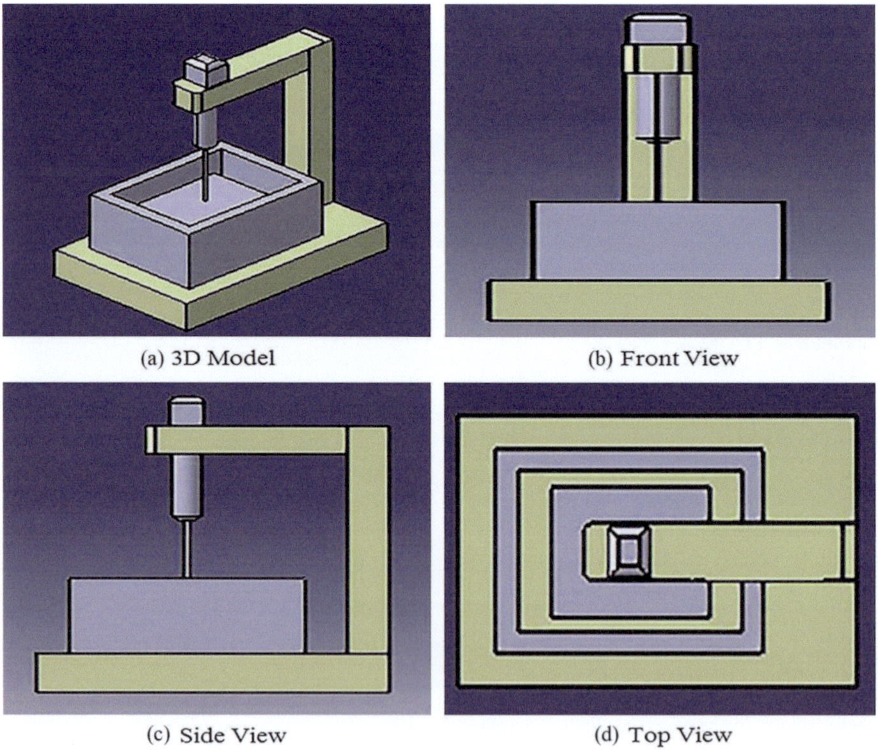

Fig. 5 **a** 3D model of ECDM, **b** front view of ECDM, **c** side view of ECDM, and **d** top view of ECDM

the four supporting round bars, base plate, column, horizontal beam, tool holder, and tool electrode are considered as a rigid body. While the workpiece is considered as a flexible body. In the explicit dynamics conditions for the analysis 1 N force and 1 mm displacement of the tool is applied toward the workpiece body. The distance between the tool and workpiece is initially 0.0001 mm and for analysis condition set end time is 0.0001 s. The start time is 0 s.

2.4 *Results and Discussion*

1 N force and 1 mm displacement were applied to the workpiece of each material, i.e., Al6061T6, magnesium alloy, nickel alloy, silicon carbide, and titanium alloy material. Each workpiece has supporting four round bars attached to the base plate.

a. **Analysis of Al6061-T6**

Figure 6 shows equivalent (von Mises) stress and time history graph of Al6061-T6. From the obtained results of explicit dynamics, the graph for equivalent von Mises stress versus time was constructed. It starts from 0 to 0.0001 s, the increase in stress is due to the increased load applied on it. The graph shows the relationship between the load applied and stress occurs. The results show that the maximum equivalent von Mises stress is 242.87 MPa at time 0.0001 s. It shows that, the stress is directly proportional to the time until reach the maximum stress. Here, due to the physical properties of the Al6061T6 material, the stress is occurred irregular throughout the surface. Figure 7 shows equivalent elastic strain and time history graph of Al6061-T6. The graph shows the maximum equivalent elastic strain is 0.0076122 mm/mm in 0.0001 s which is occurred toward the center and near to the tool tip point. The total deformation and time history graph of Al6061-T6 are presented in Fig. 8. The maximum total deformation occurs at around the tool tip point and it is 1.095 mm at 0.0001 s.

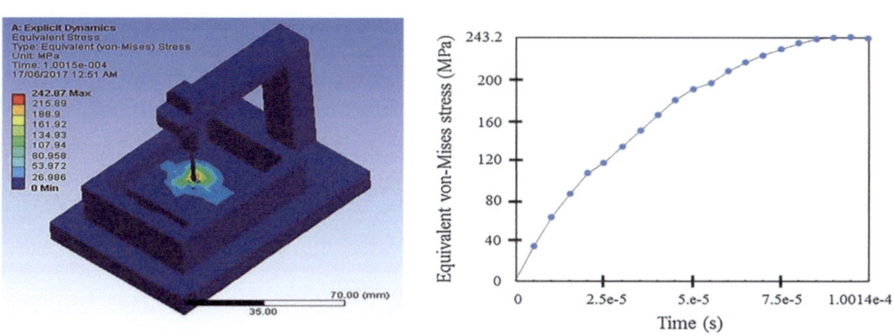

Fig. 6 Equivalent (von Mises) stress and time history graph of Al6061-T6

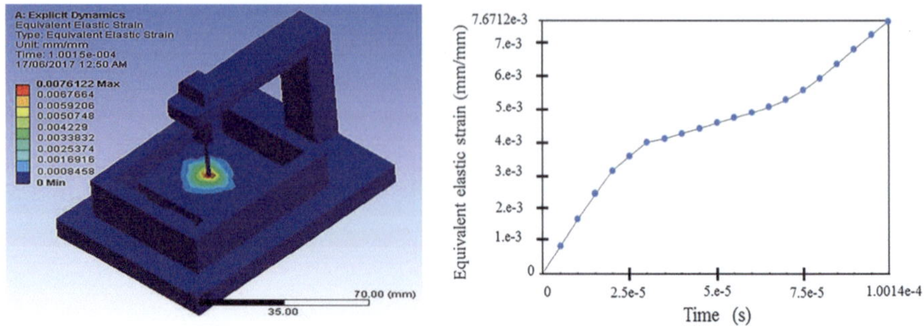

Fig. 7 Equivalent elastic strain and time history graph of Al6061-T6

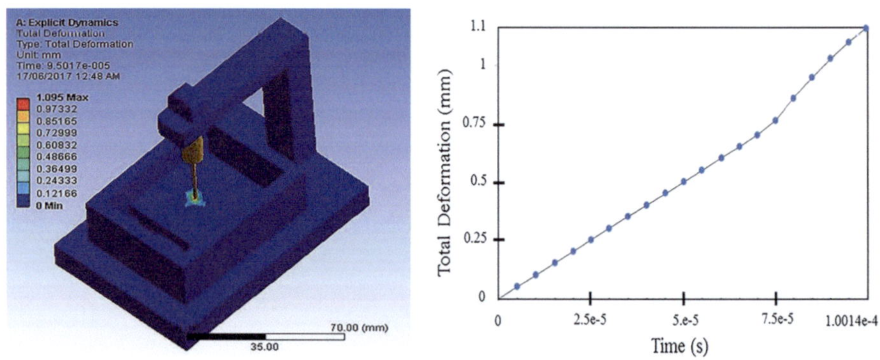

Fig. 8 Total deformation and time history graph of Al6061-T6

b. **Analysis of magnesium alloy**

The equivalent (von Mises) stress and time history graph of magnesium alloy is shown in Fig. 9. The results indicate that the maximum equivalent von Mises stress is 370.93 MPa at time 0.0001 s which is occurred irregular throughout the surface. Figure 10 shows the equivalent elastic strain and time history graph of magnesium alloy. The graph shows maximum equivalent elastic strain is 0.020645 mm/mm in 0.0001 s which is observed toward the center and near to the tool tip point and it is more than the Al6061T6. The total deformation and time history graph of magnesium alloy are presented in Fig. 11. The maximum total deformation is less than the Al6061T6 occurred around the tool tip point and it is 0.90293 mm at 0.0001 s.

c. **Analysis of nickel alloy**

Figure 12 presents equivalent (von Mises) stress and time history graph of nickel alloy. The results indicate that the maximum equivalent von Mises stress is 682.09 MPa at time 0.0001 s. For the given boundary conditions, because of physical properties of nickel alloy, the stresses occur irregularly throughout the

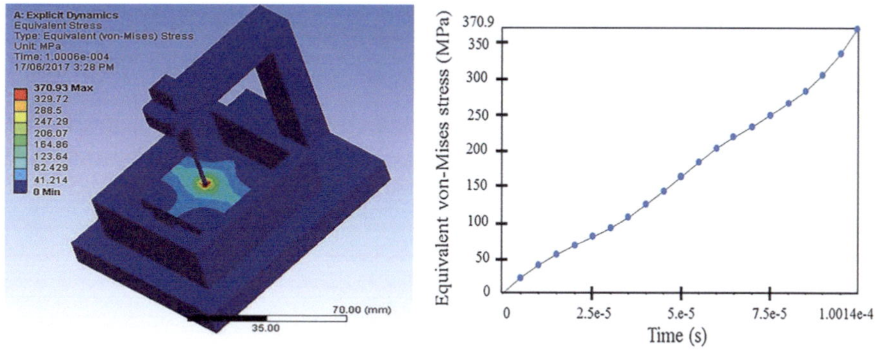

Fig. 9 Equivalent (von Mises) stress and time history graph of magnesium alloy

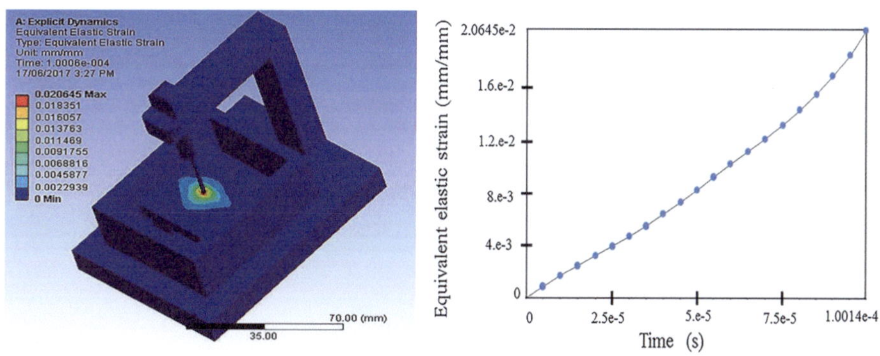

Fig. 10 Equivalent elastic strain and time history graph of magnesium alloy

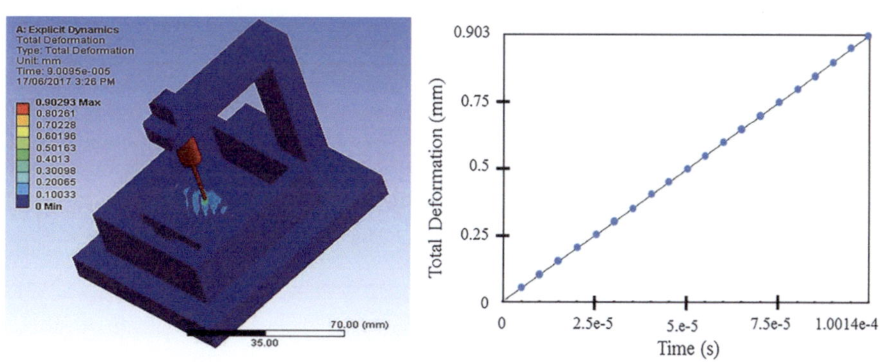

Fig. 11 Total deformation and time history graph of magnesium alloy

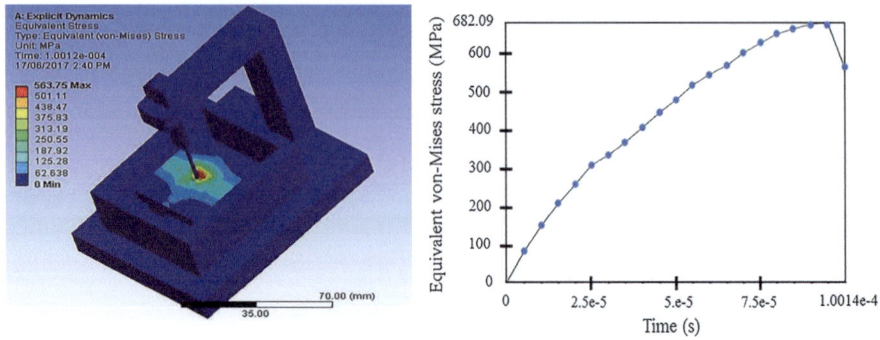

Fig. 12 Equivalent (von Mises) stress and time history graph of nickel alloy

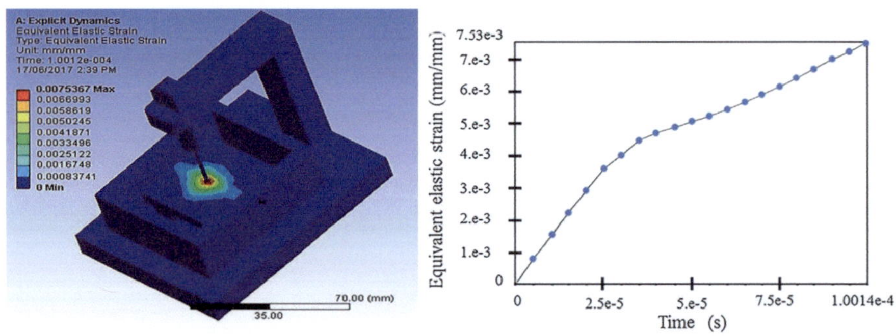

Fig. 13 Equivalent elastic strain and time history graph of nickel alloy

surface and they are greater than the magnesium alloy material. After 682.09 MPa, the equivalent (von Mises) stress drops down up to 563.75 MPa. This means after 682.09 MPa the failure is initiated and these maximum stresses occur on the expected failure region where the failure initiated. Figure 13 shows the equivalent elastic strain and time history graph of nickel alloy. The graph shows maximum equivalent elastic strain is 0.0075367 mm/mm in 0.0001 s which is observed at the center and near to the tool tip point and similar to the strain results of Al6061T6 material. The total deformation and time history graph of nickel alloy are presented in Fig. 14. The maximum total deformation occurs around the tool tip point which is more than magnesium alloy material and it is 0.94388 mm at 0.0001 s.

d. **Analysis of silicon carbide**

Figure 15 shows equivalent (von Mises) stress and time history graph of silicon carbide. The results show that the maximum equivalent von Mises stress is 2566 MPa at time 0.0001 s. After 2566 MPa, the equivalent (von Mises) stress drop down up to 779.89 MPa. Hence, 2566 MPa is the expected failure region where the failure is initiated and it is higher than the nickel alloy. Figure 16 shows

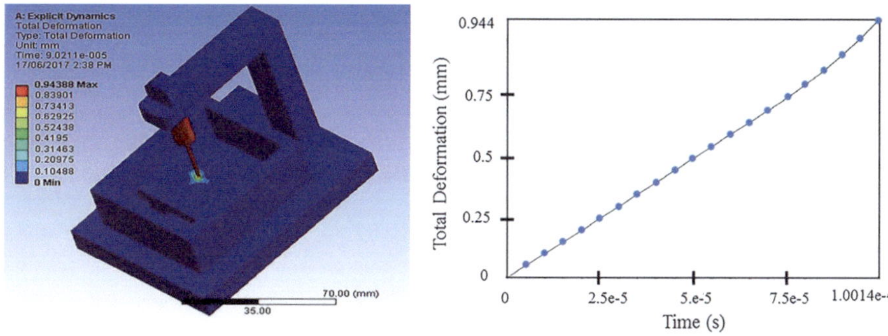

Fig. 14 Total deformation and time history graph of nickel alloy

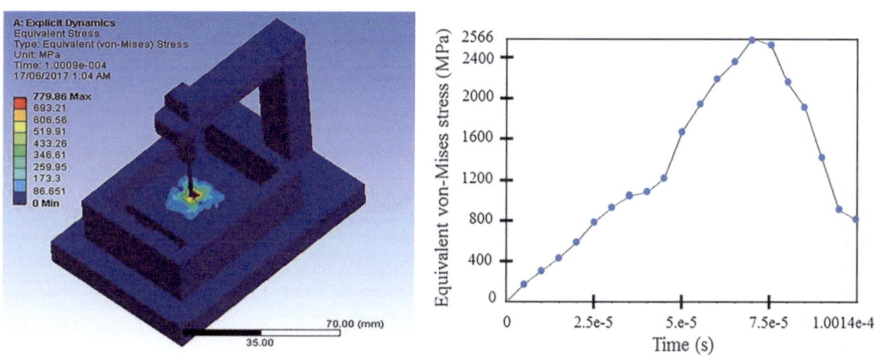

Fig. 15 Equivalent (von Mises) stress and time history graph of silicon carbide

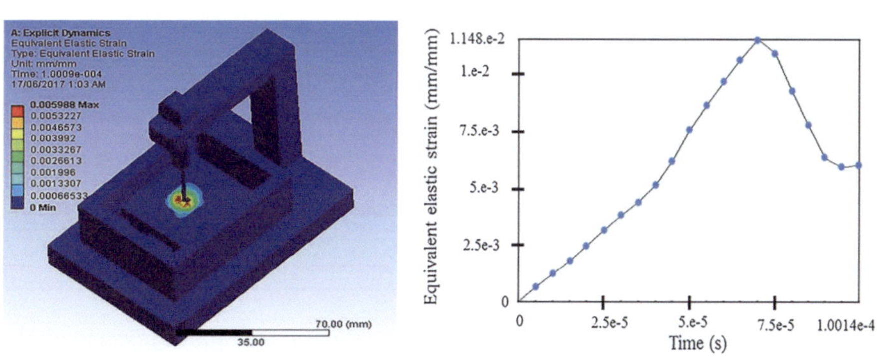

Fig. 16 Equivalent elastic strain and time history graph of silicon carbide

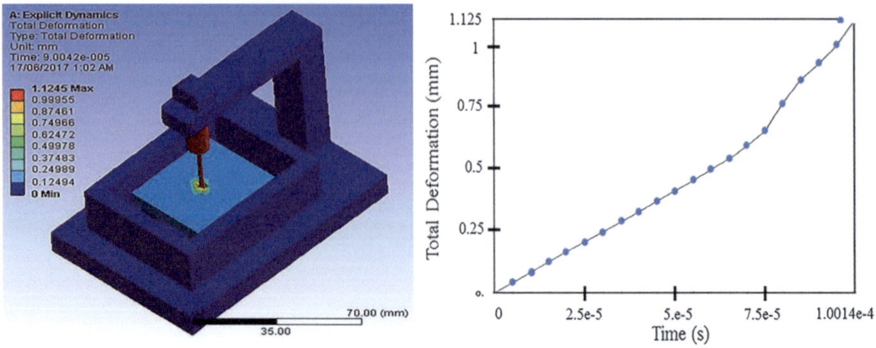

Fig. 17 Total deformation and time history graph of silicon carbide

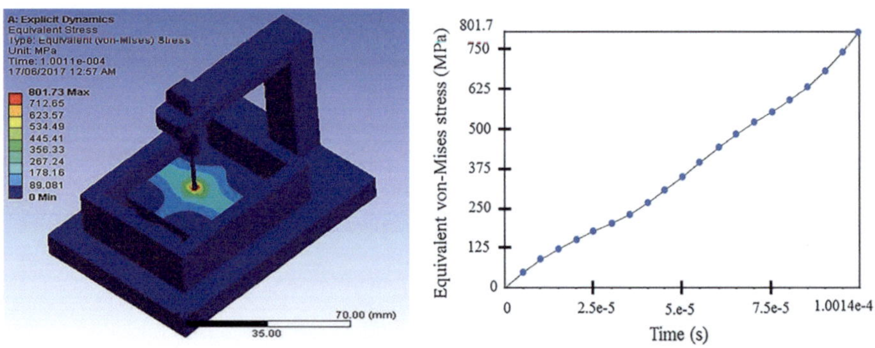

Fig. 18 Equivalent (von Mises) stress and time history graph of titanium alloy

the equivalent elastic strain and time history graph of silicon carbide. The graph shows maximum equivalent elastic strain is $1.148e^{-2}$ mm/mm in 0.0001 s. The total deformation and time history graph of silicon carbide are presented in Fig. 17. The maximum total deformation occurs at around the tool tip point and it is 1.125 mm at 0.0001 s. The uneven and irregular equivalent von Mises stress, equivalent elastic strain and total deformation are obtained during this analysis. It may be due to the extreme hardness properties of this material.

e. **Analysis of titanium alloy**

The equivalent (von Mises) stress and time history graph of titanium alloy is presented in Fig. 18. The results show the maximum equivalent von Mises stress is 801.73 MPa at time 0.0001 s. Figure 19 shows the equivalent elastic strain and time history graph of titanium alloy. The graph shows maximum equivalent elastic strain is 0.021551 mm/mm in 0.0001 s. The total deformation and time history graph of titanium alloy are presented in Fig. 20. The maximum total deformation is 0.91194 mm at 0.0001 s. The equivalent von Mises stress, equivalent elastic strain, and maximum total deformation are quite similar to the magnesium alloy material.

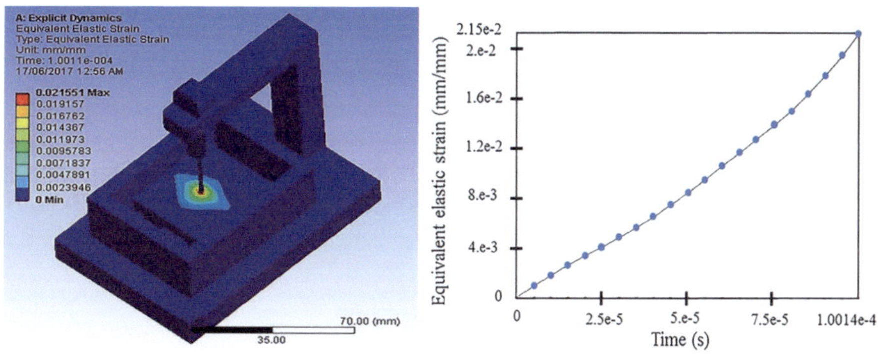

Fig. 19 Equivalent elastic strain and time history graph of titanium alloy

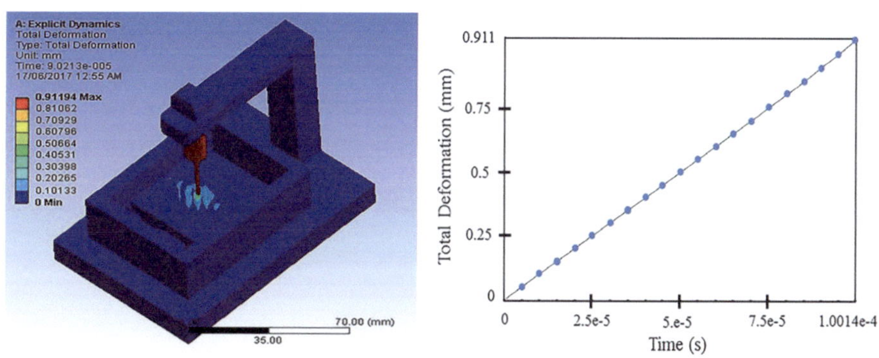

Fig. 20 Total deformation and time history graph of titanium alloy

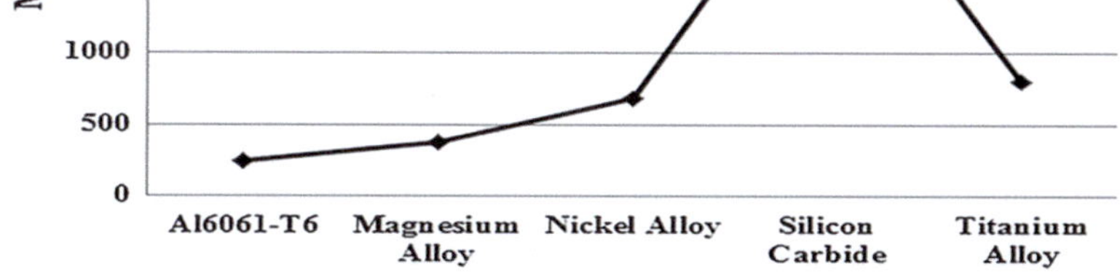

Fig. 21 Graph of maximum equivalent (von Mises) stress versus different materials

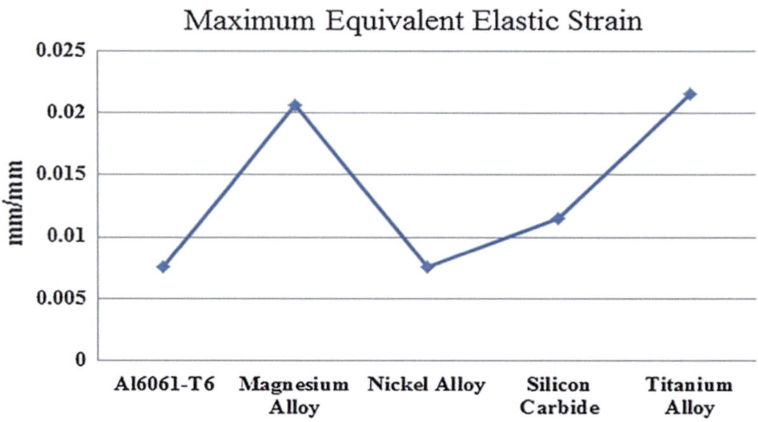

Fig. 22 Graph of maximum equivalent elastic strain versus different materials

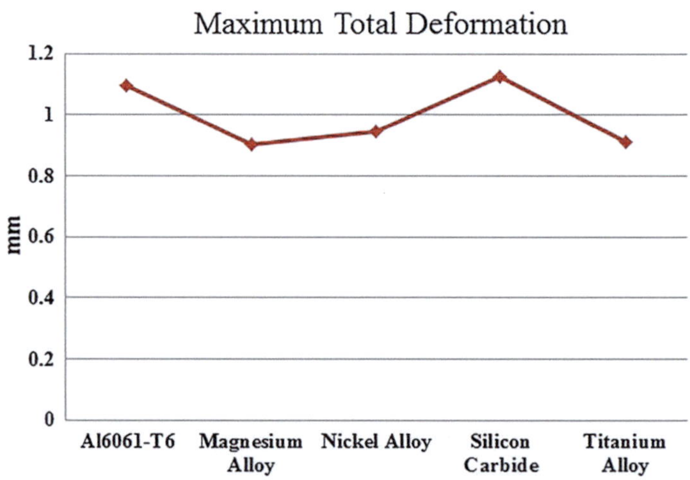

Fig. 23 Graph of maximum total deformation versus different materials

Thus, from the above observations, the line graphs are plotted. Figure 21 shows the graph of maximum equivalent (von Mises) stress *vs.* different materials. The silicon carbide shows highest equivalent (von Mises) stress than other materials. The graph of maximum equivalent elastic strain *vs.* different materials is shown in Fig. 22, it reveals that the highest equivalent elastic strain is obtained for titanium alloy as compared to the other materials. Figure 23 presents the graph of maximum total deformation versus different materials and the highest total deformation is observed for silicon carbide.

3 Conclusions

This chapter presents the basic mechanism of ECDM, past to present progress of ECDM, and also the finite element analysis of ECDM machine structure by using explicit dynamics ANSYS is carried out. The materials such as Al6061T6, magnesium alloy, nickel alloy, silicon carbide, and titanium alloy are used for the present investigation as they have a wide range of applications in different fields. The three-dimensional model is built in ANSYS workbench then explicit dynamics analysis conditions are applied to the tool electrode, i.e., 1 N force and 1 mm displacement toward the workpiece of different materials. The equivalent elastic strain, equivalent von Mises stress and total deformation were solved and recorded by ANSYS solver. From the obtained results, it can be concluded that the uneven and irregular equivalent von Mises stress, equivalent elastic strain, and total deformation are generated during analysis of silicon carbide. It may be due to the high hardness properties of this material due to which it shows higher failure region than other materials. Whereas, as compared to the other materials except silicon carbide, the titanium alloy shows maximum equivalent von Mises stress and equivalent elastic strain because it has higher bulk and Young's modulus which are the measures of material elasticity. The Al6061-T6 shows high total deformation than the other materials because it has high strength but low ductility.

Acknowledgements This chapter is a revised and extended version of the paper titled "Finite Element Method Analysis of Electrochemical Discharge Machine using Explicit Dynamics ANSYS", presented in 6th International and 27th All India Manufacturing Technology, Design and Research (AIMTDR) Conference held at College of Engineering, Pune, Maharashtra, India, December 16–18, 2016.

References

1. Dixit, P. M., & Dixit, U. S. (2008). *Modeling of metal forming and machining processes: by finite element and soft computing methods*. Springer Science & Business Media.
2. Jain, V. K. (2008). Advanced (non-traditional) machining processes. *Machining: Fundamentals and recent advances* (pp. 299–327). London: Springer.
3. Singh, T., & Dvivedi, A. (2016). Developments in electrochemical discharge machining: A review on electrochemical discharge machining, process variants and their hybrid methods. *International Journal of Machine Tools and Manufacture, 105,* 1–13.
4. Wuthrich, R. (2009). Machining with electrochemical discharges—An overview. *Micro and nano technologies, micromachining using electrochemical discharge phenomenon* (pp. 1–9). Boston: William Andrew Publishing.
5. Kurafuji, H., & Suda, K. (1968). Electrical discharge drilling of glass. *Annals CIRP, 16,* 415–419.
6. Crichton, I. M., & McGeough, J. A. (1985). Studies of the discharge mechanisms in electrochemical arc machining. *Journal of Applied Electrochemistry, 15*(1), 113–119.
7. Gupta, K., Perveen, A., & Molardi, C. (2017). Machining of glass materials: An overview. In *Advanced manufacturing technologies: Modern machining, advanced joining, sustainable manufacturing* (pp. 23–47). Springer International Publishing.

8. Wuthrich, R., & Fascio, V. (2005). Machining of non-conducting materials using electrochemical discharge phenomenon—An overview. *International Journal of Machine Tools and Manufacture, 45*(9), 1095–1108.
9. Paul, L., & Hiremath, S. S. (2016). Experimental and theoretical investigations in ECDM process—An overview. *Procedia Technology, 25,* 1242–1249.
10. Goud, M., Sharma, A. K., & Jawalkar, C. (2016). A review on material removal mechanism in electrochemical discharge machining (ECDM) and possibilities to enhance the material removal rate. *Precision Engineering, 45,* 1–17.
11. Pawar, P., Ballav, R., & Kumar, A. (2015). Revolutionary developments in ECDM process: An overview. *Materials Today: Proceedings, 2*(4–5), 3188–3195.
12. Celik, H. K. (2017). Determination of bruise susceptibility of pears (Ankara variety) to impact load by means of FEM-based explicit dynamics simulation. *Postharvest Biology and Technology, 128,* 83–97.
13. Callister, W. D., & Rethwisch, D. G. (2014). *Materials science and engineering: An introduction* (9th ed.). New York: Wiley.
14. Kulkarni, A., Sharan, R., & Lal, G. K. (2002). An experimental study of discharge mechanism in electrochemical discharge machining. *International Journal of Machine Tools and Manufacture, 42*(10), 1121–1127.
15. Singh, Y. P., Jain, V. K., Kumar, P., & Agrawal, D. C. (1996). Machining piezoelectric (PZT) ceramics using an electrochemical spark machining (ECSM) process. *Journal of Materials Processing Technology, 58*(1), 24–31.
16. Lim, H. J., Lim, Y. M., Kim, S. H., & Kwak, Y. K. (2001). Self-aligned micro tool and electrochemical discharge machining (ECDM) for ceramic materials. *Proceedings of SPIE, 4416,* 348–353.
17. Zheng, Z. P., Cheng, W. H., Huang, F. Y., & Yan, B. H. (2007). 3D microstructuring of Pyrex glass using the electrochemical discharge machining process. *Journal of Micromechanics and Microengineering, 17*(5), 960–966.
18. Wuthrich, R., Spaelter, U., & Bleuler, H. (2006). The current signal in spark-assisted chemical engraving (SACE): What does it tell us? *Journal of Micromechanics and Microengineering, 16*(4), 779–785.
19. Ziki, J. D. A., & Wuthrich, R. (2015). Nature of drilling forces during spark assisted chemical engraving. *Manufacturing Letters, 4,* 10–13.
20. Allagui, A., & Wuthrich, R. (2009). Gas film formation time and gas film life time during electrochemical discharge phenomenon. *Electrochimica Acta, 54*(23), 5336–5343.
21. Cheng, C. P., Wu, K. L., Mai, C. C., Yang, C. K., Hsu, Y. S., & Yan, B. H. (2010). Study of gas film quality in electrochemical discharge machining. *International Journal of Machine Tools and Manufacture, 50*(8), 689–697.
22. Wuthrich, R., & Hof, L. A. (2006). The gas film in spark assisted chemical engraving (SACE) —A key element or micro-machining applications. *International Journal of Machine Tools and Manufacture, 46*(7), 828–835.
23. Yang, C. K., Cheng, C. P., Mai, C. C., Wang, A. C., Hung, J. C., & Yan, B. H. (2010). Effect of surface roughness of tool electrode materials in ECDM performance. *International Journal of Machine Tools and Manufacture, 50*(12), 1088–1096.
24. Ziki, J. D. A., Hof, L. A., & Wuthrich, R. (2015). The machining temperature during spark assisted chemical engraving of glass. *Manufacturing Letters, 3,* 9–13.
25. Didar, T. F., Dolatabadi, A., & Wuthrich, R. (2009). Local hardness and density variation in glass substrates machined with spark assisted chemical engraving (SACE). *Materials Letters, 63*(1), 51–53.
26. Bhattacharyya, B., Doloi, B. N., & Sorkhel, S. K. (1999). Experimental investigations into electrochemical discharge machining (ECDM) of non-conductive ceramic materials. *Journal of Materials Processing Technology, 95*(1), 145–154.
27. Basak, I., & Ghosh, A. (1996). Mechanism of spark generation during electrochemical discharge machining: A theoretical model and experimental verification. *Journal of Materials Processing Technology, 62*(1–3), 46–53.

28. Liu, J. W., Yue, T. M., & Guo, Z. N. (2010). An analysis of the discharge mechanism in electrochemical discharge machining of particulate reinforced metal matrix composites. *International Journal of Machine Tools and Manufacture, 50*(1), 86–96.
29. Fascio, V., Wuthrich, R., & Bleuler, H. (2004). Spark assisted chemical engraving in the light of electrochemistry. *Electrochimica Acta, 49*(22), 3997–4003.
30. Kamaraj, A. B., Jui, S. K., Cai, Z., & Sundaram, M. M. (2015). A mathematical model to predict overcut during electrochemical discharge machining. *The International Journal of Advanced Manufacturing Technology, 81*(1–4), 685–691.
31. Bhondwe, K. L., Yadava, V., & Kathiresan, G. (2006). Finite element prediction of material removal rate due to electro-chemical spark machining. *International Journal of Machine Tools and Manufacture, 46*(14), 1699–1706.
32. Wei, C., Xu, K., Ni, J., Brzezinski, A. J., & Hu, D. (2011). A finite element based model for electrochemical discharge machining in discharge regime. *The International Journal of Advanced Manufacturing Technology, 54*(9), 987–995.
33. Skrabalak, G., Zybura-Skrabalak, M., & Ruszaj, A. (2004). Building of rules base for fuzzy-logic control of the ECDM process. *Journal of Materials Processing Technology, 149* (1), 530–535.
34. Coteata, M., Pop, N., Schulze, H. P., Slatineanu, L., & Dodun, O. (2016). Investigation on hybrid electrochemical discharge drilling process using passivating electrolyte. *Procedia CIRP, 42,* 778–782.
35. Gautam, N., & Jain, V. K. (1998). Experimental investigations into ECSD process using various tool kinematics. *International Journal of Machine Tools and Manufacture, 38*(1–2), 15–27.
36. Jui, S. K., Kamaraj, A. B., & Sundaram, M. M. (2013). High aspect ratio micromachining of glass by electrochemical discharge machining (ECDM). *Journal of Manufacturing Processes, 15*(4), 460–466.
37. Huang, S. F., Liu, Y., Li, J., Hu, H. X., & Sun, L. Y. (2014). Electrochemical discharge machining micro-hole in stainless steel with tool electrode high-speed rotating. *Materials and Manufacturing Processes, 29*(5), 634–637.
38. Dhanvijay, M. R., & Ahuja, B. B. (2014). Micromachining of ceramics by electrochemical discharge process considering stagnant and electrolyte flow method. *Procedia Technology, 14,* 165–172.
39. Zhang, Y., Xu, Z., Xing, J., & Zhu, D. (2015). Enhanced machining performance of micro holes using electrochemical discharge machining with super-high-pressure interior flushing. *International Journal of Electrochemical Science, 10,* 8465–8483.
40. Han, M. S., Min, B. K., & Lee, S. J. (2007). Improvement of surface integrity of electro-chemical discharge machining process using powder-mixed electrolyte. *Journal of Materials Processing Technology, 191*(1), 224–227.
41. Pawar, P., Sinha, S., Kumar, A., & Ballav, R. (2014). Review on research trends in electrochemical discharge machining. In *4th National Conference on Recent in Advances in Manufacturing (RAM-2014)* (pp. 132–137).
42. Jain, V. K., Choudhury, S. K., & Ramesh, K. M. (2002). On the machining of alumina and glass. *International Journal of Machine Tools and Manufacture, 42*(11), 1269–1276.
43. Coteata, M., Ciofu, C., Slatineanu, L., Munteanu, A., & Dodun, O. (2009). Establishing the electrical discharges weight in electrochemical discharge drilling. *International Journal of Material Forming, 2,* 673–676.
44. Abou Ziki, J. D., & Wuthrich, R. (2012). Tool wear and tool thermal expansion during micro-machining by spark assisted chemical engraving. *The International Journal of Advanced Manufacturing Technology, 61*(5), 481–486.
45. Han, M. S., Min, B. K., & Lee, S. J. (2008). Modeling gas film formation in electrochemical discharge machining processes using a side-insulated electrode. *Journal of Micromechanics and Microengineering, 18*(4), 045019.

46. Jiang, B., Lan, S., Ni, J., & Zhang, Z. (2014). Experimental investigation of spark generation in electrochemical discharge machining of non-conducting materials. *Journal of Materials Processing Technology, 214*(4), 892–898.
47. Yang, C. K., Wu, K. L., Hung, J. C., Lee, S. M., Lin, J. C., & Yan, B. H. (2011). Enhancement of ECDM efficiency and accuracy by spherical tool electrode. *International Journal of Machine Tools and Manufacture, 51*(6), 528–535.
48. Jain, V. K., & Adhikary, S. (2008). On the mechanism of material removal in electrochemical spark machining of quartz under different polarity conditions. *Journal of Materials Processing Technology, 200*(1), 460–470.
49. Mediliyegedara, T. K. K. R., De Silva, A. K. M., Harrison, D. K., & McGeough, J. A. (2005). New developments in the process control of the hybrid electro chemical discharge machining (ECDM) process. *Journal of Materials Processing Technology, 167*(2), 338–343.
50. Gao, C., Liu, Z., & Li, A. (2014). Study of micro drilling on Pyrex glass using spark assisted chemical engraving. *Micro and Nanosystems, 6*(1), 26–33.
51. Tandon, S., Jain, V. K., Kumar, P., & Rajurkar, K. P. (1990). Investigations into machining of composites. *Precision Engineering, 12*(4), 227–238.
52. Manna, A., & Narang, V. (2012). A study on micro machining of e-glass–fibre–epoxy composite by ECSM process. *International Journal of Advanced Manufacturing Technology, 61,* 1191–1197.
53. Madhavi, B. J., & Hiremath, S. S. (2016). Investigation on machining of holes and channels on borosilicate and sodalime glass using μ-ECDM setup. *Procedia Technology, 25,* 1257–1264.
54. Sahasrabudhe, A. H., & Ahuja, B. B. (2014). Experimental investigations for Al_2O_3 ceramic micro machining using ECDM process. In *5th International & 26th All India Manufacturing Technology, Design and Research Conference (AIMTDR 2014)*, December 12–14, 2014 (pp. 182-1–182-5), IIT Guwahati, Assam, India.
55. Sarkar, B. R., Doloi, B., & Bhattacharyya, B. (2006). Parametric analysis on electrochemical discharge machining of silicon nitride ceramics. *International Journal of Advanced Manufacturing Technology, 28*(9), 873–881.
56. Doloi, B., Bhattacharyya, B., & Sorkhel, S. K. (1999). Electrochemical discharge machining of non-conducting ceramics. *Defence Science Journal, 49*(4), 331–338.
57. Didar, T. F., Dolatabadi, A., & Wuthrich, R. (2008). Characterization and modeling of 2D-glass micro-machining by spark-assisted chemical engraving (SACE) with constant velocity. *Journal of Micromechanics and Microengineering, 18*(6), 065016.
58. Fascio, V., Wuthrich, R., Viquerat, D., & Langen, H. (1999). 3D microstructuring of glass using electrochemical discharge machining (ECDM). In *International Symposium on Micromechatronics and Human Science*, IEEE (pp. 179–183).
59. Cao, X. D., Kim, B. H., & Chu, C. N. (2009). Micro-structuring of glass with features less than 100 μm by electrochemical discharge machining. *Precision Engineering, 33*(4), 459–465.
60. Lee, E. S., Howard, D., Liang, E., Collins, S. D., & Smith, R. L. (2004). Removable tubing interconnects for glass-based micro-fluidic systems made using ECDM. *Journal of Micromechanics and Microengineering, 14*(4), 535–541.
61. Khairy, A. B. E., & McGeough, J. A. (1990). Die-sinking by electroerosion-dissolution machining. *CIRP Annals-Manufacturing Technology, 39*(1), 191–195.
62. Pawar, P., Ballav, R., & Kumar, A. (2015). Measurement analysis in electrochemical discharge machining (ECDM) process: A literature review. *Journal of Chemistry and Chemical Engineering, 9,* 140–144.
63. Wuthrich, R., Spaelter, U., Wu, Y., & Bleuler, H. (2006). A systematic characterization method for gravity-feed micro-hole drilling in glass with spark assisted chemical engraving (SACE). *Journal of Micromechanics and Microengineering, 16*(9), 1891–1896.
64. Maillard, P., Despont, B., Bleuler, H., & Wuthrich, R. (2007). Geometrical characterization of micro-holes drilled in glass by gravity-feed with spark assisted chemical engraving (SACE). *Journal of Micromechanics and Microengineering, 17*(7), 1343–1349.

65. Mousa, M., Allagui, A., Ng, H. D., & Wuthrich, R. (2008). The effect of thermal conductivity of the tool electrode in spark-assisted chemical engraving gravity-feed micro-drilling. *Journal of Micromechanics and Microengineering, 19*(1), 015010.
66. Mallick, B., Ali, M.N., Sarkar, B.R., Doloi, B., & Bhattacharyya, B. (2014). Parametric analysis of electrochemical discharge micro-machining process during profile generation on glass. In *5th International & 26th All India Manufacturing Technology, Design and Research Conference (AIMTDR 2014)*, December 12–14 (pp. 540-1–540-6), IIT Guwahati, Assam, India.
67. Paul, L., & Hiremath, S. S. (2013). Response surface modelling of micro holes in electrochemical discharge machining process. *Procedia Engineering, 64,* 1395–1404.
68. Coteata, M., Slatineanu, L., Dodun, O., & Ciofu, C. (2008). Electrochemical discharge machining of small diameter holes. *International Journal of Material Forming, 1,* 1327–1330.
69. Abou Ziki, J. D., & Wuthrich, R. (2013). Forces exerted on the tool-electrode during constant-feed glass micro-drilling by spark assisted chemical engraving. *International Journal of Machine Tools and Manufacture, 73,* 47–54.
70. Han, M. S., Min, B. K., & Lee, S. J. (2009). Geometric improvement of electrochemical discharge micro-drilling using an ultrasonic-vibrated electrolyte. *Journal of Micromechanics and Microengineering, 19*(6), 065004.
71. Rusli, M., & Furutani, K. (2012). Performance of micro-hole drilling by ultrasonic-assisted electro-chemical discharge machining. *Advanced Materials Research, 445,* 865–870.
72. Wuthrich, R., Despont, B., Maillard, P., & Bleuler, H. (2006). Improving the material removal rate in spark-assisted chemical engraving (SACE) gravity-feed micro-hole drilling by tool vibration. *Journal of Micromechanics and Microengineering, 16*(11), N28–N31.
73. Elhami, S., & Razfar, M. R. (2017). Study of the current signal and material removal during ultrasonic-assisted electrochemical discharge machining. *The International Journal of Advanced Manufacturing Technology*, 1–9.
74. Razfar, M. R., Behroozfar, A., & Ni, J. (2014). Study of the effects of tool longitudinal oscillation on the machining speed of electrochemical discharge drilling of glass. *Precision Engineering, 38*(4), 885–892.
75. Cheng, C. P., Wu, K. L., Mai, C. C., Hsu, Y. S., & Yan, B. H. (2010). Magnetic field-assisted electrochemical discharge machining. *Journal of Micromechanics and Microengineering, 20* (7), 075019.
76. Hajian, M., Razfar, M. R., & Movahed, S. (2016). An experimental study on the effect of magnetic field orientations and electrolyte concentrations on ECDM milling performance of glass. *Precision Engineering, 45,* 322–331.
77. Chak, S. K., & Rao, P. V. (2007). Trepanning of Al_2O_3 by electro-chemical discharge machining (ECDM) process using abrasive electrode with pulsed DC supply. *International Journal of Machine Tools and Manufacture, 47*(14), 2061–2070.
78. Schopf, M., Beltrami, I., Boccadoro, M., Kramer, D., & Schumacher, B. (2001). ECDM (electro chemical discharge machining), a new method for trueing and dressing of metal bonded diamond grinding tools. *CIRP Annals-Manufacturing Technology, 50*(1), 125–128.
79. Kulkarni, A. V., Jain, V. K., & Misra, K. A. (2011). Electrochemical spark micromachining: Present scenario. *International Journal of Automation Technology, 5*(1), 52–59.
80. Wuthrich, R., Fujisaki, K., Couthy, P., Hof, L. A., & Bleuler, H. (2005). Spark assisted chemical engraving (SACE) in microfactory. *Journal of Micromechanics and Microengineering, 15*(10), S276–S280.
81. Abou Ziki, J. D., Didar, T. F., & Wuthrich, R. (2012). Micro-texturing channel surfaces on glass with spark assisted chemical engraving. *International Journal of Machine Tools and Manufacture, 57,* 66–72.
82. Kulkarni, A. V. (2007). Electrochemical discharge machining process. *Defence Science Journal, 57*(5), 765–770.
83. Kabir, M. A., Lovell, M. R., & Higgs, C. F. (2008). Utilizing the explicit finite element method for studying granular flows. *Tribology Letters, 29*(2), 85–94.

84. Mitrovic, S. (2015). Three-dimensional nonlinear dynamic time history analysis of seismic site and structure response. Ph.D. Thesis, Faculty of Civil Engineering, University of Rijeka.
85. Lee, K. M., & Liu, C. H. (2012). Explicit dynamic finite element analysis of an automated grasping process using highly damped compliant fingers. *Computers & Mathematics with Applications, 64*(5), 965–977.
86. Liu, Y. (2008). ANSYS and LS-DYNA used for structural analysis. *International Journal of Computer Aided Engineering and Technology, 1*(1), 31–44.
87. Jung, D. W. (1998). Study of dynamic explicit analysis in sheet metal forming processes using faster punch velocity and mass scaling scheme. *Journal of Materials Engineering and Performance, 7*(4), 479–490.
88. Hu, Y., Shao, Y., Chen, Z., & Zuo, M. J. (2011). Transient meshing performance of gears with different modification coefficients and helical angles using explicit dynamic FEA. *Mechanical Systems and Signal Processing, 25*(5), 1786–1802.

Molecular Dynamics Simulation Study of Neck Growth in Micro-selective Laser Sintering of Copper Nanoparticles

Srijan Paul, Nagahanumaiah, Souren Mitra and Debabrata Roy

Nomenclature

F	Resultant force
m	Mass of atom
a or $\ddot{r}$	Acceleration of the atom
v or $\dot{r}$	Velocity of atom
$\dddot{r}$ and $\ddddot{r}$	Third-order and fourth-order derivative of position w.r.t. time
t	Time
i, j and k	Position of atom
r_{ij}	Interatomic centre distance of atom i and j
ϕ	Potential between atom i and j
ε	LJ well depth parameter
σ	Interatomic distance at which interatomic potential is zero
E	Energy
U	Embedding energy

S. Paul (✉)
Mechanical Engineering Department, Asansol Engineering College, Asansol, India
e-mail: srijanpaul@yahoo.co.in

Nagahanumaiah
Micro Systems Technology Lab, CSIR-Central Mechanical Engineering Research Institute, Durgapur, India
e-mail: naga@cmeri.res.in

S. Mitra
Production Engineering Department, Jadavpur University, Kolkata, India
e-mail: souren.mitra@gmail.com

D. Roy
Mechanical Engineering Department, NSHM Knowledge Campus, Durgapur Group of Institutions, Durgapur, India
e-mail: debabrata.roy@nshm.com

U. S. Dixit and R. Kant (eds.), *Simulations for Design and Manufacturing*,
Lecture Notes on Multidisciplinary Industrial Engineering,
https://doi.org/10.1007/978-981-10-8518-5_10

ρ_i	Total density of electron at atom i due to the neighbouring atoms of the atom i in the system
D_e	Potential well depth in Morse potential
α	Constant determined from the physical properties of material
A and B	Empirically determined range and softness parameters of the material
A', B', λ_1, λ_2, λ_3, α, β, n, c, d, h	Fitting parameters of the Tersoff potential
N	Number of atoms
V	Volume
T	Temperature
P	Pressure
D1	Diameter of spherical particle
ρ	Radius of curvature of neck
x	Radius of disc of contact of two particles

1 Introduction

Sintering is the process of forming a solid object by the application of heat and/or pressure without reaching its melting point. Sintering of minerals like ceramic particles is widely used to produce various artefacts and in late 1900, metal sintering is evident [2]. In recent years, the research has grown historically towards micro-nanoscale manufacturing as requirement for micro-parts with sub-micrometre and nanometer scale features is growing in many applications [3]. However, engineering of metal micro-parts that too in bulk quantity with acceptable repeatability and functional properties is still a daunting task. Laser sintering is the process that consolidates particles through laser heating without reaching their melting temperature is found to be one of promising metal additive manufacturing processes. However, the selective laser sintering processing capability of current commercial systems is in the range of metal particle size of 10 μm or more, leaves much from the desired capability to use in micro-nano systems.

It is advantageous to have better performance when laser radiation is employed to sinter selectively the submicron-sized particles [4]. In selective laser sintering (SLS), the particles are scanned selectively by laser beam layer-by-layer. The heated granules consolidate through sintering, but not melting. The previous researchers reveal that the SLS of macro- to mesoscale powder is accepted widely. Regenfuss et al. [5] carried out various sintering operations on different materials like metal powder or ceramic using various laser sources either in continuous or q-switched pulsed mode. He established the effect on sintered product by using different laser energy sources. He found that it has a significant effect on rapid heating and rapid cooling which helps in high geometric resolution of sintered metal parts [6]. In view of micro-nano systems engineering the demands for batch scale manufacturing of micro-parts, interest for using submicron-sized particles has been increased in the field of SLS, but this is still in research and development stage as in many cases laser beam spot diameter is larger than the particle size. Understanding the process of joining metal particle at molecular level using practical experiments in the laboratory is difficult and expensive. However, this process mechanism could be modelled and simulated using computer simulations like molecular dynamics (MD) or Monte Carlo (MC) simulations for better understanding and process improvement. MD simulation method is preferred over Monte Carlo (MC) simulations as it carried to investigate dynamic properties of the sintering process.

A few attempts at modelling the sintering process have been made in recent years. In order to simulate at atomistic levels, Daw and Baskes [7] proposed a model know as embedded-atom method (EAM) derived as an expression based on density-functional theory which deals with total energy obtained from embedded energy of the metal. This model finds out ground state properties of realistic metals which have conformity with experimentally investigated data. Richardson and Clancy [8] studied the ability of EAM model to simulate the equilibrium and non-equilibrium behaviour of metals to produce similar values of regrowth velocity as experimentally obtained. He found that on ignoring the free electron contribution towards transport property results obtained are quantitatively accurate.

Pan et al. [9] carried MD simulations to compare neck growth process in continuous laser heating. He carried out classical MD simulations on predicting neck growth for gold nanoparticles and compared it with analytical neck growth model by Johnson [10] for continuous laser heating. He found closer agreement with particles as large as 10 nm or more. Later Yang et al. [11] carried MD simulations of gold nanoparticles of size 4–20 nm to find out effect on neck growth on changing laser intensity and particle size. He observed that slowing the rate of heating causes more time to reach the final temperature, this makes the neck wider but on increasing heat rate causes melting which subsequently increases the neck width. On increasing the size of one of the particles, mean square displacement (MSD) decreases having a reverse effect on the radius of gyration which shows no effect of heating rate as the final stage of sintering reaches to melting point of the particle. Jiang et al. studied and investigated the structure of sintered hollow silver nanoparticles [12]. So far, researchers have successfully created models of sintering process for face-centred cubic (FCC) metals such as copper, gold, aluminium, silver and carried out MD simulations to effectively correlate between conventional theories and experimental results. It is revealed that the researchers have employed MD simulation to understand the laser–material interaction and successfully simulated one of the important phenomenon neck growths, which is formed during the initial stage of sintering and subsequently diminishes on the completion of the sintering.

In this present study MD simulations are employed and conducted using LAMMPS [1] to understand the conventional neck growth model during Selective laser sintering (SLS) of sub-micron sized copper (Cu) nanoparticles based on embedded atom method (EAM). In this simulation model identical spherical pairs of nanoparticles are investigated for sintering operation considering isothermal and nonisothermal effect on the neck growth of the coalition.

2 Selective Laser Sintering Process

Sintering is the process by which the powder materials are compacted and formed a solid mass or porous mass of material by means of heating and/or pressure at maintained sintering temperature lower than the melting point. During this process, atoms across the boundaries of the particles diffuse together and create one solid piece after solidification. During fusion, the neck is initiated at the contact points of the neighbour powder particles and later the neck grownup. Sintering can be possible to be used for metals and non-metals. In the conventional sintering process, sintering furnace is used in which whole compacted powder material is sintered inside it. Later various difficulties are faced by the researchers to develop complex shapes adopting this method of sintering.

In the past, several technologies were developed based on additive manufacturing concepts (bottom-up). Selective laser sintering of powder layer-by-layer is one such process widely accepted for ceramic and metal prototypes. This is a liquid phase sintering process; the powder particles are converted into a coherent solid mass by laser heating without through melting. In this process of sintering, laser is used as the heat source. Laser sintering is a process that uses a laser beam to selectively heat powder. It is done layer-by-layer with some depth increments, with a laser build line. The un-sintered powder acts as the support structure for the sintered part. To fabricate micro-parts, selective laser sintering is widely accepted.

The selective laser sintering (SLS) is a typical rapid prototyping (RP) technique, in which laser scans over the layer of metal powder sinters over previous layers to build 3D parts [5, 6, 13, 14]. In the process of SLS, the particles are scanned selectively by laser beam spot. The particles are heated through by the laser beam and they consolidate through sintering or melting and solidification reactions. The powder size, laser parameters, layer thickness play a vital role in porosity, minimum and maximum size limits of sintered part.

There are some limitations with respect to particle size during SLS. The powder grains are considered as spherical. During sintering, the upper layer of spherical grains is melted and creates a bond with the adjacent spherical grain after solidification. During sintering, the porosity is created. If grain size of powder particle is more, the sintered part becomes more porous. In order to control the above limitations driven by powder particle size, several alternatives have been adopted by the researchers. These include the use of multi-materials, mixture of powder of different particle sizes and use of submicron size of powder. The use of submicron-sized powder reduces both porosity and minimum feature size that could be built by laser sintering layer-by-layer.

2.1 Sintering Mechanisms

During sintering, atoms across the boundaries of the powder particles diffuse. The diffusion of atoms across the boundaries takes place by a gradient of interatomic potential. The major phenomena of sintering are densification and contact formation or neck formation [10, 14]. The atoms follow different paths to initialize and grow the neck. Fig. 1 shows six common mechanisms of sintering [15]. They are surface diffusion, vapour transport, lattice diffusion from surface and from grain boundary, grain boundary diffusion and plastic deformation (Table 1). Neck is sink of diffused matter for all six mechanisms.

In some of the sintering mechanisms, atoms displace from one position to another position following different paths. Atoms from the surface move to another surface or another portion of the same surface. Due to rearrangement of atoms of grain boundary, porosity arises between particles. In some other mechanisms, bulk number of atoms move to the surface of pores. Due to bulk movement of atoms, porosity may be eliminated and the density of the sample is increased. During sintering, centre of mass of particles changes continuously. Densification and shrinkage in the sintered part occur during condensation and the centres of mass of the particles approach towards each another. The paths of atom movement during formation of the neck are shown in Fig. 1.

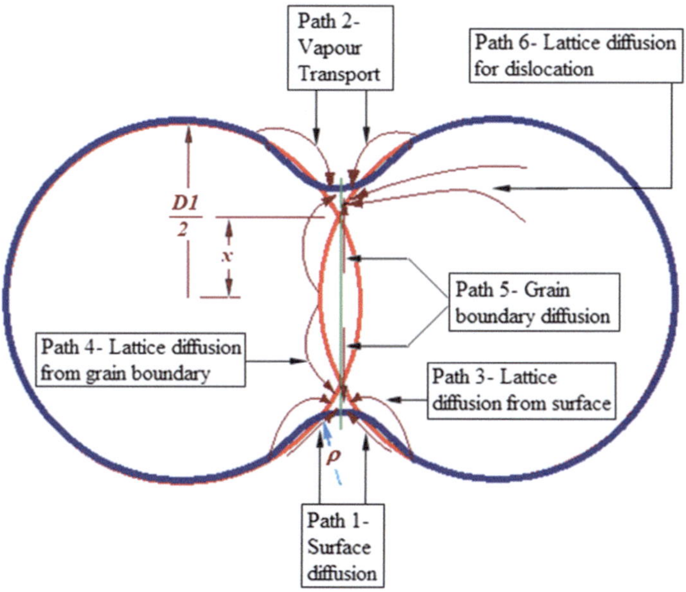

Fig. 1 Mechanism of sintering

Table 1 Common mechanisms of sintering

Source of matter	Path followed by atoms	Path of transport
Surface	Path-1	Surface diffusion
	Path-2	Vapour transport
	Path-3	Lattice diffusion from surface
Grain boundary	Path-4	Lattice diffusion from grain boundary
	Path-5	Grain boundary diffusion
Bulk dislocation of atoms	Path-6	Lattice diffusion

2.2 *Neck Growth Mechanism*

In selective laser sintering of powder, neck is initialized a short time after heating is started and the last stage of neck growth takes place on a long time. The neck grows according to the theory of sintering stated by Frenkel's [16]. The neck growth is shown in Fig. 2. Across grain boundary, rearrangement of atoms takes place; due to this high temperature, the solid materials behave like viscous flow which is caused due to surface stresses. Due to the diffusion of material atom, vacancies are created which directs to transportation of atoms.

Conventionally it was supposed that there are three stages of sintering. Later one more stage called stage zero was introduced by M. F. Ashby. In this stage, instantaneous neck formation occurs when powder particles come into contact [15]. The stages of neck growth during sintering are as follows:

Stage 0: Instantaneous neck formation.
Stage 1: Early stage of neck growth.

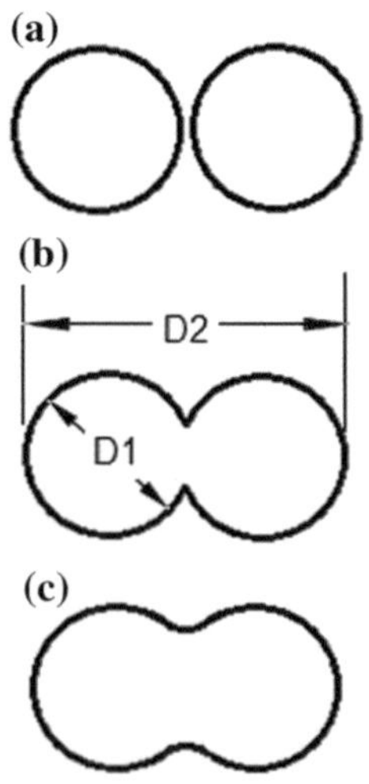

Fig. 2 Schematic diagram of neck growth of two particles during SLS

Stage 2: Intermediate stage: In this stage neck diameter is quite large and the pores are roughly cylindrical.
Stage 3: Final stage: In this stage, neck diameter reaches maximum, and the pores become isolated and spherical.

In Fig. 1, the mechanisms of sintering responsible for the neck growth are shown. From Fig. 2, it can be seen that all the mechanisms are responsible for neck growth, and mechanisms 4, 5 and 6 are also responsible for densification.

Initially, a small gap has been maintained between the two particles (a) after heating started the neck is initialized (b). Later due to heating for a long duration, the neck diameter is increased due to viscous flow of atoms from the particles (c). However, this mechanism is not very well understood for selective laser sintering of submicron-sized metal powder. In the following section, molecular dynamics (MD) based simulation has been used for micro-laser sintering of sub-micrometric metal powder.

3 Basic Theories of Molecular Dynamics

The basic problems of molecular dynamics (MD) simulation are to be solved using the equations of Newton's second law of motion. Each molecule interacts with its neighbour molecules by means of interactive forces. The interactive forces can be mathematically expressed as

$$F_{ij} = m_i a_i \tag{1}$$

where

F_{ij} resultant force acting between atom i and j
m_i mass of atom i and
a_i acceleration of the atom i $\left(a_i = \frac{d^2 r_i}{dt^2}\right)$

These resultant forces acting between atom i and j can be expressed as the interatomic potential energy gradient between neighbour atoms. MD simulations require to select an appropriate interatomic or intermolecular potential. Initial positions and masses of particles are defined according to lattice structure and type of material, respectively. A potential function is used to define the inter-particle interaction $\phi(r_1, r_2, r_3, \ldots, r_n)$, where $(r_1, r_2, r_3, \ldots, r_n)$ are denoted as positional coordinates of atoms. The resultant forces acting between each atom are to be computed by differentiating the potential function w.r.t. the distance between them as shown below:

$$F_i = -\frac{\partial}{\partial r_i}\phi(r_1, r_2, r_3, \ldots, r_n) \tag{2}$$

A number of interatomic or intermolecular potential are used in MD simulations. Among them, a popularly used interatomic or intermolecular potential is Lennard-Jones (LJ) potential and embedded-atom method (EAM) potential form metallic materials. LJ and EAM potential are used in present work. The basic theory of LJ and EAM potential are discussed briefly later. After selecting an appropriate interatomic or intermolecular potential, MD simulations require a suitable computation method to integrate Newton's equations of motion for which many algorithms have been established. Among them, one of the most recommended algorithms is Verlet algorithm.

3.1 Interatomic/Intermolecular Potential

3.1.1 Lennard-Jones (LJ) Potential

This proposed mathematical model of approximation demonstrates the interaction between two neutral atoms or molecules. F_{ij} is resultant force acting between atom i and j. We can replace F_{ij} with gradient of potential energy between atom i and j. Mathematically [17],

$$F_{ij} = -\frac{\partial \phi_{ij}}{\partial r_{ij}} \tag{3}$$

where ϕ_{ij} is LJ potential between atom i and j. Common expression of LJ potential [18] is

$$\phi_{ij} = 4\varepsilon\left[\left(\frac{\sigma}{r_{ij}}\right)^{12} - \left(\frac{\sigma}{r_{ij}}\right)^{6}\right] \tag{4}$$

where

ϕ_{ij} LJ Potential. This is the interatomic potential between the two atoms.

ε LJ well depth parameter. This is a measure of strength of attraction between the two molecules.

σ interatomic distance at which the interatomic potential between the two atoms become zero (shown in Fig. 3a). This is the equilibrium separation parameter (shown in Fig. 3b). It is equal to the half of the interatomic distance between nonbonding atoms.

r_{ij} separating distance between two atoms (it is the centre to centre distance).

Minimum value of ϕ_{ij} at r = $R_{\min}$ (Fig. 3b).

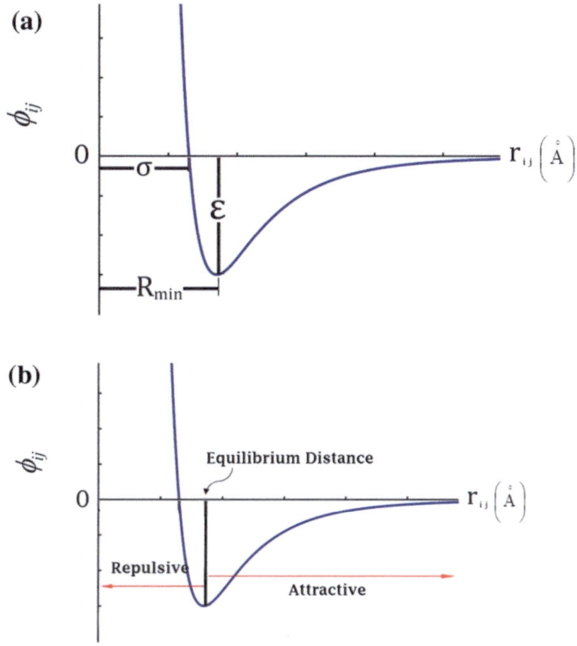

Fig. 3 LJ Potential energy versus intermolecular separation curve

3.1.2 Embedded-Atom Method (EAM) Potential

The embedded-atom method (EAM) [7, 19] is one of the approximation methods of interatomic potential energy between multiple atoms or bodies [20]. E_{tot} is a sum of energies of separation between multiple atoms and its neighbouring atoms [8, 9]. EAM potential mathematical model is suitable for MD simulations of metallic materials. EAM potential can be expressed mathematically as [20].

$$E_{tot} = \sum_{i=1}^{n} E_i \tag{5}$$

where E_{tot} is the total energy of the system and E_i is energy of atom i.

For each atom, the energy function can be broken up into two different terms. The first term is interatomic potential between atom i and j. The later one is embedding energy function which is necessary to embed atom into the electron cloud due to neighbour atoms of the atom i. Mathematical expression of the energy of atom i is [19, 21]

$$\sum_{i=1}^{n} E_i = \frac{1}{2} \sum_{\substack{i,j \\ j \neq i}} \phi_{ij}(r_{ij}) + \sum_{i} U(\rho_i) \tag{6}$$

where r_{ij} is the distance between atoms i and j; $\phi_{ij}(r_{ij})$ is a standard interatomic potential between atoms i and j; U is the embedding energy; ρ_i is the total electron density at position r_i due to neighbouring atoms of atom i.

Both summations in the formula are applicable for overall neighbour atom i within the cut-off distance. Cut-off distance is the distance between atom i and j when infinitesimal small potential will work between them. Atoms or molecules or particles can interact when the distance between them will be less than cut-off distance.

3.1.3 Morse Potential

The Morse potential model expresses interatomic interactions for creating bonds between two neighbour atoms. Between the pair of atoms, there will be interatomic forces that will create a diatomic molecule when they approach each other closely. The Morse potential [22] energy can be mathematically expressed as [17, 23]

$$\phi(r_{ij}) = D_e\left(1 - \mathrm{e}^{-\alpha\left(r_{ij}-r_e\right)}\right)^2 \tag{7}$$

where r_{ij} is instantaneous interatomic separation or interatomic distance, r_e is the equilibrium interatomic separation D_e is the potential well depth in the curve (Fig. 4), α is a constant determined from the physical properties of the material.

It can be noted that Morse function is minimum and zero at the equilibrium intermolecular separation and it tends to dissociation energy D_e when intermolecular separation tends to infinite (Fig. 4a). Hence, subtracting

D_e from ϕ_{ij}, we have [17, 23]

$$\phi_{ij}^* = \phi_{ij} - D_e = D_e\left[\mathrm{e}^{-2\alpha\left(r_{ij}-r_e\right)} - 2\mathrm{e}^{-\alpha\left(r_{ij}-r_e\right)}\right] \tag{8}$$

where ϕ_{ij}^* is having a minimum value, i.e. $\phi_{ij}^* = -D_e$ at $r_{ij} = r_e$ and $\phi_{ij}^* \to 0$ at $r_{ij} = \infty$.

3.1.4 Born–Mayer Potential

Born–Mayer have proposed that intermolecular repulsion as exponential repulsive form [17],

$$\phi(r) = A \cdot \mathrm{e}^{-\left(\frac{r}{B}\right)} \tag{9}$$

where $\phi(r)$ is the Born–Mayer potential, r is intermolecular separation and A and B are empirically determined range and softness parameters of the material. It does not produce potential well.

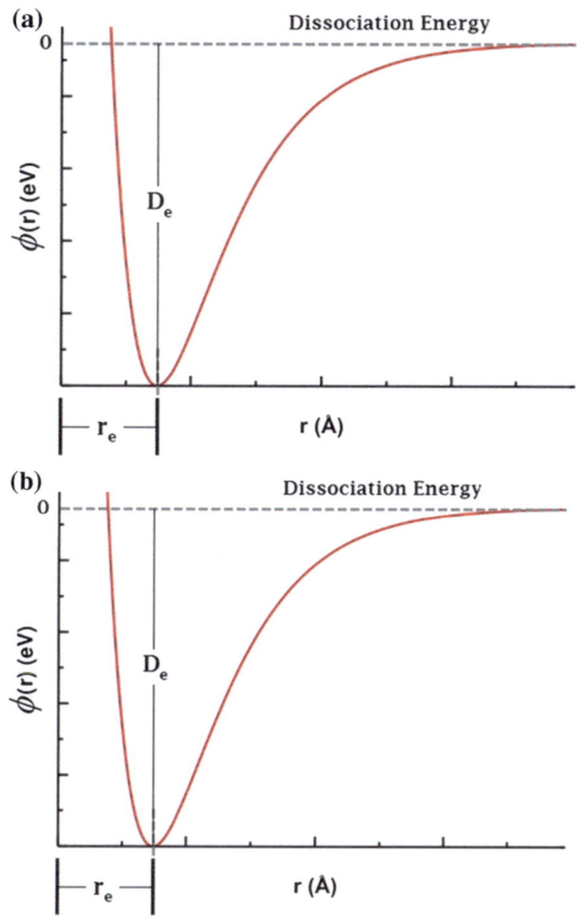

Fig. 4 Morse potential energy versus intermolecular separation curve

3.1.5 Tersoff Potential

According to the proposal of Tersoff [24], the total energy of the pair of atoms can be modelled as the summation of interatomic interactions. The energy can be mathematically expressed as the function of the atom positions [17]

$$E = \sum_i E_i = \frac{1}{2}\sum_i \sum_{i \neq j} \phi_{ij} \tag{10}$$

where

$$\phi_{ij} = f_C(r_{ij})\left[a'_{ij} f_R(r_{ij}) + b'_{ij} f_A(r_{ij})\right]$$

$$f_R(r) = A' \cdot e^{(-\lambda_1 r)}$$

$$f_A(r) = -B' \cdot e^{(-\lambda_2 r)}$$

$$f_C(r) = \begin{cases} 1, & r < R - D \\ \frac{1}{2} - \frac{1}{2}\sin\left(\frac{\pi}{2}(r - R)/D\right), & R - D < r < R + D \\ 0, & r > R + D \end{cases}$$

$$b'_{ij} = \left(1 + \beta^n \zeta_{ij}^n\right)^{-1/2n}$$

$$\zeta_{ij} = \sum_{k(\neq i,j)} f_C\left(r_{ij}\right) g\left(\theta_{i,j,k}\right).e^{\left[\lambda_3^3\left(r_{ij} - r_{jk}\right)^3\right]}$$

$$g(\theta) = 1 + \frac{c^2}{d^2} - \frac{c^2}{\left[d^2 + (h - \cos\theta)^2\right]}$$

$$a'_{ij} = \left(1 + \alpha^n \eta_{ij}^n\right)^{-1/2n}$$

$$\eta_{ij} = \sum_{k(\neq i,j)} f_c\left(r_{ij}\right).e^{\left[\lambda_3^3\left(r_{ij} - r_{jk}\right)^3\right]}$$

Here, i, j and k label the atoms of the system, r_{ij} is the length of the ij bond and $\theta_{i,j,k}$ is the bond angle between bonds ij and ik. Where R and D are cut-off parameters; A', B', λ_1, λ_2, λ_3, α, β, n, c, d, h are fitting parameters of the Tersoff potential.

3.2 Integration Algorithm

In Eq. (1), a is acceleration. Assuming simple case where acceleration is constant,

$$a = \frac{dv}{dt} \tag{11}$$

Integrating Eq. (11) expression of velocity v can be obtained

$$v = at + v_0 \tag{12}$$

We know that $v = \frac{d_r}{d_t}$ and after integration, the following expression can be obtained:

$$r = vt + x_0 \tag{13}$$

Combining Eqs. (12) and (13) expression of distance between particles or atoms or molecules, r at time t can be obtained

$$r = \frac{1}{2}at^2 + v_0 t + r_0 \tag{14}$$

where r is a function of time t, acceleration a, initial velocity $v_{0,}$ initial position x_0.

For integration of Newton's equations of motion and approximation of the positions, velocities and accelerations, Taylor's series expansion is applied in the algorithms [21].

$$\left.\begin{aligned} r(t+\Delta t) &= r(t) + \dot{r}(t)\Delta t + \frac{1}{2}\ddot{r}(t)\Delta t^2 + \cdots \\ v(t+\Delta t) &= \dot{r}(t) + \ddot{r}(t)\Delta t + \frac{1}{2}\dddot{r}(t)\Delta t^2 + \cdots \\ a(t+\Delta t) &= \ddot{r}(t) + \dddot{r}(t)\Delta t + \frac{1}{2}\ddddot{r}(t)\Delta t^2 + \cdots \end{aligned}\right\rangle \tag{15}$$

where Δt is a predetermined or restricted timestep, r is the position, $\dot{r}$ is the velocity (v), $\ddot{r}$ is the acceleration (a), $\dddot{r}$ and $\ddddot{r}$ are the third- and fourth-order derivative of position w.r.t. time. The above equations can be rewritten as [21]

$$\left.\begin{aligned} r(t+\Delta t) &= r(t) + v(t)\Delta t + \frac{1}{2}a(t)\Delta t^2 + \cdots \\ v(t+\Delta t) &= v(t) + a(t)\Delta t + \frac{1}{2}\dddot{r}(t)\Delta t^2 + \cdots \\ a(t+\Delta t) &= a(t) + \dddot{r}(t)\Delta t + \frac{1}{2}\ddddot{r}(t)\Delta t^2 + \cdots \end{aligned}\right\rangle \tag{16}$$

3.2.1 Basic Verlet Algorithm

Verlet Algorithm [25] is the most recommended algorithm and it is based simply on the position of the molecule or atom or particle, i.e. employing this algorithm the positions can be found. To find the next step, the following equation can be used [21]:

$$r(t+\Delta t) = r(t) + v(t)\Delta t + \frac{1}{2}a(t)\Delta t^2 + \cdots \tag{17}$$

Similarly, to find the previous step following equation can be used,

$$r(t - \Delta t) = r(t) - v(t)\Delta t + \frac{1}{2}a(t)\Delta t^2 + \cdots \tag{18}$$

Summing the Eqs. (17) and (18), we get [21],

$$r(t + \Delta t) \approx 2r(t) - r(t - \Delta t) + a(t)\Delta t^2 \tag{19}$$

To estimate the new position of atom contains a little error due to approximation. From Newton's laws of motion, we can rewrite acceleration as

$$\ddot{r}(t) = a = \frac{F(t)}{m}$$

where m is mass and $F(t)$ is force.

Equation (19) can be rewritten as [21],

$$r(t + \Delta t) \approx 2r(t) - r(t - \Delta t) + \frac{F(t)}{m}\Delta t^2 \tag{20}$$

In basic Verlet algorithm, velocity is not used to find the new position of the atom.

There are some similar algorithms of Verlet algorithm. They are Verlet Leapfrog algorithm and velocity Verlet algorithm.

3.2.2 Verlet Leapfrog Algorithm

Verlet Leapfrog algorithm uses the velocities to find the new position. By taking the average value half-integer timesteps, velocities can be calculated. The equations defining the velocities at half-integer timesteps can be written as [21]

$$v\left(t - \frac{\Delta t}{2}\right) = \frac{r(t) - r(t - \Delta t)}{\Delta t} \tag{21}$$

and

$$v\left(t + \frac{\Delta t}{2}\right) = \frac{r(t + \Delta t) - r(t)}{\Delta t} \tag{22}$$

From Eq. (22), the expression of new position can be obtained based on the old positions and velocities [21]

$$r(t + \Delta t) = r(t) + v\left(t + \frac{\Delta t}{2}\right)\Delta t \tag{23}$$

From Verlet algorithm, the expression for the updated velocity can be written as [21]

$$v\left(t+\frac{\Delta t}{2}\right)=v\left(t-\frac{\Delta t}{2}\right)+\frac{F(t)}{m}\Delta t \tag{24}$$

Positions and velocities can be determined applying this algorithm, in which positions and velocities leap over one another.

3.2.3 Velocity Verlet Algorithm

Unlike Verlet Leapfrog algorithm for finding velocities and positions, velocity Verlet algorithm determines velocity, position and acceleration for the same value of time variable t. It incorporates velocity in solving Basic Verlet algorithm as shown [21]

$$r(t+\Delta t)=r(t)+v(t)\Delta t+\frac{1}{2}a(t)\Delta t^2 \tag{25}$$

$$v(t+\Delta t)=v(t)+\frac{1}{2}[a(t)+a(t+\Delta t)]\Delta t \tag{26}$$

It can be noted that low memory is required for computation for velocity Verlet algorithm, because during simulation, this algorithm does not keep track of the velocity at every time step. The algorithm follows the following steps [21]:

Step-1 → Compute velocity at halfway timestep, $v\left(t+\frac{1}{2}\Delta t\right)=v(t)+\frac{1}{2}a(t)\Delta t$.
Step-2 → Compute position at same timestep, $r(t+\Delta t)=r(t)+v\left(t+\frac{1}{2}\Delta t\right)\Delta t$.
Step-3 → Derive $a(t+\Delta t)$ from the interaction potential using $r(t+\Delta t)$.

Eliminating the halfway timestep velocity, calculations may be done as follows [21]:

Step-1 → Compute the position, $r(t+\Delta t)=r(t)+v(t)\Delta t+\frac{1}{2}a(t)\Delta t^2$.
Step-2 → Derive $a(t+\Delta t)$ from the interaction potential using $r(t+\Delta t)$.
Step-3 → Compute velocity at same timestep, $v(t+\Delta t)=v(t)+\frac{1}{2}[a(t)+a(t+\Delta t)]\Delta t$.

This is noted that this algorithm assumes that acceleration $a(t+\Delta t)$ only depends on position $r(t+\Delta t)$, and does not depend on velocity $v(t+\Delta t)$

4 Thermodynamic Ensembles

In MD simulations, our approach is very much atomistic so it becomes very difficult to examine each and every atom for its properties as various atoms exhibit different states at microscopic level. Statistical mechanics is required to determine atomistic position and moment, but if the system is taken as a whole, i.e. macroscopic level, they tend to exhibit same thermodynamic states which can be defined in terms of thermodynamic properties. So, a system at a given macrostate can be defined in terms of basic thermodynamic properties such as pressure (*P*), volume (*V*), temperature (*T*) and entropy (*S*), etc. which are grouped to form basic thermodynamic ensembles as NVE, NVT, NPT and μVT [26] keeping some of the properties as constant like temperature which is controlled using thermostat and rest as variable or not accounted in observation (Table 2).

5 Molecular Dynamics (MD) Simulation for SLS of Nanoparticles

5.1 *Process of Simulation Study*

In this simulation-based study, the following sequence of process has been followed to study neck growth mechanism in selective laser sintering of nanoparticles (Fig. 5):

Step 1: Model selection.
Step 2: Select an interatomic potential appropriately.
Step 3: Selection of algorithm for the integrating equations of motion.
Step 4: Model initialisation.
Step 5: Moderating the model's initial state to its dynamically equilibrium state.
Step 6: Simulating the model and analysing the results.

Table 2 Summary of variables in basic thermodynamic ensembles

Ensembles	No of atoms, *N*	Volume, *V*	Pressure, *P*	Temperature, *T*	Overall energy, *E* (PE + KE)
Microcanonical (NVE)	Constant	Constant	Variable	Variable	Constant
Canonical (NVT)	Constant	Constant	Variable	Constant	
Isobaric isothermal (NPT)	Constant		Constant	Constant	
Grand canonical (μVT)				Constant	

5.2 MD Simulation of Laser Sintering of 1.82 nm Diameter Nanoparticles

An MD simulation program has been developed for the study of sintering of copper nanoparticles. In this MD simulation of sintering, periodical boundary conditions are used. The steps of MD simulations are as follows.

At first, a cuboid-shaped three-dimensional simulation box of 4.26×2.34 $2.34\ nm^3$ has been created. Inside the two spherical nanoparticles of 1.82 nm diameter have been created. Using Verlet algorithm the atomic velocities are defined to achieve temperature control. Timestep has been used as 1.0 fs and atom neighbour list is updated. For simulation of the laser sintering, two identical particles are kept side by side and the centre distance of the two particles of diameter D1 (=1.84 nm) are kept D1 + 0.02 nm (D1 is diameter of each spherical nanoparticles) at 300 K temperature. The neck growth initializes in a short time and the last stage of neck growth happens for a long time. The steps of neck growth according to Frenkel's theory of sintering are shown in Fig. 2. Initially, a small gap has been maintained between the two particles (a) After heating started the neck is

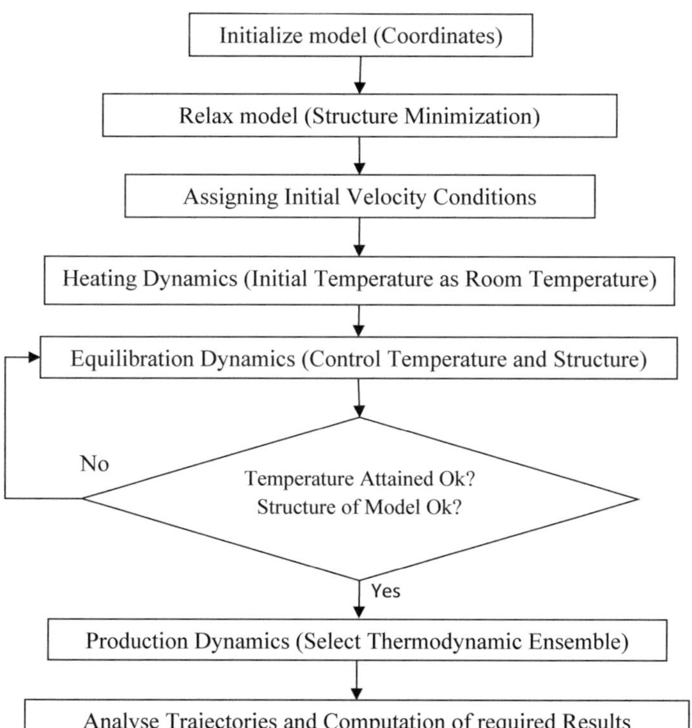

Fig. 5 Flow chart for molecular dynamics simulation study

initialized (b). Later due to heating for a long duration, the neck diameter is increased due to viscous flow of atoms from the particles (c).

5.2.1 Atomistic Model

Two identical spherical copper nanoparticles are created inside a simulation box. The simulation box size is $4.26 \times 2.34 \times 2.34$ nm^3. Each of the copper nanoparticles is of diameter 1.82 nm and the centre distance is kept 1.84 nm. A gap of 0.02 nm is maintained between two spherical copper nanoparticles. The atomistic models of copper nanoparticles are shown in Fig. 6. The number of the atoms in the particles is 12,646 and 12,630, respectively, and sufficient to simulate thermal behaviour.

Initial configuration is prepared by randomizing the velocities of atoms in the nano-sized powders according to Maxwell–Boltzmann distribution at the room temperature.

5.2.2 Sintering Simulation Model

To simulate the SLS the copper nanoparticles pair is subjected to continuous laser heat. Also, it is considered that no heat loss is occurred during the simulation study [9]. The periodic boundary conditions are enforced in all directions during creating atomistic models of copper nanoparticles.

A. *Isothermal Sintering*: After development of atomistic model, the model is simulated at temperature set ($\sim$300, $\sim$600, $\sim$900, $\sim$1200 and $\sim$1500 K) with NVT considerations in which Nose–Hoover thermostat is used. The atomistic model is initially kept at above-mentioned temperature set and the atoms

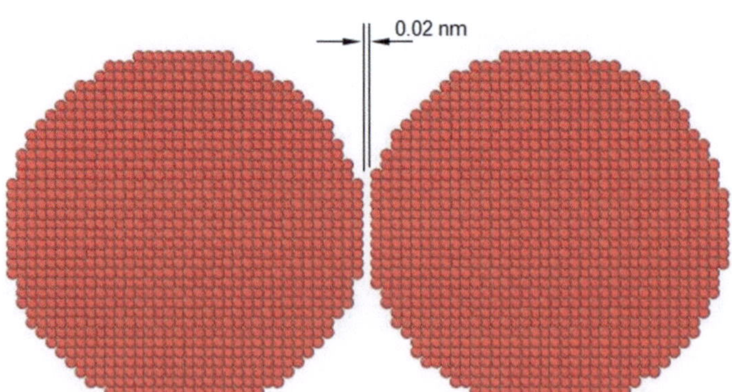

Fig. 6 Atomistic models of copper nanoparticles (each particle diameter 1.82 nm) with a gap of 0.02 nm

Fig. 7 Comparison temperature versus timestep graph for isothermal sintering at different temperatures (300, 600, 900, 1200 and 1500 K)

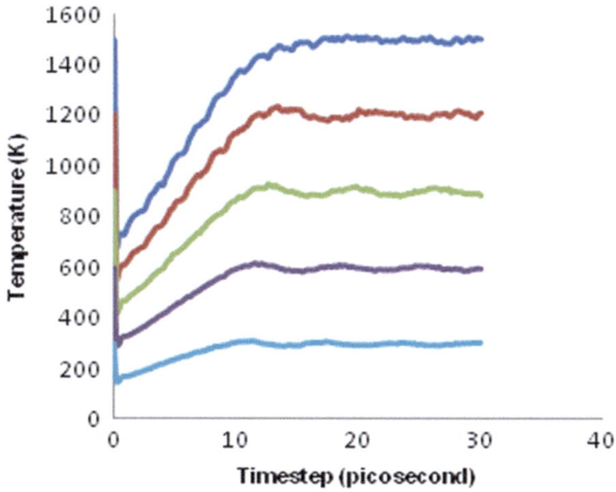

are randomized with a velocity following Gaussian distribution. The atoms are equilibrated at the temperature above-mentioned temperatures and the sintering time is set to 30 picoseconds (ps) with a timestep of 1.0 femtosecond (fs). During the equilibration process, the neck growth initialized with the EAM potential [7, 19] gradients. Neck diameters at various timesteps are measured. Temperature vs. timestep graph has been shown in Fig. 7.

B. *Non-isothermal Sintering*: The external heat is added to the particles. The particles are heated for 30 ps. During heating, the temperature is increased from 300 to 1500 K. The particles are heated at the rate of 4.00×10^{13} K/s with intensity of laser around 10^{10} W/m^2 which is in the range of traditional heating rate. It is observed that the temperature is increased linearly to the model of laser sintering. Neck diameters at various timesteps are measured. temperature versus timestep graph has been shown in Fig. 8.

5.2.3 Results and Discussion

- *Neck growth during copper–copper nanoparticle pair isothermal sintering*

From MD simulation in isothermal sintering of copper–copper nanoparticle pair it is observed that temperature of copper nanoparticle pair is decreased initially up to 0.2 ps and then increased to the considered temperature. The temperature profile is shown in Fig. 7. In isothermal sintering, the atomic model is initially kept in the considered temperature set and the atoms are randomized with a velocity following Gaussian distribution. The total simulation time is 30 ps with a timestep of 1.0 fs.

The dimensional change of neck growth during the simulation is shown in the graph of neck diameter vs. timestep in Fig. 9. In every case of isothermal sintering,

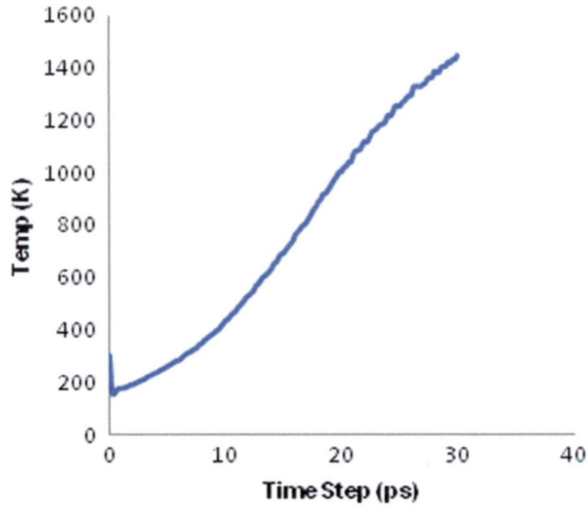

Fig. 8 Temperature versus timestep graph for non-isothermal sintering from temperatures 300 to 1500 K

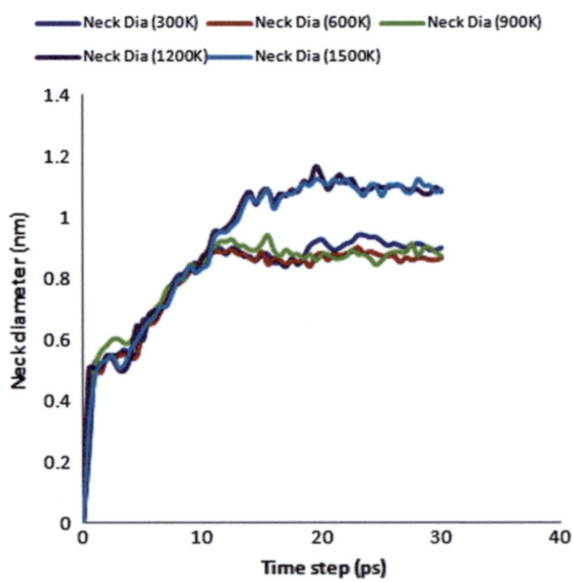

Fig. 9 Comparison neck diameter versus timestep graph for isothermal sintering at different temperatures (300, 600, 900, 1200 and 1500 K)

the rapid neck growth is observed initially. In cases of 300, 600 and 900 K after 10 ps, neck diameter is getting stable. And in cases of 1200 and 1500 K after 15 ps, neck diameter is getting stabled.

- *Neck growth during copper–copper nanoparticle pair non-isothermal sintering*

It has been observed that the temperature of copper nanoparticle pair increases from 300 to 1444 K in 30,000 timesteps, i.e. 30 ps with each step of 1.0

femtosecond (fs). Figure 8 shows the temperature profile during sintering simulation. Initially, particle pair temperature is set at 300 K. Just after starting of the simulation (i.e. before 400 fs), slight oscillations have been observed. Afterwards, the temperature increased almost linearly from temperature 300 to 1444 K. Total time of simulation is 30 ps.

Figure 10 shows the change in configuration of the pair of copper nanoparticle during the sintering simulation process. Initially, the two nanoparticles are at a centre distance of D1 + 0.02 nm. Just after starting of simulation, the pair of nanoparticles comes in contact with each other. Afterwards, temperature increases and the atoms close to the boundary of each particle started to travel together in order to minimize the surface tension of the system. This phenomenon leads to neck formation between two particles. Growth of neck occurs due to the dislocation of nearby atoms of neck. All the surface atoms lose crystalline structure and melting initializes at the outer layer of the particles.

The easier method to investigate the configuration changes during sintering is to measure the dimensional change of the two particles. The total length D2 and

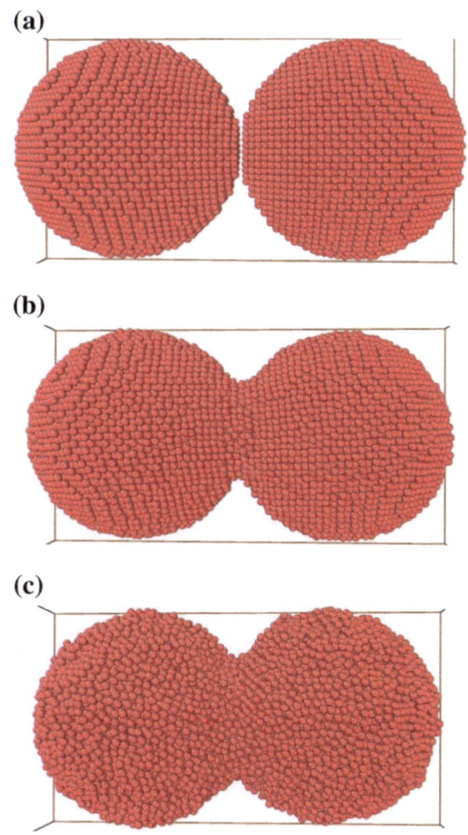

Fig. 10 Change in configuration of nanoparticle pair of copper throughout sintering

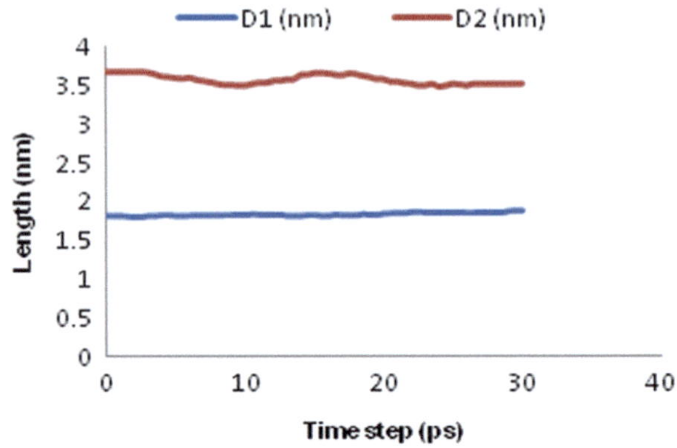

Fig. 11 Dimensional change of D1 and D2 during sintering

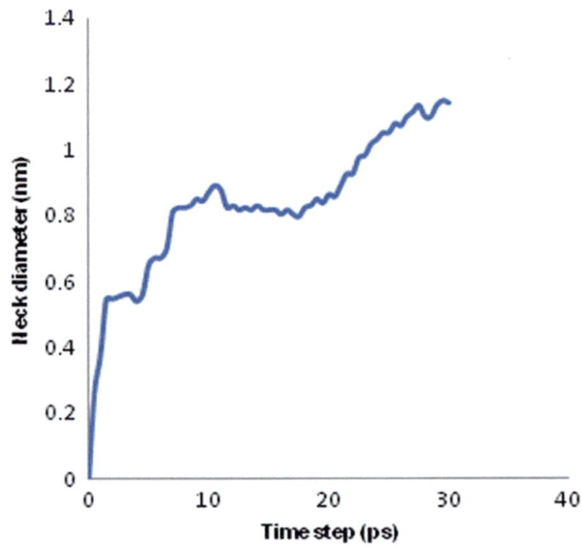

Fig. 12 Dimensional change of neck diameter during sintering

particle diameter D1 are measured (Fig. 2) and the measured neck diameters are shown in Figs. 11 and 12.

Figure 11 shows the dimensional change of the copper nanoparticles pair model, including total length (D2) and particle diameter (D1). The neck diameter shows the total length decreased due to sintering with increasing sintering time. This is because the atoms near the neck moved to the neck plane from the initial contact point. One can also observe that the diameter of each individual particle does not change significantly during the sintering process. Thus, the neck diameter keeps increasing during the 30 ps simulation.

5.3 MD Simulation-Based Virtual Experiments

In order to study and establish relation between the neck growth in sintering of different sized nanoparticle pairs undergoing similar heating rates, MD simulation-based virtual experiments were conducted. In this case, sintering of spherical identical copper nanoparticles, three separate models are created. In this MD simulation of sintering, periodical boundary conditions are used.

At first, a cuboid-shaped three-dimensional simulation box is created. Inside the simulation box pair of spherical identical nanoparticles has been created. The centres of gravity of both the particles are present horizontally at equal *y*-coordinate and *z*-coordinate. The centres of gravity are at a distance of the particle diameter, i.e. there is no gap between two particles. Initially, the particles are kept at room temperature, i.e. 300 K. The atoms are randomized to achieve temperature control using Verlet algorithm. Due timesteps have been used as 1 fs. The neck growth in each case has been observed during sintering. The steps of neck growth according to Frenkel's theory [16] of sintering are shown in Fig. 2.

5.3.1 Development of Atomistic Model

In the present simulation, the atomistic models of pairs of identical spherical copper nanoparticles are developed to investigate neck growth and other mechanisms. Particle diameters are considered as 1, 2 and 3 nm. For sintering, three models are developed. In the first model, two identical spherical copper nanoparticles of diameter 1 nm are created inside a three-dimensional simulation box. In the second model, the identical spherical copper nanoparticles of diameter 2 nm are created inside a three-dimensional simulation box. In the third model, the identical spherical copper nanoparticles of diameter 3 nm are created inside a three-dimensional simulation box. In the first second and third models, there are two particles, each particle containing 2103, 16,779, and 56,606 atoms, respectively. The total number of atoms in the above models is 4206, 33,558 and 113,212, respectively. The three models are shown in Fig. 13. The neck growth of these models and the sintering mechanism are studied using these models.

Maxwell–Boltzmann distribution is considered to randomize the atoms in the nano-sized particles to prepare the initial configuration at room temperature.

5.3.2 Sintering Simulation Model

To simulate the SLS the copper nanoparticle pairs are subjected to laser beam irradiation continuously. The rate of heating is maintained at 4×10^{13} K/ps. Also, it is considered that no heat loss is occurred during the simulation study [9]. The periodic boundary conditions are enforced in all directions during creating atomistic

Fig. 13 Atomistic models of spherical copper nanoparticles pair of (figures are not in scale)

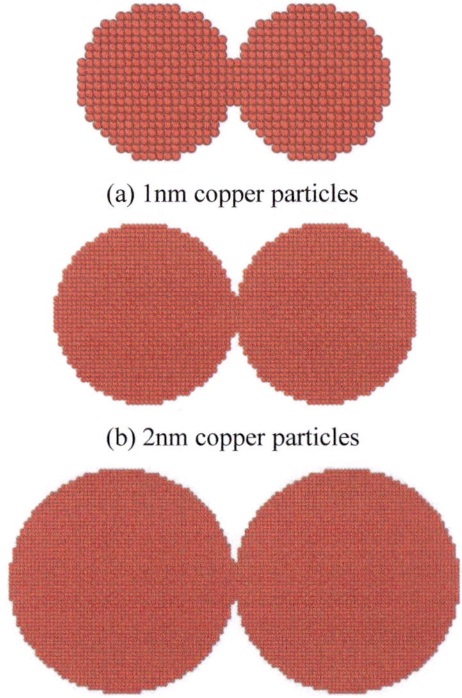

(a) 1nm copper particles

(b) 2nm copper particles

(c) 3nm copper nanopartciles

models of copper nanoparticles. In every model, the pairs of nanoparticles are heated for 30 ps with due timesteps of 1.0 fs and atom neighbour list is updated.

5.3.3 Results and Discussion

- *Snapshots and Vector plots of Laser sintering of copper–copper nanoparticle pair*

In every case of laser sintering, a pair of copper nanoparticles is heated according to a linear equation from temperature 300 to 1500 K, and the rate of heating is kept 4×10^{13} K/ps. The pair of copper nanoparticles is heated for 30 ps of time with timestep of 1.0 fs. Figures 14, 15 and 16 show the snapshots for 1, 2 and 3 nm copper particles pairs, respectively, at five timesteps 5.6, 12, 18.4, 24.8 and 30 ps. Comparison of the three models shows that after 30 ps of time of sintering, they reach in different temperatures. 1, 2 and 3 nm copper nanoparticles pairs reach the temperature of 1426.3587, 1437.0025 and 1391.1457 K after 30 ps heating.

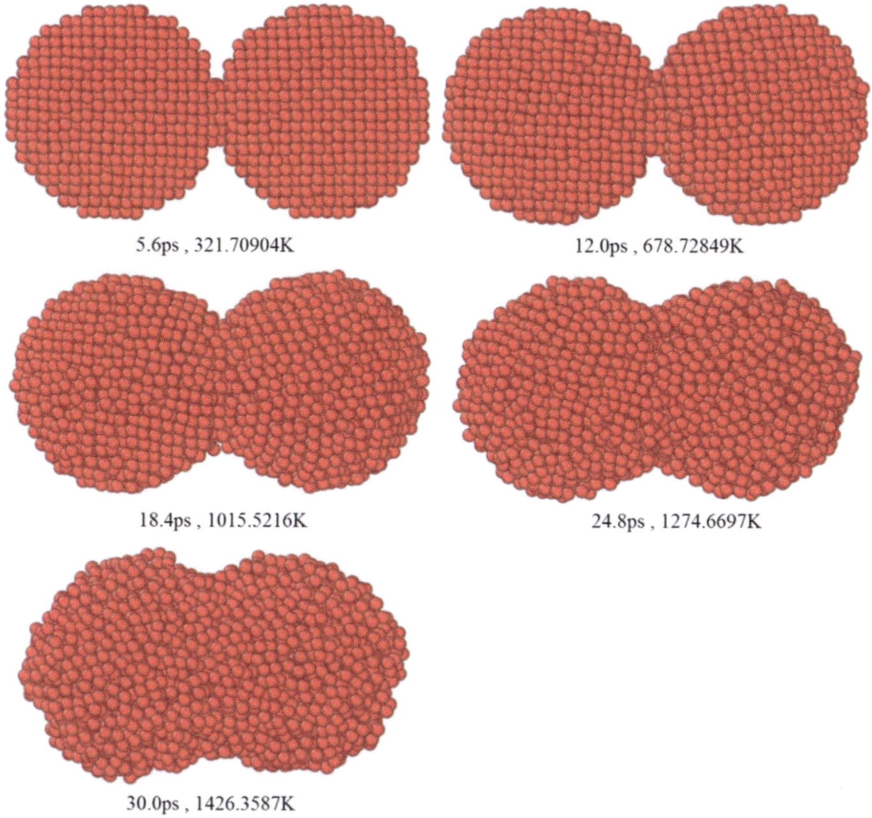

Fig. 14 Sintering snapshots of the pairs of nanoparticles of 3 nm at different timesteps (5.6, 12, 18.4, 24.8 and 30 ps)

The snapshots at 30 ps timestep exhibit the sintered shape of the nanoparticles. In comparison with the snapshots of nanoparticle pairs of 1, 2 and 3 nm, it is observed that at the same heating rate the neck growth is better in lower size particles. The change in neck size is significantly more in 1 nm nanoparticles pair. Whereas, in 3 nm nanoparticles, the pair neck size is almost similar throughout the duration of sintering. In 2 nm nanoparticles pair, the neck growth is also significant. The 1 nm nanoparticles pair achieved the more sphere-like shape. At the same time duration, i.e. 30 ps, sintering is almost complete in the 1 nm nanoparticles pair. This phenomenon is mainly happening due to the diffusion of atoms and ultimately melting. It is also observed that neck formation is initiated at the timestep of 2.4 ps after sintering starts.

The insights during sintering of nanoparticle pairs of different particle sizes are shown in Figs. 14, 15 and 16, at five timesteps 5.6, 12, 18.4, 24.8 and 30 ps. The directions of the atom movements are shown in those vector plots. The vectors of atom movement at different timesteps are illustrated in Figs. 17, 18 and 19 during

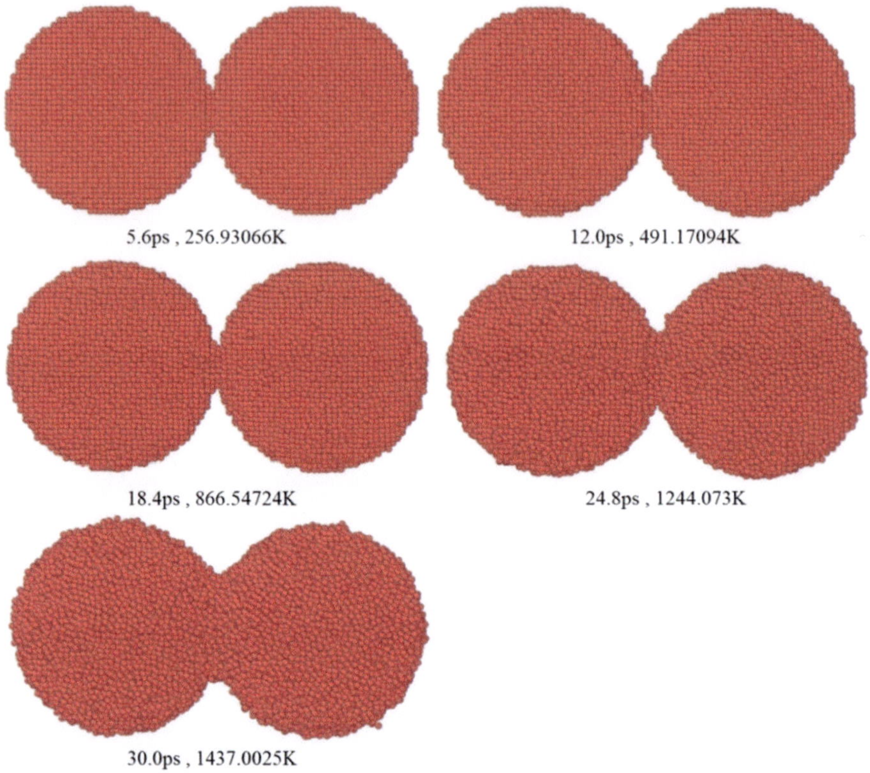

Fig. 15 Sintering snapshots of the pairs of nanoparticles of 2 nm at different timesteps (5.6, 12, 18.4, 24.8 and 30 ps)

sintering the pairs of copper nanoparticles, i.e. 1, 2 and 3 nm sized pairs of nanoparticles, respectively. The arrows in Figs. 17, 18 and 19 represent magnitude and direction of the displacement vectors of the atoms with reference to the initial coordinates of the same atom.

Comparison of three cases, during sintering the pair of copper nanoparticle pair it is observed that at 5.6 ps, the directions of movement in 1 nm particles are significant, but in 2 and 3 nm particles, it is hardly distinguished. This phenomenon has also affected the neck growth. The neck growth history is shown in Figs. 21 and 22. In 2 nm particle pair, the movement is quite significant at 18.4 ps whereas in 3 nm particle pair it is at 24.8 ps. In three of the cases, it is observed that the movement of the atoms is in a whirling way. The atoms whirl and are moving towards the neck. The direction of whirling is just opposite in the particles. This phenomenon indicates that the melting of nanoparticles arises and tends to propel the atoms to form a single spherical particle. At the outer surface of the nanoparticle, the displacement vectors are clearly visible. From the displacement vector plots (Figs. 17, 18 and 19), it is observed that the atoms of the left-sided particle are whirling in anticlockwise

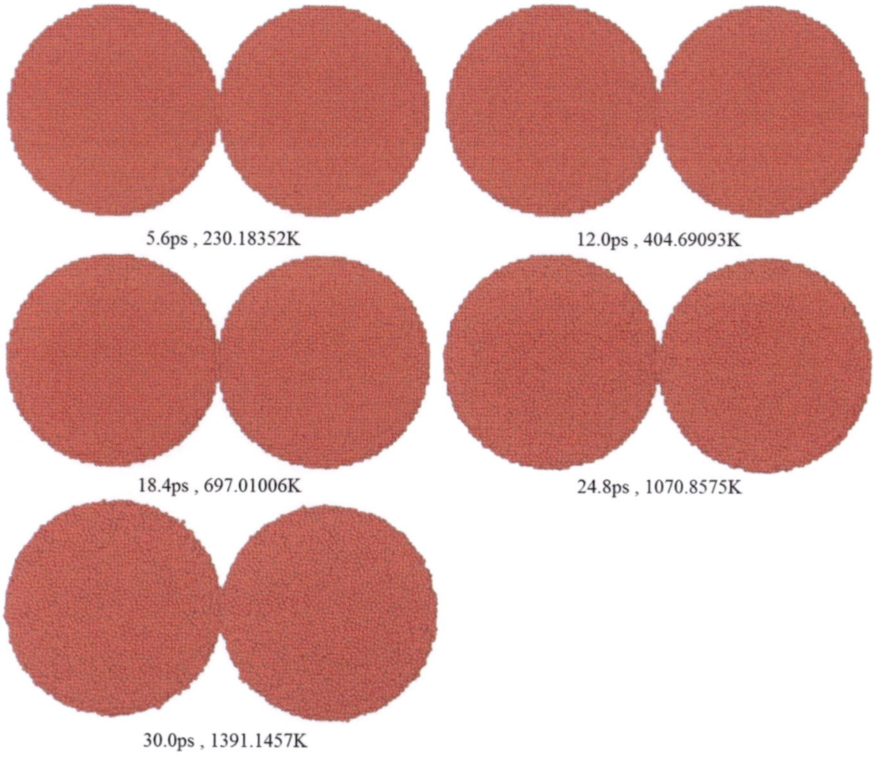

Fig. 16 Sintering snapshots of the pairs of nanoparticles of 3 nm at different timesteps (5.6, 12, 18.4, 24.8 and 30 ps)

direction and in the right-sided particle in clockwise direction. The displacement vectors of the atoms inside the particle are not visible properly. But from the vector plots it is clearly understood that melting of nanoparticles continues and they are driven to merge to a single particle. It is found that at higher temperature atom movements are random. At the same heating rate, the development of sintering is more thorough in the case of smaller particle size.

- *Neck growth during Laser sintering of copper–copper nanoparticle pair*

It has been observed that the temperature increases from 300 to 1426 K, 300 to 1437 K and 300 to 1391 K of the copper nanoparticle pairs of 1, 2 and 3 nm, respectively, in 30,000 timesteps, i.e. 30 ps with each step of 1.0 fs. Figure 20 shows the temperature profile during sintering simulation. The initial particle pair temperature is set at 300 K. It has been observed that at the beginning of the simulation, the temperature drops and after that, the temperature increased linearly during 30 ps of the simulation time.

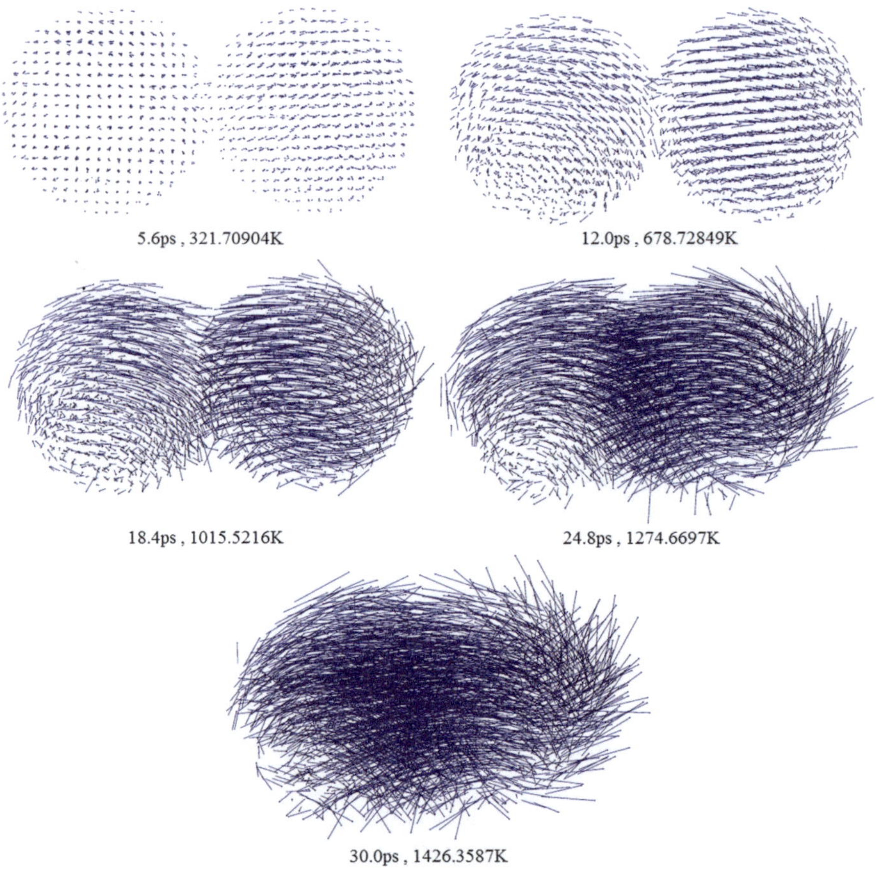

Fig. 17 Vector plots of the pairs of nanoparticles of 1 nm at different timesteps (5.6, 12, 18.4, 24.8 and 30 ps)

The effect of particle size on neck growth is illustrated in Figs. 21 and 22. The neck growth is quite smooth and rapid in the 1 nm copper particle pair. It is clearly observed that the neck growth in the 3 nm copper particle pair is almost stagnant. A very slow neck growth is observed in the case of 3 nm particles pair. In 2 nm particles pair, the neck growth is slow but quite smooth.

Therefore, it is observed that in same heating rate in the 1 nm particles pair neck formation is smooth, stable and rapid. In 1 nm particle pair, the neck formation started rapidly after 10 ps of sintering and it is smooth and stable. In 2 nm particle pair, the neck formation started rapidly after 20 ps of sintering and it is quite smooth and stable. But in 3 nm particles pair, the neck formation has become rapid after 15 ps of sintering. After approximately 17 ps time, the rate of neck formation is again very less and almost stagnant up to the end of sintering time.

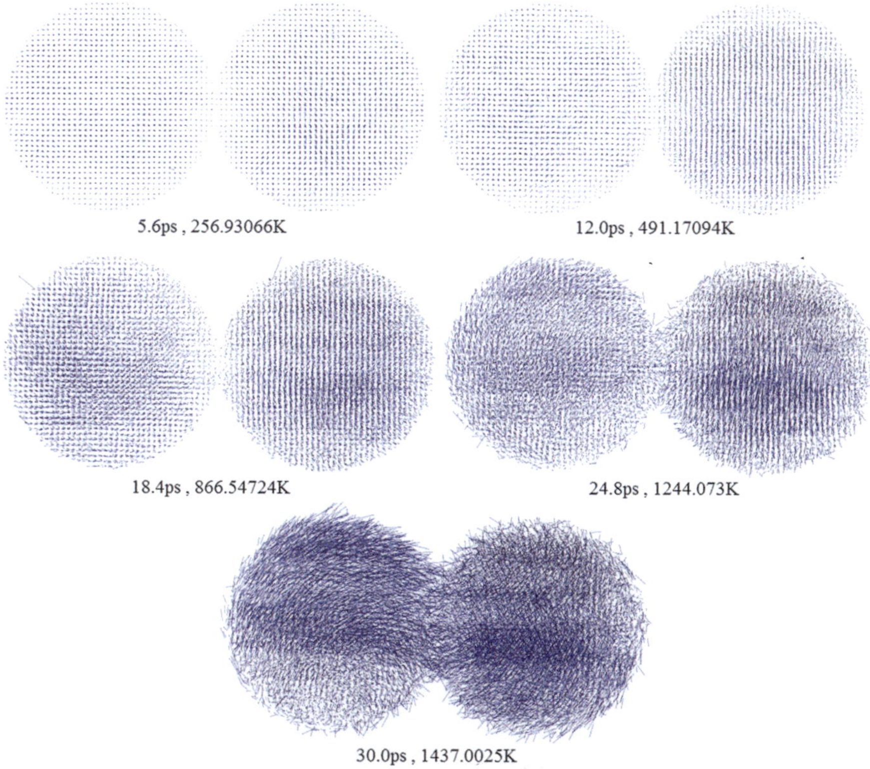

Fig. 18 Vector plots of the pairs of nanoparticles of 2 nm at different timesteps (5.6, 12, 18.4, 24.8 and 30 ps)

- *Mean square displacement (MSD) during Laser sintering of copper–copper nanoparticle pair*

The mean square displacement (MSD) is the distance of the position that an atom dislocates during laser heating process with respect to a reference position [26]:

$$\mathrm{MSD} = \frac{1}{N}\sum_{i=1}^{N}[r_i(t) + r_i(0)]^2 \tag{27}$$

N is the number of sampled atoms of the nanoparticle which are to be averaged in fixed time t at the position r. A vector constituting of four quantities is considered. The three elements of the vector are squared displacements, i.e. d$x \times$ dx, d$y \times$ dy and d$z \times$ dz. The fourth element of the vector is the sum of squared displacements, i.e. (d$x \times$ dx + d$y \times$ dy + d$z \times$ dz). The displacement of an atom is computed from its reference location. MSD of atoms, in the present simulation

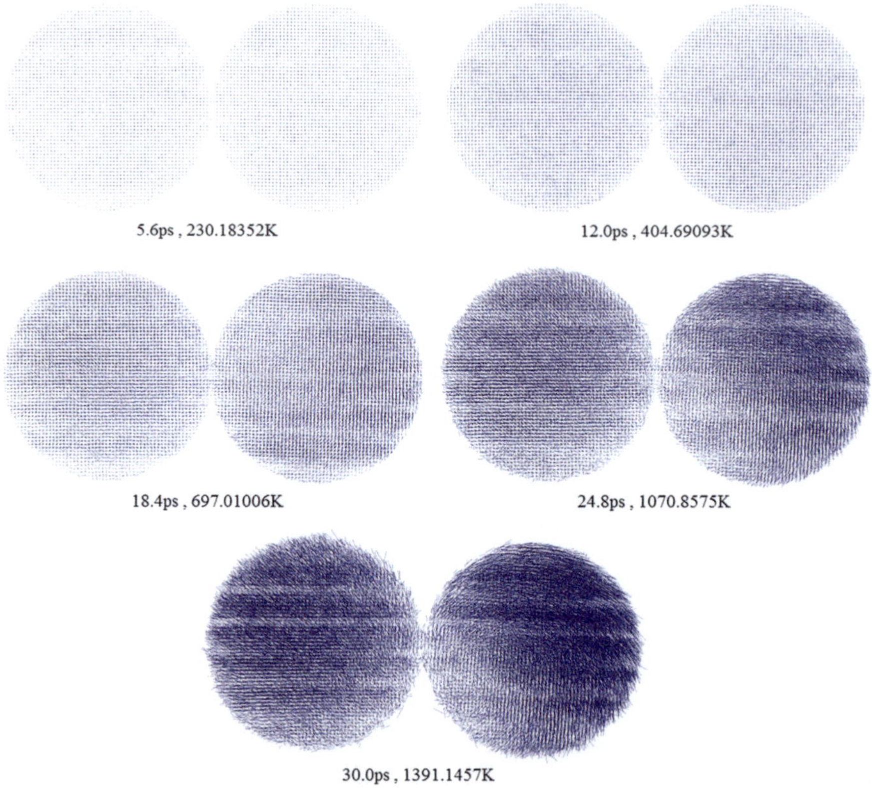

Fig. 19 Vector plots of the pairs of nanoparticles of 3 nm at different timesteps (5.6, 12, 18.4, 24.8 and 30 ps)

Fig. 20 Temperature versus time graph for different sized nanoparticles during sintering

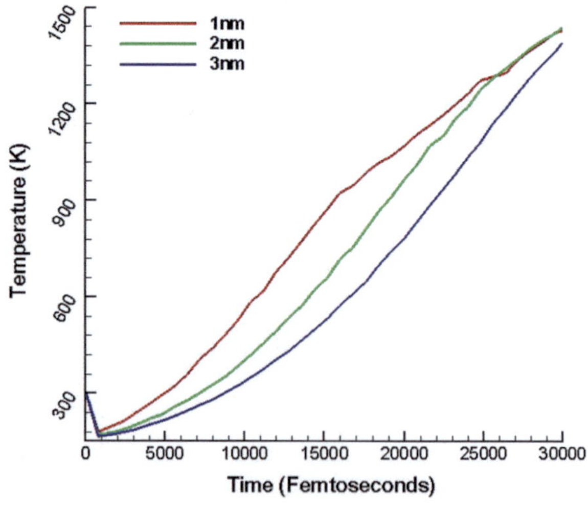

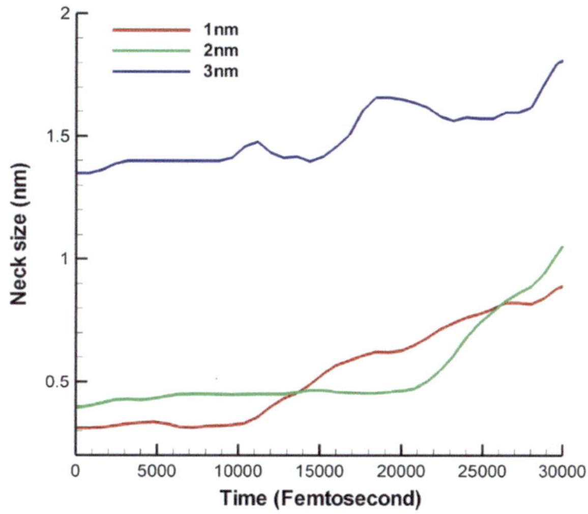

Fig. 21 Variation in neck size with time during laser sintering for different pairs of nanoparticles

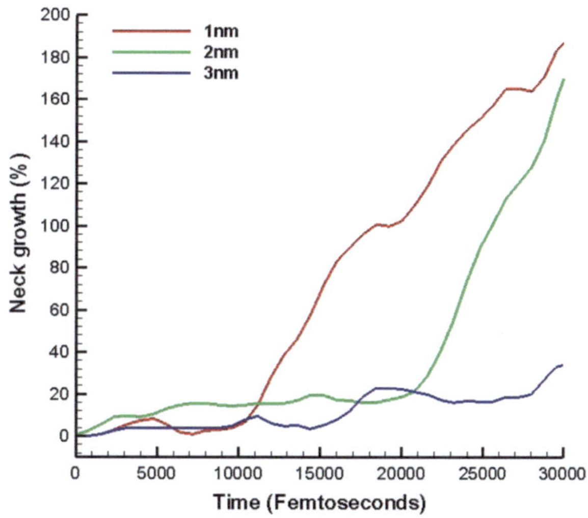

Fig. 22 Effect of time on neck growth during laser sintering for nanoparticle pairs of 1, 2 and 3 nm size

models, is computed with reference to the initial positions of atoms when laser sintering was initiated. The curve of the MSD versus time is plotted in Fig. 23. This reveals that the atom dislocation increases with the time. Therefore, it can be stated that coefficient of diffusion of atoms is directly proportional to the slope of the curve.

In Fig. 23, the MSD for the copper nanoparticles pairs is compared. The MSD is calculated w.r.t. the preliminary atom positions (reference positions) at the initiation stage of sintering. In Fig. 23, it is observed that the value of MSD initially near to zero and at the finishing point of sintering process MSD values depart from the zero

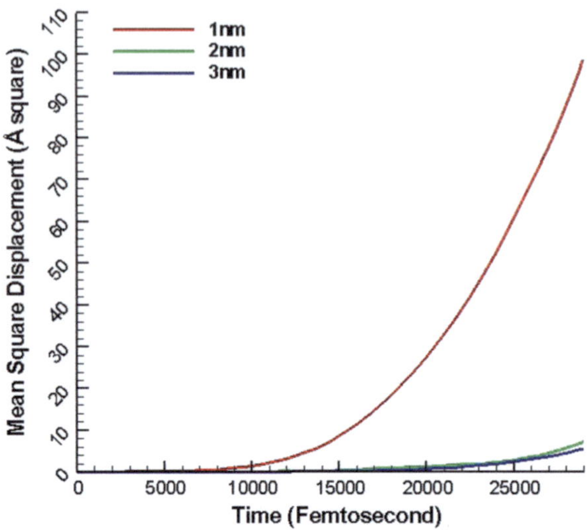

Fig. 23 Mean square displacement (Å^2) versus time (femtosecond) graph during laser sintering for different sized nanoparticle pairs

line and it increased. In the case of 1 nm copper particles pair, the change in MSD is significant and is increased rapidly after 10 ps time of sintering. The smooth MSD versus time curve for 1 nm particle pair implies that the diffusion occurs and atoms displaced from the initial positions. This implies the confirmation of thorough melting of material during the sintering of 1 nm copper nanoparticles pair. On the other hand, the MSD versus time curve of 2 and 3 nm copper nanoparticles pairs are smooth but both the curve is closer to zero MSD value throughout the sintering. The graph indicates that the displacements of the atoms are very less during sintering. This implies that the melting in 2 and 3 nm copper nanoparticles pairs is very slow, and less amount of material has been melted. Due to this reason, the neck growth rate is small in the 2 and 3 nm particle pairs.

6 Conclusion

MD-based simulation study in this chapter shows that neck formation between two adjacent nanoparticles as a result of ultrashort-pulsed laser irradiation plays a key role in optimizing selective laser sintering of nanoparticles. The laser irradiation over 1.82 run diameter copper powder resulted in a rapid heating with increase in temperature from 330 to 1444 K in 30 ps. It resulted in a linear increase of neck diameter up to the maximum of 1.20 nm, which is about 70% of particle size and formed strong bonding between sintered particles. Three different atomistic models of pairs of nanoparticles of 1, 2 and 3 nm sizes were prepared. From the vector plots, it is clearly visible that the atoms' displacement is towards the contact point of two particles. It is also noted that during sintering the neck growth of lower sized nanoparticles is thorough and consistent as compared to the larger nanoparticles pair under the same heating rate. While this study established the neck formation mechanism between the particles of the same layer, it needs to be extended to compute neck size between the particles laying one over the other to understand micro-sintering process mechanism and also to compute porous size of built micro-part.

Acknowledgements This work is supported under CSIR Network projects NWP-30 and ESC0112 at CSIR-Central Mechanical Engineering Research Institute (CSIR-CMERI), Durgapur, India.

References

1. LAMMPS MD Simulator. http://lammps.sandia.gov/index.html.
2. German, R. M. (2013). History of sintering: Empirical phase. *Powder Metallurgy, 56,* 117–123.
3. Li, L., Hong, M., Schmidt, M., Zhong, M., Malshe, A., Huis, B., et al. (2011). Laser nano-manufacturing—State of the art and challenges. *CIRP Annals–Manufacturing Technology, 60,* 735–755.
4. Ko, S. H., Pan, H., Grigoropoulos, C. P., Luscombe, C. K., Fréchet, J. M. J., & Poulikakos, D. (2007). All-inkjet-printed flexible electronics fabrication on a polymer substrate by low-temperature high-resolution selective laser sintering of metal nanoparticles. *Nanotechnology, 18,* 345202.
5. Regenfuss, P., Streek, A., Hartwig, L., Klötzer, S., Brabant, T., Horn, M., et al. (2007). Principles of laser micro sintering. *Rapid Prototyping Journal, 13,* 204–212.
6. Regenfuß, P., Ebert, R., & Exner, H. (2007). Laser micro sintering—A versatile instrument for the generation of microparts. *Laser Technik Journal, 4,* 26–31.
7. Daw, M. S., & Baskes, M. I. (1984). Embedded-atom method: Derivation and application to impurities, surfaces, and other defects in metals. *Physical Review B, 29,* 6443–6453.
8. Richardson, C. F., & Clancy, P. (1991). Picosecond laser processing of copper and gold: A computer simulation study. *Molecular Simulation, 7,* 335–355.
9. Pan, H., Ko, S. H., & Grigoropoulos, C. P. (2008). The solid-state neck growth mechanisms in low energy laser sintering of gold nanoparticles: A molecular dynamics simulation study. *Journal of Heat Transfer, 130,* 092404.
10. Johnson, D. L. (1969). New method of obtaining volume, grain-boundary, and surface diffusion coefficients from sintering data. *Journal of Applied Physics, 40,* 192–200.
11. Yang, L., Gan, Y., Zhang, Y., & Chen, J. (2012). Molecular dynamics simulation of neck growth in laser sintering of different-sized gold nanoparticles under different heating rates. *Applied Physics A, 106,* 725–735.
12. Jiang, S., Zhang, Y., Gan, Y., Chen, Z., & Peng, H. (2013). Molecular dynamics study of neck growth in laser sintering of hollow silver nanoparticles with different heating rates. *Journal of Physics. D. Applied Physics, 46,* 335302.
13. Kathuria, Y. (1999). Microstructuring by selective laser sintering of metallic powder. *Surface & Coatings Technology, 116,* 643–647.
14. Dai, C., Zhu, H., Ke, L., Lei, W., & Chen, B. (2011). Development a Cu-based metal powder for selective laser micro sintering. *Journal of Physics: Conference Series*, 012016.
15. Ristic, M. (1975). *Science of sintering and its future.*
16. Ristić, M. M., & Milosević, S. (2006). Frenkel's theory of sintering. *Science of Sintering, 38,* 7–11.

17. Oluwajobi, A. (2012). Molecular dynamics simulation of nanoscale machining, in molecular dynamics. In P.L. Wang (Ed.), *Studies of synthetic and biological macromolecules*. InTech.
18. Johnson, J. K., Zollweg, J. A., & Gubbins, K. E. (1993). The Lennard-Jones equation of state revisited. *Molecular Physics, 78,* 591–618.
19. Foiles, S., Baskes, M., & Daw, M. (1986). Embedded-atom-method functions for the fcc metals Cu, Ag, Au, Ni, Pd, Pt, and their alloys. *Physical Review B, 33,* 7983.
20. Bonny, G., Pasianot, R. C., & Malerba, L. (2009). Fe–Ni many-body potential for metallurgical applications. *Modelling and Simulation in Materials Science and Engineering, 17*.
21. Frenkel, D., & Smit, B. (2002). *Understanding molecular simulation: From algorithms to applications* (vol. 1, pp. 1–638). Amsterdam: Elsevier (formerly published by Academic Press).
22. Morse, P. M. (1929). Diatomic molecules according to the wave mechanics. II. Vibrational levels. *Physical Review, 34,* 57–64.
23. Lim, T.-C. (2003). The relationship between Lennard-Jones (12-6) and Morse potential functions. *Zeitschrift für Naturforschung A, 58,* 615–617.
24. Tersoff, J. (1988). Empirical interatomic potential for silicon with improved elastic properties. *Physical Review B, 38,* 9902–9905.
25. Verlet, L. (1967). Computer "experiments" on classical fluids. I. Thermodynamical properties of Lennard-Jones molecules. *Physical Review, 159,* 98–103.
26. LAMMPS-ICMS User's Manual. (2016). http://lammps.sandia.gov.

FSC
www.fsc.org

MIX
Papier aus verantwortungsvollen Quellen
Paper from responsible sources
FSC® C105338

If you have any concerns about our products, you can contact us on
ProductSafety@springernature.com

In case Publisher is established outside the EU, the EU authorized representative is:
Springer Nature Customer Service Center GmbH
Europaplatz 3, 69115 Heidelberg, Germany

Printed by Libri Plureos GmbH
in Hamburg, Germany